KEY SCIENCE

Biology

David Applin BSc MSc PhD FRES
Head of Science, Chigwell School

Stanley Thornes (Publishers) Limited

First published in 1994 by:
Stanley Thornes (Publishers) Ltd
Ellenborough House
Wellington Street
CHELTENHAM GL50 1YD
England

Reprinted 1995

A catalogue record of this book is available from the British Library.

ISBN 0-7487-1676-9

Related titles
Key Science: Chemistry (0-7487-1675-0)
Key Science: Physics (0-7487-1674-2)
Key Science: Biology Teacher's Guide (0-7487-1722-6)
Key Science Resource Bank (0-7487-1719-6)

Design and artwork by Cauldron Design Studio, Auchencrow, Berwickshire. Additional artwork by Barking Dog Art, Nailsworth, Gloucestershire and Mark Dunn.
Typeset by Word Power, Auchencrow, Berwickshire and Tech-Set, Gateshead, Tyne & Wear.
Printed and bound in Spain by Mateu Cromo.

Acknowledgements

I would like to express my thanks to Mr Gareth Williams who has offered invaluable guidance and original ideas for the Ecology, Populations, Soil and Farming topics and for the questions which he has contributed.

Dr Jerry Wellington and Mr Jon Scaife have provided some excellent material on information technology, and I thank them for their contributions to the text.

I thank the following organisations and people who have supplied photographs.
Allsport: p. 30 (athlete; Gary Mortimore), p. 169 (Bob Martin), 11.2D (Gary Mortimore)
Andrea Silvester/Institute of Biology: ix (bottom)
Ardea: 8.6C, 21.5N(e) (Francois Gahier), 12.2A(a) (John Clegg), 12.2A(b) (John Mason), 17.2D (Valerie Taylor), 21.1D(b) (Jack Swedburg), 21.2A, 21.5A, 21.5I, 21.5J (Adrian Warren), 21.5M (J-P Ferrero), 21.5N(b) (Peter Steyn)
Barr-Brown: 10.5H
Biofotos: 2.1B(d) (S Summerhays)
Biophoto Associates: 1.4A(b), (c), p. 12 (cocci, vibrios), p. 13 (spirilla, bacilli), p. 14 (amoeba), p. 15 (spirogyra, euglena), p. 16 (field mushroom), p. 18 (moss), p. 21 (scot's pine), p. 22 (hydra), p. 23 (coral, jellyfish), p. 24 (planaria, fluke, tapeworm), p. 26 (snail, chiton), p. 27 (common mussel), p. 29 (spider, centipede, crayfish), 3.7A(e), (f), (g), 3.7B, 3.7D, 4.1A (blue tit), 4.2B (larva), 9.1H, 9.3A(a), (b), 9.4F, 10.2E, 10.2F, 10.2G, 10.3B, 11.1D(a), 11.5G, 11.6E, 11.6G, 12.1C, 12.2C, 12.2K(a), (b), 12.2N(a), (b), 14.1B, 14.1E, 14.2B, 14.2G, 14.4C, 15.2B(a), 18.3C, 19.3A(a), (b), 19.3C(b)
British Diabetic Association, London: 20.2B
Bruce Coleman Ltd: 1.3 (Hans Reinhard), 1.3B(a) (Scott Nielson), 1.3B(b) (Joe Van Warmer), 10.2H (Jane Burton)
Cheltenham General Hospital Medical Photography: 15.5M
Department of Military Entomology, Royal Army Medical College: 4.3A
Dr Eileen Ramsden: 6.7C, 6.8B, 9.2A
Dr Jeremy Rayner, University of Bristol: 17.2N
Dr John Clark, Roslin Institute, Edinburgh: 20.3E
Health and Diet Co. Ltd, Manchester: 4.4F (royal jelly, propolis)
Heather Angel: 21.1B(b), (c), 4.1A (Aphidoletes larva and adult), 15.2C, 15.3B(a), (b), 15.3C, 16.4A
Holt Studios International: 4.1A (earwig, ground beetle, lacewing larva and adult, hoverfly larva and adult, Anthocoris nymph and adult, ladybird larva and adult, aphid), 4.2B (adult), 4.3B, 4.4B (frame), 11.4C(a), 20.1A, 20.5E
Hulton-Deutsch Collection: 21.1F

ICI Chemicals and Polymers: p. 28
International Centre for Conservation Education: 2.1C(b), 3.2A (WWF/Rautkari), 3.7A(a) (Manfred Kage), 3.7A(b) (Cath Wadforth), 3.7A(c) (Eric Grove), 3.7A(d) (EM Unit, CVL Webridge), 5.1E, 6.1A, 6.3A, 6.4A, 19.5C(a), (b) (Mark Boulton), 8.4A (Andy Purcell), 8.6D (Mark Tasker), 8.6E (Don Hinrichsen), 11.1A (Chris Rose)
John Innes Institute: 19.5A, 20.3C
Mark Boulton: 2.1B(a), (e), 2.2A, 3.5B(b), 15.1A(b), 15.5D, 15.5Q, 15.5R, 16.1A(b), 19.1A, 19.2A, 21.5N(a), (c), (d)
Marlow Foods: 20.4A
Martyn Chillmaid: 4.4B(a), 20.1B, 20.2C
Mary Evans Picture Library: 16.2G, p. 366, 21.1A, 21.1C, 21.1G, 21.2D, 21.3A
Mike Peters: p. 380
NASA: p. 1, 1.1A
National Medical Slide Bank: 4.2B (person), 15.5B, 15.5T, 16.5B
Natural History Museum: 1.2A (Dr N E Stork)
Natural History Photographic Agency: 10.3A (Peter Johnson), 15.1A(a) (Karl Switak), 21.1E(b) (Philippa Scott), 21.5K(a) (Dave Watts/ANT), 21.5K(b) (ANT)
Oxford Scientific Films Ltd: p. 25 (mason worm; G I Bernard) (medical leech; Mike Birkhead), p. 27 (cuttlefish; Rodger Jackman), p. 31 (carp, pigeon; G I Bernard), 2.1B(c) (Ronald Templeton), 2.1B(f), 2.1C(a) (Stephen Mills), 2.1C(d), 2.5A (G I Bernard), 3.1A(c) (Waina Cheng), 3.5A (Stan Osolinski), 3.6C (Len Zell), 3.6F (P and W Ward), 4.4A, 4.4B(c), (d), 5.3E (Michael Fogden), 6.7B(a) (Stan Osolinski), 6.7B(b) (Deni Brown), 9.4A, 9.4D(a), 9.4G (Carolina Biological Supply Co.), 11.5E (London Scientific Films), 15.1B(d), 15.1H (Barrie Watts), 15.1I, 15.1J, 15.1K, 15.1L, 15.2B(b) (Gordon Maclean), 14.2A, 14.2C, 14.2F, 14.2H, 15.6B, 15.6C(b) (Val Cooke, Marcia W Griffen/Animals Animals), 15.6E(a) (David Thompson, London Scientific Films), 15.6F(a) (Rudie H Kuiter, Scott Camazine), 15.6F(b) (Breck P Kent/Animals Animals, London Scientific Films), 16.1B (Breck P Kent/Earth Scenes), 16.2J, 17.1A(a), 17.2C(a), (c) (London Scientific Films), 17.2C(b) (C G Gardener), 19.4B(b) (David Fox), 19.5C(c) (Edward Parker), 21.1D(a) (Len Rue Jr./Animals Animals), 21.1E(a) (Peter Ryley), 21.4C (Peter Parks), p. 17 (bread mould; J A L Cooke)
Plant Breeding International, Cambridge: 20.3D
Popperfoto: 10.5G, 10.5J (Erik Hill)
Science Photo Library: cover (Charlotte Raymond), B p. vii (Dr Gopal Murti), p. 13 (Dr Steve Patterson), 4.2B (eggs; George Bernard), 4.2B (pupa; John Burbage), 5.2A (Susan Leavines), 6.1B (Peter Menzel), 6.2A (Peter Ryan), 7.9B, p. 139 (G F Gennard), 9.1C (Larry Mulvehill), 10.4B (David Scharf), 10.4C, 10.4D, 10.5E (Oxford Molecular Biophysics Laboratory), 10.5I (Dr Arthur Lesk, Laboratory of Molecular Biology, MRC), 11.1C (Simon Fraser), 11.1F (NASA), 11.2G (Biophoto Associates), 11.2H (St Mary's Hospital Medical School), 11.4C(b) (P Hawtin, PHLS, University of Southampton), 11.6I, 13.1A (Simon Fraser, Royal Victoria Infirmary, Newcastle Upon Tyne), 15.1B(c) (Phillipe Plailly), 15.1B(e), 15.2H (Dr Jeremy Burgess), 15.5A(e) (Andrew Syred), 15.5H (Petit Format/CSI), 15.5N (SIU), p. 298 (Will and Deni McIntyre), p. 314 (Science Source, 18.2A(CNRI), 19.3C(a) (Richard Hutchings), 19.3E(a) (Peter Menzel), 19.3E(b) (Biophoto Associates), 20.5C, 20.5I (Erie Grave), 20.5G (Department of Medical Photography, St Stephen's Hospital, London), 20.6B (effluent; John Walsh), 20.6B (Vorticella, Andrew Syred), 21.3B
Sporting Pictures (UK) Ltd: 16.1A(c), 16.5D, 19.4B(b)
St Mary's Hospital Medical School, London: 20.5D
Sue Boulton: 19.4B(a)
The Environmental Picture Library: 5.4A, 5.4B, 5.4C, 5.4D, 7.8A (David Townend), 8.1A (Alan Greig)
The Forensic Science Service: 20.2A
The Image Bank: ix (top)
Weidenfeld (Publishers) Ltd: 10.5F
Zoological Society of London: 20.3F, 20.3G

I thank the following examining groups for permission to reproduce questions from examination papers:
University of London Examinations and Assessment Council, Midland Examining Group, Northern Examinations and Assessment Board, Northern Ireland Schools Examinations and Assessment Council, Southern Examining Group, Welsh Joint Education Committee.

The examining groups bear no responsibility for the answers to questions taken from their past question papers contained in this publication.

The production of a science text book from manuscripts involves considerable effort, energy and expertise from many people, and I wish to acknowledge the work of all the members of the publishing team. I am particularly grateful to Penelope Barber (Senior Publisher), Adrian Wheaton (Science Publisher), Lorna Godson, Helen Roberts and John Hepburn (Editors), and Malcolm Tomlin (Senior Editor), for their advice and support.

Finally I thank my family for the tolerance which it has shown and the encouragement which it has given during the preparation of this book.

David Applin

Contents

Preface

Key Science: Biology has been written for GCSE Science: Biology and, with two other books, for all science syllabuses under the National Curriculum at GCSE. The related books are:
Key Science: Chemistry, Eileen Ramsden
Key Science: Physics, Jim Breithaupt.

Topics are differentiated into core material for Double/Single Science and extension material for Science: Biology (yellow margin). The examining groups have chosen different extension topics in addition to the Programme of Study for Key Stage 4; therefore, teachers and their students will need to familiarise themselves with the specific requirements of the syllabus which they are following.

Key Science: Biology has features, often located in the margin, such as **It's a fact**, **Science at work** and **Who's behind the science?**, which enhance the text and improve the accessibility of the text to all students.

• **Look at Links** identify related topics in this book and in *Key Science: Chemistry* and *Key Science: Physics*. Their role is to help students to integrate the different areas of biology, to appreciate the structure of the subject and its relationship to other branches of science. For the teacher, they have another benefit in that they make it easy to depart from the order of topics and take a different route through the text. The links are optional allowing each book to stand alone.

• **Summaries** and **Checkpoints** at regular intervals in the text help students to review their work and test their knowledge and understanding. More taxing questions which appear at the end of each theme will exercise their ability to apply their knowledge to novel situations, to analyse and interpret data and to solve problems. These longer questions co-ordinate the topics in the preceding theme and link up with other themes.

• **Information technology** 'IT' appears in the National Curriculum both as a cross-curricular skill (analogous to literacy and numeracy) and as an element of the separate subjects. The value of information technology, however, extends through all of the attainment targets in science. The 'IT' entries in the book reflect the wide range of applications of information technology. These 'IT' features show how computer-assisted learning (perhaps of a direct tutorial nature), simulations using 'IT', databases, videodisc, datalogging and other educational uses of information technology can be used throughout science at Key Stage 4.

Consultants in information technology, Dr Jerry Wellington and Mr Jon Scaife, both of the Educational Research Centre, University of Sheffield, have provided expert guidance for students (alongside the text) and for teachers (in the teacher's guides).

Key Science: Biology is supported by a detailed Teacher's Guide and Resource Bank of photocopiable activities and assignments – some of which can be used for teacher-based assessment of Attainment target 1 for Science, Sc1. Details are available from the publisher.

Student's Note

I hope that this book will provide all the information you need and enough practice to enable you to do well in the tests and examinations in your GCSE Biology course. However, examinations are not the most important part of your course. I hope also that you will develop a real interest in science, which will continue after your course is over. Finally, I hope that you enjoy using the book. If you do, all the effort will have been worth while!

David Applin

Adventures in science

The Mirror of Galadriel

In the book *Lord of the Rings*, amazing things happen to Gandalf and the hobbits. They risk their lives on a perilous journey to save their world. They journey through secret tunnels and unknown lands and they witness astonishing events. In the early part of their travels, they are invited to look into a magic mirror – a silver bowl filled with water. It shows 'things that were, and things that are, the things that yet may be.' But the viewer can't tell which it is showing!

Imagine you have travelled in time from two centuries ago to the present day. What would you make of television – a silver bowl showing events from the past and the present? Television is a product of the scientific age in which we live.

Science has many more amazing things to reveal. Imagine food 'grown' in a factory or round-the-world flight in a few hours or 'intelligent' computers that need no programming. These and many more projects are the subject of intense scientific research now. Fact is stranger than fiction.

How science works

Science involves finding out how and why things happen. Natural curiosity sometimes provides the starting point for a scientific investigation. Another investigation might set out to solve a certain problem and therefore have a definite aim. Even then, unexpected results may emerge to stimulate our natural curiosity.

The history of science has many examples where an important discovery or invention came about unexpectedly. The electric motor was invented when an electric generator was wired up wrongly at the Vienna Exhibition of 1873. X-rays, radioactivity and penicillin are further examples of unexpected scientific discoveries.

Discoveries in science are unpredictable. No-one can predict which research projects will lead to exciting new developments because no-one knows what new areas of science lie waiting to be discovered beyond the frontiers of knowledge.

Biology as a key branch of science

Biology is explained as 'The science of life'. Why do many scientists say that 'Biology is the key to all our futures'? On a June day in 1953, James Watson and Bernard Crick rushed from the Cavendish Laboratory in the centre of Cambridge, England into 'The Eagle' nearby and excitedly announced to the surprised lunch-time gathering that they had discovered 'the secret of life'. The two scientists had worked out the structure of DNA: they had 'cracked' the genetic code (Figure B). Their discovery has been the motor for innumerable exciting developments in medicine, agriculture and pharmaceuticals: it has spawned biotechnology, which promises to be the industry of the future, producing the countless goods we need and want – from plastics to fuel, drugs to food.

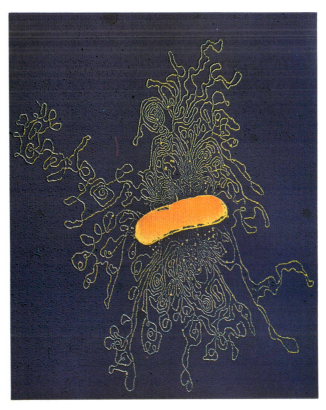

Figure B ● DNA of a bacterium – enzyme treatment which weakened the cell wall was followed by immersion in water which burst the bacterium, spilling its DNA.

Studying biology also helps us look at two sides of the same coin: investigating life tells us something about ourselves – who, what and why we are. Physics and Chemistry support Biology, but Biology is more than the application of chemical and physical laws. It investigates living processes at different levels of organisation (Figure C).

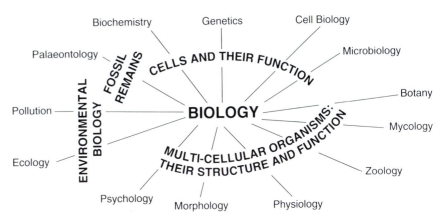

Figure C ● Some of Biology's –ologys! *Can you briefly explain the type of Biology each area of study covers?*

All of the –ologys shown in Figure C, and many more, are fields for investigation and research. In this way, biologists (scientists who study biology) explore the living world using the methods of science. Their work helps us make sense of the scientific laws that govern our existence.

The next steps after GCSE

Read this section carefully, bearing in mind that your working life will probably be about forty years. If you want your life ahead to be interesting; if you want to make your own decisions about what you do; if you want to make the most of your talents; then you should continue your studies. After GCSE, you can continue in full-time education at school or at college, or you can train in a job through part-time study. If you take a job without training, you will soon find that your friends who stayed on have much better prospects.

Most students aiming for a career in science or technology continue full-time study for two years, taking either GCE A- and AS-levels or a GNVQ course in Science. Successful completion of a suitable combination of these courses can then lead to a degree course at a university or a college of higher education or, alternatively, straight into employment.

The A-level route to higher education requires successful completion of a two-year full-time course, usually consisting of three or more A-level subjects. Students taking A-level Biology usually also take A- or AS-level Chemistry. A-levels in Biology and Chemistry and one other subject, such as Mathematics or Physics or Geography, lead to a wide range of degree courses including environmental science, medicine, agriculture, veterinary science, pharmacy, pharmacology, nutrition, animal and plant biology, genetics and management studies.

The GNVQ science route is also a two-year full-time course at advanced level, leading to a qualification equivalent to double award A-level Science. All students study some biology, chemistry and physics and can then specialise. The course is assessed continuously through tests and laboratory work. Students who do not have grade CC in science to enter advanced level GNVQ can take the one-year GNVQ science intermediate course as a preparation for the advanced course.

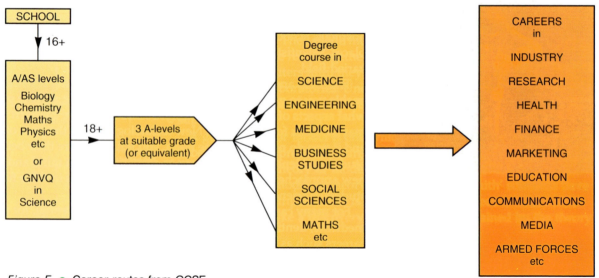

Figure E ● Career routes from GCSE

THEME A
Life on Earth

Throughout the universe we know of only one planet where there is life – planet Earth. Scientists have described about 5 million types of living thing. Millions more await discovery and millions once lived on Earth but are now extinct. The phrase 'The balance of nature' refers to the delicate relationships that exist between living things and the environments in which they live. People upset this balance with their need for food, homes, hospitals, schools and their demands for a wide range of manufactured goods. Now people are becoming aware that the human race is responsible for preserving the environment and that each one of us can play a part in protecting planet Earth.

TOPIC 1

VARIETY AND CLASSIFICATION

1.1 Earth as a place to live

> Earth is home to millions of different kinds of living things. It seems likely that none of the other planets in the solar system support life.

IT'S A FACT

> The first living things on Earth were probably simple bacteria like organisms. They originated about 4000 million years ago. The human race appeared about 2.5 million years ago.

What is it about Earth that enables it to support life? One factor is the distance between Earth and the Sun. Earth is close enough to the Sun for its surface temperature to be in the range in which life can exist. The energy of sunlight is converted by living things into the energy which they need for their life processes.

The size of Earth is a factor. It is massive enough to have a sufficiently large gravity to hold down an atmosphere. The gases oxygen, nitrogen, carbon dioxide and water vapour are essential for living organisms and are present in the air on Earth's surface.

Several factors combine to keep the temperature at the surface of Earth at an average of 30 °C.
- The distance of Earth from the Sun.
- The layer of ozone which surrounds Earth prevents too much ultraviolet light from the Sun reaching Earth. (Too much ultraviolet light destroys living organisms.)
- The layer of carbon dioxide and water vapour which blankets Earth prevents too much heat radiating from Earth into space. Without it, the temperature at the surface of Earth would be –30 °C.

There are at least 35 million different kinds of living organism on Earth. Their variety of appearance is enormous, as you will see in later sections of this topic. All living things, however, share certain characteristics, which non-living things do not. These are the **characteristics of life**.

LOOK AT LINKS
The characteristics of living organisms are:
1. They **move**; for **movement** see Topic 17.
2. They **feed**; for **nutrition** see Topic 11.
3. They **respire**; for **respiration** see Topic 12.
4. They **grow**; for **growth** see Topics 11 and 15.
5. They **excrete** waste products; for **excretion** see Topic 13.
6. They **reproduce**; for **reproduction** see Topic 15.
7. They are **sensitive to stimuli**; for **sensitivity** see Topic 16.

Figure 1.1A ● The Earth from space – 71% of the planet's surface is covered by water!

❶ What would happen to Earth's water if Earth were (a) nearer to and (b) further from the Sun?

❷ The Moon is about the same distance from the Sun as Earth is. The Moon is not massive enough to hold down an atmosphere. How does this affect the possibility of finding life on the Moon?

❸ **Mrs Gren** will help you remember the characteristics of living organisms. Complete the names of these processes which occur in living things.
M _____   R _____   S _____
G _____   R _____   E _____   N _____

❹ An aeroplane moves, it has to be 'fed' with fuel, and it gives out waste products. Do these characteristics prove that an aeroplane is alive? Explain your answer.

FIRST THOUGHTS

1.2 Variety of life

Have you ever thought about how many different living things there are in a garden? As you read about the variety of life think about:
- the large number of species of living things that we have named,
- the even greater estimate of the number of species awaiting discovery,
- using biological keys to identify living things.

Will we ever know all the species (a species is a particular type of living thing) of life on Earth? So far we know of about 5 million species but this figure is only a fraction of the millions of species that biologists estimate are awaiting discovery.

For example, techniques of collecting insects from the tops of trees in tropical rain forests lead us to believe that over 30 million new insect species are living in the leaves, branches, flowers and fruit (Figure 1.2A).

IT'S A FACT

Over 600 new species of beetle have been discovered living in just one type of tree that grows in the rainforests of Costa Rica.

Figure 1.2A ● Machines make a fog of insecticide that knocks down insects living in the trees. They fall down the funnels into the collecting jars beneath. The catch is examined later on in the laboratory. After careful work new discoveries are given names and added to the list of species already known.

Figure 1.2B ● A deciduous wood

Using keys

How many kinds of living thing can you see in Figure 1.2B? You can see from the picture that leaves from trees litter the ground with a thick carpet of dead organic material. Leaf litter is a rich source of food and home for many small animals. Specimens collected from leaf litter can be grouped according to their appearance and structure. Each description is a clue that helps to identify a particular group of animals found in leaf litter (Figure 1.2C). We call a set of clues a **key**.

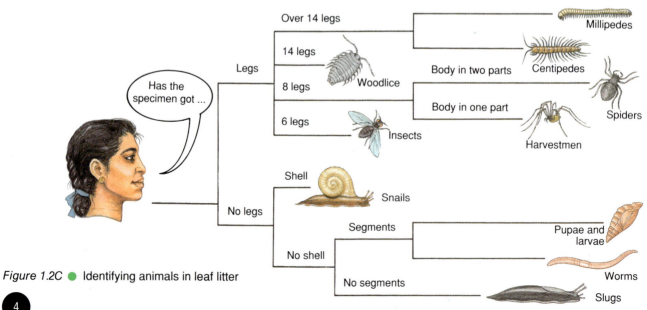

Figure 1.2C ● Identifying animals in leaf litter

IT'S A FACT

False scorpions are common in leaf litter but they are well hidden and very small (about 1–2.5 mm long) and little is known about them. False scorpions are relatives of true scorpions (together with spiders, harvestmen and mites) but they do not have a sting in the tail.

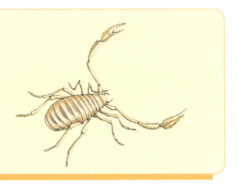

The easiest type of key to use is called a dichotomous key. 'Dichotomous' means branching into two. Each time the key branches you have to choose between the two statements. Eventually you will track down the identity of your specimen.

TRY THIS

Naming names

Amphibians are **vertebrates** (animals with backbones) that spend part of their lives on land and the rest of the time in water. Identify each of the illustrated specimens using the key on the next page, then draw a table by putting the correct letter against each name for all of the specimens.

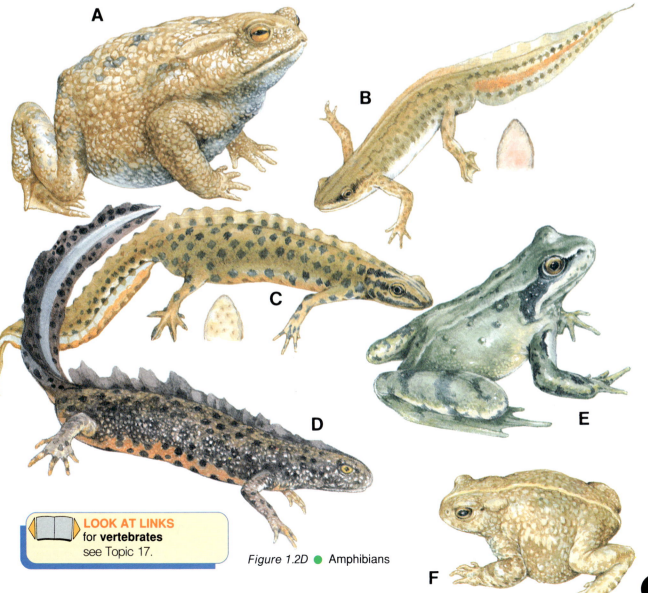

LOOK AT LINKS
for **vertebrates**
see Topic 17.

Figure 1.2D ● Amphibians

● Key to Amphibians

1 The animal has a tail	**Newts**	go to 2
The animal has no tail	**Frogs and toads**	go to 3
2 The newt has a rough warty skin, is dark brown with dark spots on top and has a yellow or orange belly blotched with black markings	**Great crested (warty) newt**	
The newt has a smooth skin, is green or brownish in colour with or without dark spots; belly is yellow or orange and may be spotted; 10 cm or less in length	**Smooth or palmate newt**	go to 4
3 The animal has a smooth, moist skin and a dark flash behind the eye	**Common frog**	
The animal has a warty skin and no dark flash behind the eye	**Toads**	go to 5
4 The throat is whitish and spotted	**Smooth newt**	
The throat is pinkish and unspotted	**Palmate newt**	
5 The toad has a yellow stripe running down the middle of the back	**Natterjack toad**	
There is no yellow stripe running down the middle of the back	**Common toad**	

A key in biology, therefore, is a guide to a name. Different keys are used to name different living things.

SUMMARY

Although about 5 million species of living thing have been described, millions more are estimated to await discovery, especially in tropical rain forests. Making a collection of living things in the environment will give an idea of life's variety. Biological keys help us to identify them.

CHECKPOINT

❶ Look at Figure 1.2A. Briefly describe the method being used to investigate the wildlife living in tropical treetops.

❷ What is leaf litter?

❸ (a) What is a biological key used for?
(b) Suggest why certain characteristics like size, exact colour and mass are not suitable for keys.

1.3 What's in a name?

FIRST THOUGHTS

What's in a name?
This section emphasises the importance of biological names. As you read it think about:
• the confusion caused by having the same name for different living things and different names for the same living thing,
• Carolus Linnaeus who cleared up the confusion by defining each type of living thing with a species name,
• the meaning of 'species'.

Sometimes a living thing has several different names which describe it. For example, the plant shown in Figure 1.3A is called 'cuckoo pint', 'lords and ladies', 'parson-in-the-pulpit' and 'wake-robin' in different parts of the country.

On the other hand different living things are sometimes given the same name. For example, the robin in the USA is different from the robin in the UK. They are shown in Figure 1.3B. *List some of the differences between the two birds.*

If living things were known only by their everyday names, you can imagine the confusion there would be when British and American ornithologists (an ornithologist is someone who studies birds) spoke to each other about 'robins'.

Figure 1.3B ●
a) The North American robin

b) The British robin

Figure 1.3A ● Cuckoo pint

The Swedish naturalist Carl von Linné (better known as Carolus Linnaeus which is the Latin version of his name) was born in 1707. In 1741 he was made Professor of Medicine and Botany at Uppsala University. His book *Systema Naturae* (published in 1735) laid the foundations for the system of naming organisms that we use today.

IT'S A FACT

A mule is the hybrid offspring from a mating between a horse and a donkey.

SUMMARY

The Swedish naturalist Carolus Linnaeus introduced the idea of the species. Each type of living thing is given a species name which distinguishes it from all other types of organism.

How can people describe living things to each other and know that they are talking about the same organism?

The Swedish scientist Linnaeus tackled this question in 1735. Linnaeus' system was to give each living organism a name consisting of two parts. The first part of the name is the name of the **genus** to which the organism belongs. The second part of the name is the name of the **species** to which the organism belongs. Since each name has two parts, the Linnaeus method of naming is called the **binominal system** (from the Latin: *bis* – twice, *nomen* – a name). The genus name begins with a capital letter; the species name begins with a small letter.

To illustrate how the binominal system works look at the biological name for human beings – *Homo sapiens*. Our species name is *sapiens* (Latin for 'wise'). Our genus name is *Homo*. The other species of our genus are now extinct.

So the binominal system clears up confusion over names. For example, ornithologists can easily distinguish between the robins mentioned earlier when the birds are called by their binominal names which are *Turdus migratorius* (common name North American robin) and *Erithacus rubecula* (common name British robin).

What, then, is a species? Most biologists agree that if individuals can sexually reproduce offspring which are themselves able to reproduce, then they belong to the same species. Sometimes members of closely related species mate and produce offspring called hybrids. However, animal hybrids are usually sterile, that is they cannot reproduce.

Different species of plant also form hybrids between themselves. Unlike most animal hybrids, hybrid plants are often fertile and able to produce offspring. This throws open the question 'What is a species?', to which there is no entirely satisfactory answer.

CHECKPOINT

❶ Who was Linnaeus? Explain how his binominal system of biological names works.

❷ What is a hybrid? Use a named example in your explanation.

1.4 Classification

Why is it important to classify living things? Here are some points to think about:
- classification as a way of organising living things into groups,
- groups within groups,
- classification making sense of the variety of life.

Linnaeus' system is more than just a list of names cataloguing the variety of life on Earth. Linnaeus believed that as many characteristics as possible should be used to describe species, rather like the way we use characteristics to describe a type of car. Some characteristics are unique to the species but others are shared with other species. This means that species which have characteristics in common can be put together into groups.

Family group cats

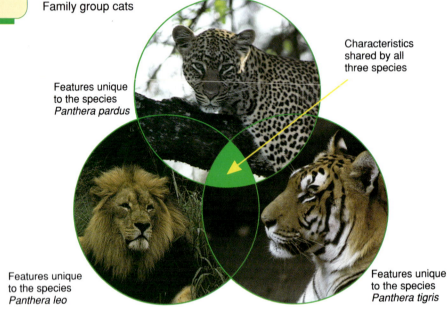

Characteristics shared by all three species

Features unique to the species *Panthera pardus*

Features unique to the species *Panthera leo*

Features unique to the species *Panthera tigris*

Figure 1.4A ● Unique features describe each species: shared features unite cats into a larger family group

Organising living things into groups means that we don't need to know everything about every species to be able to identify them. For example, we do not have to know about every insect species to know what an insect looks like!

RESOURCE ACTIVITY PACK

Groups within groups

Groups can be combined to form larger groups. The cats (Figure 1.4A) are a **family** which belong to the **order** called carnivores. Also included in the order carnivores are dog, bear, seal and sealion. The carnivores and other orders are members of the **subclass** placentals, which is one of the **class** mammals.

Organising things into groups is called **classification**. Figure 1.4B shows the classification of cats. It also shows that cats are part of a still larger group called a **phylum**. In this phylum mammals are grouped with fish, amphibians, reptiles and birds. All these animals have a feature in common – a backbone. Animals with backbones are called **vertebrates**.

Phyla (plural of phylum) come together in the largest group used in classification of all – the **kingdom**. For example, phylum 'vertebrates' along with 32 other phyla are grouped in the animal kingdom.

SUMMARY

Living things which have features in common are grouped together. The groups are named according to Linnaeus' system of classification, which creates groups within groups. This helps us to make sense of life's variety without having to know all about every living thing.

Phylum | **Class** | **Subclass** | **Order** | **Family** | **Genus** | **Species**

Species

Felis silvestris
Wildcat

Felis catus
Domestic cat

Felis lynx
Lynx

Panthera pardus
Leopard

Pathera tigris
Tiger

Panthera leo
Lion

Genus

Felis

Panthera

Family

Cats

Dogs

Bears

Seals and sealions

Plus six more families

Order

Carnivores

Primates*
Monkeys and apes

Cetaceans
Whales and dolphins

Lagomorphs
Rabbits and hares

Plus twenty-two more orders

Figure 1.4B ● Classification of cats. Groups within groups: notice monotremes and placentals belong to the larger and more inclusive group, mammals, etc. Group names increase in size indicating that more and more species are included in groups to the left of the figure. In-between groups are named by adding the prefixes sub or super to any of the main groups for a more detailed classification.

* Our order – humans are primates and close relatives of apes.

Subclass

Placentals
Mammals whose young complete their development inside the mother attached to her by a placenta. Food and oxygen pass from mother to young and wastes pass from young to mother across the placenta.

Marsupials
Mammals that give birth to young which complete their development attached to a milk nipple inside the mother's pouch (marsupium). Most species, like the kangaroo, are found in Australia though some come from South America.

Monotremes
Mammals that lay eggs. Includes the duck-billed platypus from Australia and three species of spiny anteater; one species from Australia and the other two from New Guinea

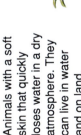

Class

Mammals
The female feeds her young with milk produced from milk glands called mammary glands which give the class its name. Many mammals have fur that helps to retain body heat.

Birds
Animals with feathers that help to retain body heat. Feathers also make it possible for birds to fly.

Reptiles
Animals with dry, horny scales that waterproof their body.

Amphibians
Animals with a soft skin that quickly loses water in a dry atmosphere. They can live in water and on land.

Fish
Animals with flat, wet scales which make the body smooth (bony fish) or with tiny, backward sloping spines (cartilagenous fish, e.g. sharks).

Phylum

Vertebrates
Backboned animals with a skeleton made of bone and/or cartilage.

Skeleton of the cat *Felis*. Notice the rib cage and the arching backbone.

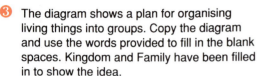

CHECKPOINT

1. With the help of Figure 1.4A explain why organising living things into groups helps us to understand them.

2. What does the word 'arthropods' mean literally?

3. The diagram shows a plan for organising living things into groups. Copy the diagram and use the words provided to fill in the blank spaces. Kingdom and Family have been filled in to show the idea.

   species   class   order   genus   phylum

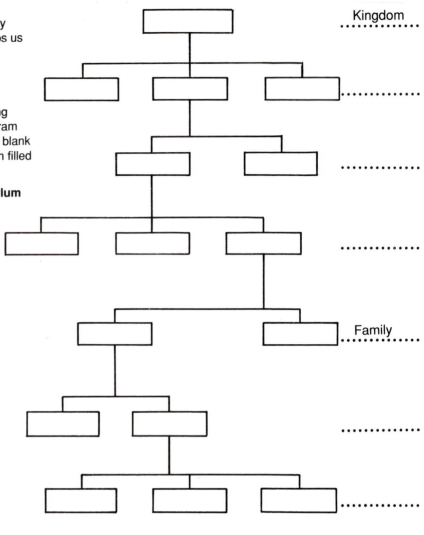

Kingdom ....

....

....

....

Family ....

....

....

1.5   Five Kingdoms

FIRST THOUGHTS

The largest grouping of animals is the kingdom. Here we review the five kingdoms. As you read, think about:
• each kingdom, representing a way of life shared by all its members,
• the size of living things, ranging from the very small to the very large,
• how living things are adapted for their way of life.

There are five kingdoms. Living things in each kingdom obtain food in different ways. Their structure and body chemistry are different. Each kingdom, therefore, represents a way of life which all of its members share. This helps us to learn about the wide variety of living things. (See Figure 1.5A).

The organisms pictured in Figure 1.5B are not drawn to scale. They vary in size from single-celled bacteria only just visible under a light microscope to the giant redwoods (*Sequoiadendron giganteum*) which tower into the sky. The wide range of size of living things, from the very small to the very large is shown.

The Five Kingdoms

		Bacteria	Protists	Fungi	Plants	Animals
Cell Structures	Nucleus	Absent.	Present.	Present.	Present.	Present.
	Cell wall	Made of different polysaccharides (not cellulose).	Of different types in different species.	Made of polysaccharide (mainly chitin).	Made of polysaccharide (mainly cellulose).	Absent.
	Chloroplasts	Absent (although chlorophyll is found in some species).	Present in some species.	Absent.	Present.	Absent.
Bodies		Single-celled (unicellular) although clusters or chains of identical cells may form.	Most species are single-celled though some are made of many cells of different types (multicellular).	Nearly all species are multicellular although a few are unicellular.	All species are multicellular.	All species are multicellular.
Food		Some species make their own food by photosynthesis. Some species are parasites; they live off other living things. Some species are saprophytes. They secrete enzymes which digest plant and animal material and then absorb the digested food. Saprophytes cause decay and decomposition.	Some species make food by photosynthesis. Some species take in food from their environment in different ways and digest it in the body. A few species make food by photosynthesis and take it in from their surroundings.	Most species are saprophytes. Hyphae secrete enzymes onto organic material which once digested is absorbed. This causes decay and decomposition. Some species are parasites.	All species make their own food by photosynthesis.	All species take in food from the environment and digest it inside the body. Exceptions are parasites like the tapeworm which absorbs the semi-digested food of its host.

Figure 1.5A ● Characteristics of the five kingdoms

(kilometre, km)	10^3 m		
	10^2 m	Giant redwood _____	
	10^1 m	Indian python _____	
(metre, m)	10^0 m	Dog _____	
	10^{-1} m	Locust _____	
	10^{-2} m	Woodlouse _____	
(millimetre, mm)	10^{-3} m	Mite _____	
	10^{-4} m	Human egg cell _____	
	10^{-5} m	Red blood cell _____	
(micrometre, (μm)	10^{-6} m	Bacterium _____	

Figure 1.5B ● The scale and size of living things

Careful study of pages 12 to 31 will help you to understand the way of life of different organisms from each kingdom. Some of the organisms will be familiar to you; others will not. Many have been selected because they are easy to obtain and study. Each of the selections on particular kingdoms and phyla contains a fact file, body file and food file on representative organisms of that group. The files list important characteristics of the organisms shown, given details of their appearance and describe how they feed.

Kingdom Bacteria

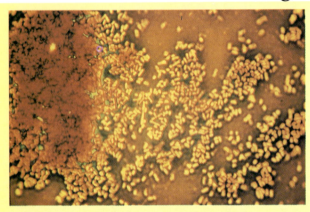

Vibrios are comma-shaped bacteria. The photograph shows *Desulfovibrio desulfuricans* which releases sulphur into the environment. If the bacteria are living in water rich in iron compounds, the sulphur they release reacts to form iron sulphide which glitters like gold (iron sulphide is called 'fool's gold').

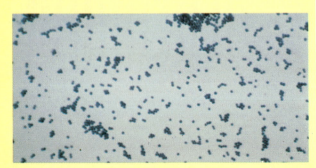

Cocci bacteria are spherical in shape and may stick together in pairs, chains or clusters as illustrated opposite. The photograph illustrates *Staphylococcus aureus* which causes boils and food poisoning.

FACT FILE

- Bacteria are single celled organisms or organisms made up of chains or clusters of similar cells.
- They are called microorganisms because they are only visible under the microscope.
- Their cell body has no distinct nucleus.
- Bacteria are found everywhere: in water, in the air and in the soil. They also live on and in other organisms.
- The cell shape helps to identify different types of bacteria.

BODY FACTS

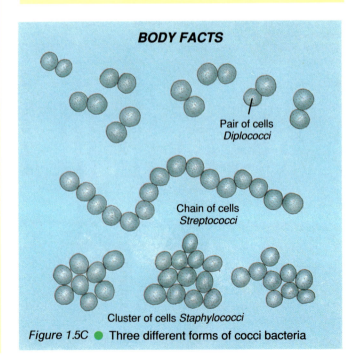

Pair of cells
Diplococci

Chain of cells
Streptococci

Cluster of cells *Staphylococci*

Figure 1.5C ● Three different forms of cocci bacteria

Viruses are not cells. They do not seem to need food as a source of energy and they cannot reproduce independently. Crystallized and stored like minerals, some viruses can remain unchanged for years until moistened and provided with a favourable environment for renewed activity. *Are viruses alive? Check p. 2 for the characteristics of living organisms and try to decide for yourself.*

Consisting of a strand of nucleic acid (DNA or RNA – see p. 162) enclosed in a coat of protein, a virus can only be seen using an electron microscope (see p. 141). Viruses infect and take over the biological machinery of living cells to reproduce new viruses. Some types of virus cause diseases in plants and animals.

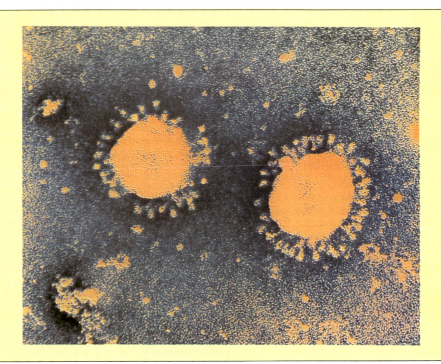

Kingdom Bacteria

FOOD FACTS

Most bacteria are **saprophytes** and are responsible for decay and decomposition. Bacterial cells secrete enzymes which convert dead organic matter and dead organisms into food which the cells absorb. Some bacteria are parasites, that is they live by feeding on the tissues of other organisms. Bacteria which cause disease belong to this group. Other bacteria can obtain their food through photosynthesis or by making use of chemicals like sulphur and hydrogen sulphide which they extract from the environment.

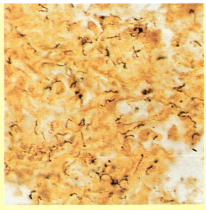

Spirilla bacteria look like tightly coiled springs. The photograph shows *Treponema pallidum* which causes the **sexually transmitted disease**, syphilis.

LOOK AT LINKS
for **saprophytes** and **fungi**
See Topic 3.5.

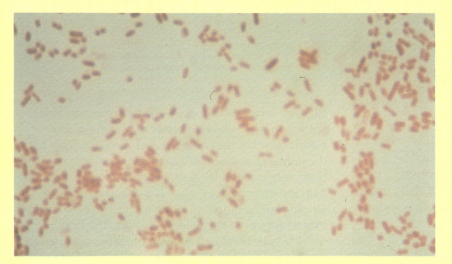

Bacilli bacteria are rod-shaped and can form into filaments of cells that look like the hyphae of fungi. The photograph shows *Escherichia coli* which is found in the human intestine. More is known about *E.coli* than any other living thing because it is used by scientists to investigate a wide range of biological problems.

LOOK AT LINKS
for **sexually transmitted disease**
See Topic 15.5.

Kingdom Protists

TRY THIS

Place some filaments of *Spirogyra* on a microscope slide and cover them with a few drops of water. Put a cover slip in place and examine the filaments under the low-power objective of a microscope. Draw what you see and label the cell wall and chloroplast.

LOOK AT LINKS
for **calcium carbonate**
The importance of the delicate shells of these amoebas is described in the carbon cycle. See Topic 2.7.

RESOURCE
ACTIVITY
PACK

FACT FILE

- Protists include both single-celled organisms and also organisms made of filaments of cells.
- Their cell bodies have distinct nuclei.
- Protists are neither animals, plants, fungi nor bacteria.

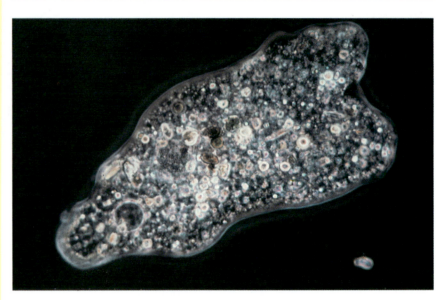

Amoeba lives in ponds and ditches. Flowing extensions (called pseudopodia: literally meaning 'false feet') of its single-celled body change its shape. Movement occurs when they flow in a particular direction. The name *Amoeba* describes all single-celled organisms that have pseudopodia. The bodies of some are surrounded by delicate shells of **calcium carbonate**. Some types of amoeba are parasites and can cause dysentery in their animal hosts.

Bladder wrack is a brown seaweed, common along the sea shore. Its body is made of branched, flat fronds. The holdfast at the end of the short stalk holds the seaweed in place on exposed rocks. Pairs of air-filled sacs (called bladders, hence the common name) help the plant to float when it is covered in water. The whole plant is limp and slimy to touch. These features protect it against the action of waves. The limp seaweed can bend and sway in the moving water and water can flow freely over the slimy surface of the fronds.

Kingdom Protists

BODY FACTS

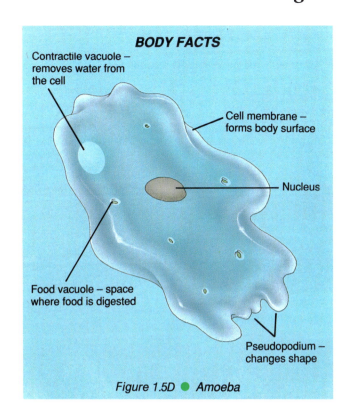

Contractile vacuole – removes water from the cell

Cell membrane – forms body surface

Nucleus

Food vacuole – space where food is digested

Pseudopodium – changes shape

Figure 1.5D ● Amoeba

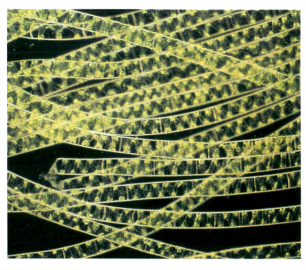

Spirogyra forms filaments of cells joined end to end. The filaments make a green tangled mat on the surface of ponds. Each cell is surrounded by a wall made of cellulose which in turn is surrounded by a layer of slimy substance called mucilage. A chloroplast winds in a spiral against the inner wall of each cell, giving *Spirogyra* its name.

FOOD FACTS

The diagram shows how *Amoeba* captures its food. Pseudopodia flow around the food particle which *Amoeba* detects by sensing the chemicals released into the water by the food. The pseudopodia join around the food particle and a food vacuole forms in the cell body. Digestion takes place in the food vacuole. The method of feeding by capturing food in this way is called phagocytosis.

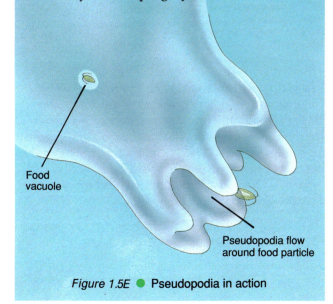

Food vacuole

Pseudopodia flow around food particle

Figure 1.5E ● Pseudopodia in action

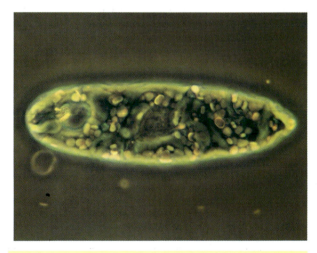

Euglena is a single-celled organism that lives in ponds and puddles, especially where there is a high level of nitrogen compounds. It has chloroplasts and therefore can photosynthesise its food. However, like *Amoeba*, it can gather food from the environment. Single cells closely related to *Euglena* do not have chloroplasts and obtain food only by feeding. Other close relatives develop chloroplasts in bright light but lose them when they are kept in the dark. *Euglena* and closely related organisms display both plant and animal characteristics. This makes it difficult to put them into either the animal or plant kingdom and they are therefore included within the protists.

Kingdom Fungi

Cut a mushroom in half longitudinally (down through the middle). Describe what you see in a few sentences. Draw the mushroom and label the cap, stalk and gills.

TRY THIS

Make a spore print. Cut the cap off an open mushroom. Place it, gills down, on a sheet of paper, fix it there with a pin and leave it for several days. When you remove the mushroom cap there will be a pattern on the paper of spores which have fallen from the gills. A spore-print is easily spoiled if touched, but they can be sealed with a spray of clear varnish and then mounted. Look at it through a hand lens and you will see the spores more clearly. Draw what you see.

FACT FILE

- Fungi are made up of slender tubes called hyphae (singular, hypha).
- The mass of hyphae which form an individual is called the mycelium (plural, mycelia).
- Fruiting bodies produce spores. Each spore can grow and develop into a new mycelium.

Field mushroom: the fruiting body is all that we see above ground. The word 'mushroom' refers to the field mushroom which can be eaten. However, many types of so-called 'toadstool' make a good meal while others are very poisonous. So, 'mushroom' and 'toadstool' have little meaning. Mushrooms and some toadstools are edible; other toadstools are not. A few species are fatal if eaten! NEVER collect fungi for the table unless you are certain of their identification - BE SAFE NOT SORRY.

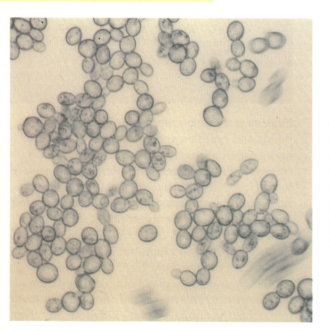

Yeasts do not grow hyphae but live as single cells. Their name *Saccharomyces* means 'sugar fungi' and they are commonly found where ever sugar occurs. For thousands of years yeast has been used to ferment the sugars in rice and barley to produce beer, and the sugar in grapes to make wine. Yeast added to dough ferments the sugar in the dough to produce carbon dioxide. This makes the dough rise.

Kingdom Fungi

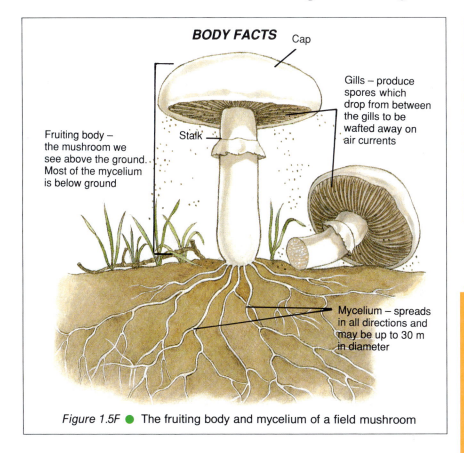

BODY FACTS

Cap

Gills – produce spores which drop from between the gills to be wafted away on air currents

Fruiting body – the mushroom we see above the ground. Most of the mycelium is below ground

Stalk

Mycelium – spreads in all directions and may be up to 30 m in diameter

Figure 1.5F ● The fruiting body and mycelium of a field mushroom

FOOD FACTS

The hyphae secrete enzymes which convert dead organic material in the soil into food which the hyphae absorb. Obtaining food in this way is called saprophytism. Most fungi are saprophytes but some are parasites, feeding on the living tissue of plants and animals.

IT'S A FACT

A lichen is a close partnership between an alga and a fungus. The algal cells grow in the fungal mycelium. They make food by photosynthesis and some of it is passed to the fungus. The fungus cannot survive without the alga. However, the alga does quite well on its own and grows even faster when free of the fungus.

Bread mould hyphae spread over and into uncovered food, spoiling it. They form a mat-like mycelium. Hyphae (called aerial hyphae) poke up into the air, each carrying a capsule-shaped fruiting body. In the picture you can see that some are black, a sign that they contain ripe spores. Soon they will burst open, releasing a cloud of spores which are carried away on the air currents. If the spores land in a suitable place they will grow into new mycelia.

Other types of mould also spoil food but some add flavour. For example, blue cheeses such as Gorgonzola, Stilton and Roquefort owe their taste to the moulds which create the blue veins in them.

The mould *Penicillium notatum* produces a substance which can destroy bacteria. The substance, called **penicillin**, is an antibiotic used to fight diseases caused by bacteria.

Plant Kingdom 1: Spore-producing plants

FACT FILE

- The plants reproduce by means of spores.
- Water is needed for the sperm to swim to the eggs.
- Spore-producing plants live on land in damp habitats.

Mosses quickly lose water in a dry atmosphere, so the plants live in damp places such as the banks of streams and the woodland floor where the air is moist. They cannot draw water from the ground as ferns and seed-producing plants do. They rely on capillary action to soak up water like a sponge. Closely fitting leaves help the capillary movement of water. This mechanism for water collection limits the plant's size. The tallest species of moss is only about 40 cm high and most are much shorter.

In summer, stalks grow from the moss plants, each carrying a capsule which contains spores. When ripe the capsule opens and the spores are shaken out to be carried away on air currents. If spores land on damp soil, they develop into new moss plants.

Plant Kingdom 1: Spore-producing plants

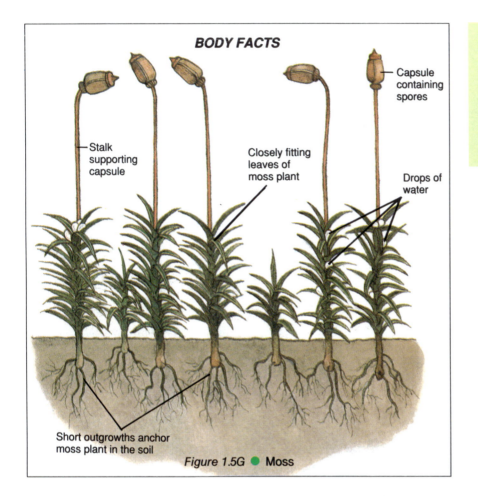

BODY FACTS

Stalk supporting capsule

Closely fitting leaves of moss plant

Capsule containing spores

Drops of water

Short outgrowths anchor moss plant in the soil

Figure 1.5G ● Moss

FOOD FACTS

Moss leaves are thin and delicate, each made up of only a few layers of cells. Chloroplasts in the cells of the leaves make food by photosynthesis.

Ferns have clumps of leaves called fronds which grow near the top of a thick stem. Roots sprout from the base of each clump, anchoring the plant firmly into the soil. Each leaf is built up of leaflets called pinnae which stand out on either side of a sturdy rib that runs through the centre of the leaf. Each leaflet is further subdivided into lobe-shaped pinnules.

A double row of greenish-white patches develops on the underside of the pinnae. Each patch consists of a group of spore-producing capsules which darken as the spores mature. When ripe, the capsules break open and the spores shoot out to be carried away on air currents.

A waxy layer waterproofs the plant's surfaces, reducing water loss. Ferns can live in a drier environment than moss, although it grows best in damp, shady places. Tissues in the roots, stem and leaves draw water from the soil and carry it to all parts of the plant. They also strengthen and support the plant which is much bigger than the moss plant, growing to a height of a metre or more.

Plant Kingdom 2: Seed-producing plants

FACT FILE

- Plants reproduce by means of seed which protects the embryo plant from drying out on land.
- The sperm of seed-producing plants is transported to the eggs by pollen grains.
- Very successful land plants are able to live in dry, hot places where there is little water.

Beech trees are widespread on gentle slopes and low lying land. They are woody plants with massive trunks covered with smooth bark and bearing a crown of thick branches which carry leaves. New wood grows every year, increasing the girth of the tree. A cross-section of a tree trunk shows rings for each year's growth. Counting the annual rings will tell you the age of the tree. The beech tree's growth is greatest in the spring. Every autumn its broad leaves are shed. Plants that lose their leaves in this way are described as **deciduous**. Beech trees are **perennial**; that is they keep on growing and producing seeds for many years. A beech tree can live for 200 years or more.

Buttercup grows in meadows and unimproved grassland. It is a non-woody **annual**; that is it grows from seed to maturity and produces new seeds all within one growing season. It then dies with the onset of the first frosts, leaving the seed to lie dormant through the winter ready to grow when conditions improve the following spring. Non-woody plants are called **herbaceous** plants.

Plant Kingdom 2: Seed-producing plants

Scots pine trees grow naturally in cool climates. They cover hill slopes and valleys with thick forests. The Scots pine is a large tree with a crown of branches covered with rough scaly bark. Short shoots growing from the branches carry clusters of needle-like leaves which are lost and replaced throughout the year. This is why the Scots pine and trees like it are called **evergreens**: they are covered with leaves all year round. Cones form on young branches that grow in spring. If the tree is damaged a sticky resin oozes out from the wound and plugs it, preventing infection by disease-carrying organisms.

BODY FACTS

Petal

Anther — male sex organ producing pollen

Carpel – female sex organ producing egg

Flower – carries the reproductive organs

Sepal – protects the flower in bud

Leaf stalk (petiole)

Node – point where leaf stem attaches to the stem

Stem – supports the leaves and flowers

Internode – distance between two succesive nodes

Root hairs – large surface area for the absorption of water from the soil

Root – anchors the plant in the soil

Figure 1.5H ● Buttercup

FOOD FACTS

Buttercup leaves are complex structures. Each leaf is made up of different types of cell. The cells nearest the upper surface receive most light and have the most chloroplasts for photosynthesis.

IT'S A FACT

Some trees live for a very long time. For example the Cedar of Lebanon lives for up to 1000 years. However, Bristlecone pines may be much older. They grow high in the mountains of eastern California and some are believed to be over 4000 years old.

FACT FILE

- The body shape shows radial symmetry: that is it has no front or rear and its parts are arranged evenly 'in the round'.
- Tentacles surround an opening which is both mouth and anus.
- Stinging cells are used to capture prey.

Hydra lives in ponds and water-filled ditches. It hangs from water plants and other firm surfaces and trails its tentacles in the water. The tentacles carry sting cells which can paralyse small organisms which touch them. *Hydra* can then feed on its defenceless prey.

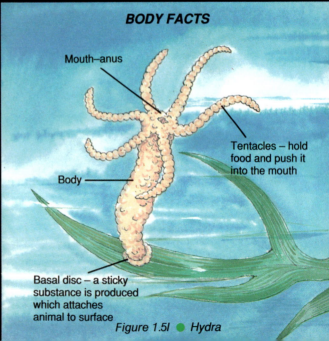

BODY FACTS

Mouth–anus

Tentacles – hold food and push it into the mouth

Body

Basal disc – a sticky substance is produced which attaches animal to surface

Figure 1.5l ● *Hydra*

FOOD FACTS

Food enters and undigested remains leave by the same opening - the mouth–anus.

Animal Kingdom 1: Hydra and its relatives

Sea anemone. The specimen in the picture is called beadlet. It lives attached to rocks in pools that are left as the tide moves down the sea-shore.

Jellyfish usually hang, bell-like, in water with the tentacles floating around the mouth. When the bell pulses in and out the water jet which is produced propels the animal through the water. Most types of jellyfish live in the sea, but a few live in ponds and streams.

Coral is made from limestone which hydra-like animals secrete around themselves. It builds up into reefs in warm sunlit seas. The Great Barrier Reef off the northeast coast of Australia is 2000 kilometres long and the world's largest reef.

IT'S A FACT

The sea blubber *Cyanea* is the largest jellyfish in the world. Its tentacles are more than 30 metres long and the bell measures over 3 metres in diameter.

IT'S A FACT

Hydra takes its name from a mythical water-snake with nine heads. An ancient Greek legend tells that the hero Hercules was sent to kill the monster. Hercules cut off each of Hydra's heads but the ninth was immortal so he rolled a huge rock over it, trapping it for ever.

TRY THIS

Looking for Hydra

Hydra lives in ponds and ditches where there are floating water lily leaves. The best time to find *Hydra* is between the months of May and October. Look at the undersides of the leaves. *Hydra* is only just visible to the naked eye, so you have to look carefully! Dab off specimens with a fine paint brush and place them in a jar filled with water from the pond.

Another way of finding *Hydra* is to scoop duckweed into a jar of pond water and leave it overnight. Look at the jar next day and specimens will probably be seen attached to the sides of the jar or hanging down from the tiny roots of the floating duckweed.

Animal Kingdom 2: Flatworms

FACT FILE

- Flatworms have flattened bodies with upper (dorsal) and lower (ventral) surfaces.
- Their body shape shows bilateral symmetry.
- Flatworm bodies have a front end (anterior) and a rear end (posterior).
- Many of the different types of flatworm are parasites.

Planaria is a flatworm which is common in ponds and streams where it lives under stones and plants. Notice the eyespots at the front end. They do not form images like our eyes but are sensitive to changes in light and shade. Other sense organs help the animal to test new environments as it moves forward.

FOOD FACTS

In Planaria the digestive cavity branches to all parts of the animal and its outline is just visible through the body wall. A part called the pharynx pokes out from the mouth–anus and captures small animals or breaks off fragments of dead larger ones. Most non-parasitic flatworms obtain their food in this way. Most parasitic flatworms feed either like the tapeworm or fluke shown here.

Fluke is a common flatworm parasite. The specimen illustrated lives in the livers of sheep and cattle. It feeds on the tissue and blood of the liver causing great damage and even death. Notice the branching digestive cavity.

Tapeworm is a common flatworm parasite that lives in the intestines of vertebrates surrounded by the host's semi-digested food. The body is long and flat and, unlike *Planaria*, made up of sections called proglottids. The front end is called the scolex and has hooks and suckers which fasten the tapeworm to the inner lining of the host's intestine. The tapeworm does not have a digestive cavity. It absorbs some of the host's semi-digested food through its body wall. Tapeworms are particular about where they live. For example, a dog tapeworm is not usually found in humans or other animals; we say that tapeworms are 'host specific'.

BODY FACTS

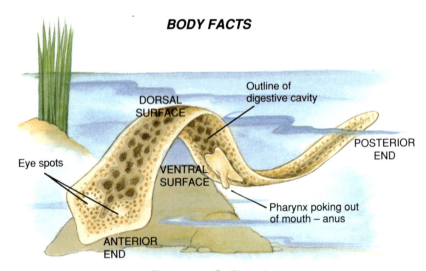

Figure 1.5J ● Planaria

Animal Kingdom 3: Larger worms

FACT FILE

- The body of the worm is long and thin with a distinct head end.
- The body is made up of many rings, called segments.
- The body is bilaterally symmetrical.

FOOD FACTS

The worm has a mouth at the head end and an anus at the rear end. The digestive cavity is an open tube that runs the length of the body. The means of taking in food and the means of passing out the undigested remains of a meal are separated. Food can therefore be digested and processed more efficiently than in flatworms which have only one opening to the digestive cavity.

BODY FACTS

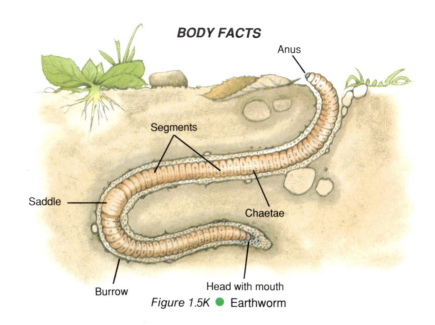

Figure 1.5K ● Earthworm

Earthworms are so-called because they live in soil. One hectare of fertile grassland may support more than seven million earthworms! A worm makes its burrow by taking in soil through its mouth. The soil is finely ground up in the intestine where any organic material is extracted for food. The remains are passed out through the anus to form a cast above ground. Notice the saddle which is formed from thickened segments. It produces envelopes of mucus. Each of these makes a protective cocoon around a fertilised egg.

IT'S A FACT

Medical leeches live in ponds and streams and feed on the blood of different vertebrate hosts. The leech has suckers which fix it to the host's body. It makes a small cut in the host's skin. Blood flows freely with the help of an anti-coagulant which the leech produces. The anticoagulant stops the host's blood from clotting. A leech may take in up to five times its own weight of blood in a single meal and may not need another one for several months. When full, it drops from the host and digests the blood. Even until the nineteenth century doctors used to think that taking blood from a person who was ill would help them to recover. They used leeches to 'let blood' from the patient. This is why doctors were given the nick-name 'leeches'.

Flatworms, probably imported in food from New Zealand, are attacking earthworms in Northern Ireland. The flatworms grow up to 15 cm long and feed on the earthworms. Scientists are trying to find ways of controlling them before the damage to earthworm populations becomes too serious.

Animal Kingdom 4: Molluscs

Although surrounded by a shell, snails are food for a variety of animals. For example, the picture shows a thrush hitting a snail against a stone to smash open the shell to get at the juicy flesh inside. The stones they use are called 'thrushes' anvils'.

FACT FILE

- The body of a mollusc is made up of a head, a foot, a mantle and a visceral mass which contains the digestive cavity and other organs.
- The mantle of a mollusc produces the shell. This is outside the body of some molluscs but is inside the body of others.
- The body shape shows bilateral symmetry.
- There is no segmentation in the body of a mollusc.

Garden snails live under stones, rotting plants and in other damp, cool places. The foot of the snail moves the animal along on a trail of slimy mucus. If disturbed, it withdraws into its protective shell. This also helps to reduce the loss of water from the body. In dry weather the snail may retreat into its shell and secrete a film of mucus (called the epiphragm) across the opening. By sealing the animal inside further water loss is avoided. When it rains the epiphragm is cast off and the snail becomes active again.

Chiton lives on the sea shore attached to rocks and stones. The shell is made of eight overlapping plates and fits over the animal. When large waves threaten a chiton's grip on its rocky platform, the animal draws the shell down around itself, sealing itself away from danger.

Animal Kingdom 4: Molluscs

BODY FACTS

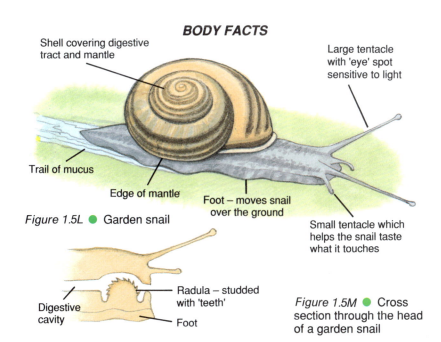

Shell covering digestive tract and mantle

Large tentacle with 'eye' spot sensitive to light

Trail of mucus

Edge of mantle

Foot – moves snail over the ground

Small tentacle which helps the snail taste what it touches

Figure 1.5L ● Garden snail

Digestive cavity

Radula – studded with 'teeth'

Foot

Figure 1.5M ● Cross section through the head of a garden snail

FOOD FACTS

Figure 1.5M is a cross-section through the head of a garden snail. It shows the radula which is a file-like tongue that the snail uses to rasp fragments of food into its mouth. All molluscs have a radula except for the molluscs that have two shells, such as mussels.

The cuttlefish is a close relative of the squid and the octopus. Its shell, which is inside the body, is called the cuttlebone. It lives in shallow water near the sea-shore, half buried in sand during the day. At night, it swims over the sea-bed in search of the crabs, fish and prawns which are its food. The siphon, or funnel, is able to produce a water jet that can propel the animal through the water in any direction at high speed. When threatened, a cuttlefish can squirt an inky liquid into the water. Under the black cloud it creates, the cuttlefish can escape unseen by predators.

Common mussels live in shallow water near the sea-shore. The mussel attaches itself by the foot to rocks and stones. Two shells close round the animal. Water is drawn through the crack where the shells meet and filtered for food. Mussels and their close relatives, oysters, are fished to provide delicacies for the dinner table.

Animal Kingdom 5: Insects and their relatives

FACT FILE

- The body of an insect is covered by a hard outer layer called the exoskeleton.
- Jointed limbs are adapted variously as legs for walking, paddles for swimming and for many other uses.
- The head and other regions of the body are formed from segments of the exoskeleton fused together.
- The insect head is well developed and has compound eyes.

The locust shows the body features typical of most insects. It has two pairs of wings and is a strong flier. Its exoskeleton has a waxy, waterproof layer which helps to reduce the loss of water from the locust's body. This feature is so successful that locusts live in hot dry places where few other types of animal survive. All insects are waterproofed in this way, which is why there are so many of them living on land.

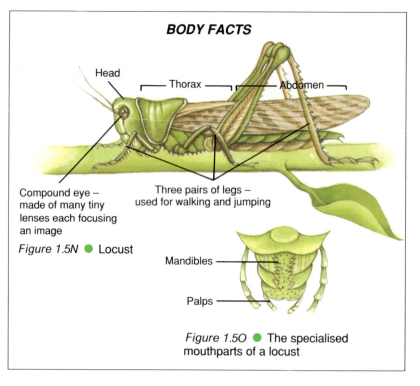

BODY FACTS

Head

Thorax

Abdomen

Compound eye – made of many tiny lenses each focusing an image

Three pairs of legs – used for walking and jumping

Figure 1.5N ● Locust

Mandibles

Palps

Figure 1.5O ● The specialised mouthparts of a locust

Animal Kingdom 5: Insects and their relatives

FOOD FACTS

Figure 1.5O shows the specialised mouthparts of a locust. The mandibles have saw-like edges which work like scissors. They cut up and chew plant food which the other mouthparts guide into the mouth. The palps help the locust to taste what it touches. Other species of insects, e.g. house fly, butterfly, have mouthparts which are specialised in different ways to utilise different food sources.

The garden spider spins sticky webs in which it catches its prey. Not all spiders spin webs; some chase their prey, others lie hidden in wait to ambush their victims. Like that of insects, the exoskeleton of the spider is waterproofed. They can live, therefore, in hot dry places. There are two regions to the body of a spider, a fused head and thorax with four pairs of legs and no wings and an abdomen.

Crayfish live in clean, clear streams. Their body is divided into two regions, an abdomen and a fused head and thorax, with pairs of limbs adapted for swimming, catching food and other uses. Crabs, shrimps and prawns are close relatives of crayfish that live in the sea. The woodlouse is another close relative that lives on land. The woodlouse does not have the waterproof layer present in the exoskeleton of insects and spiders and is therefore restricted to living under stones, rotting wood and in other damp places.

The centipede is carnivorous and lives on land. Like the woodlouse, it does not have a waterproof exoskeleton and is also restricted to living in damp places. The millipede, a close relative of the centipede feeds on dead leaves and other plant remains. The body is divided into a head and a trunk which carries many pairs of legs.

Animal Kingdom 6: Vertebrates

FACT FILE

- All vertebrates have a vertebral column (backbone) which runs along the dorsal (top) surface of the body.
- The backbone extends to form a tail.
- Vertebrates have internal skeletons, called endoskeletons which are most often made of bone.
- Limbs are variously adapted for swimming, walking and flying.

Lizards are reptiles and are fully adapted for life on land. Their skin is dry and covered with horny scales which restrict water loss from the body. All reptiles are waterproofed like this and so can live in hot, dry places. They lay eggs in which the young develop, protected by a hard shell. Water, therefore, is not necessary for breeding.

BODY FACTS

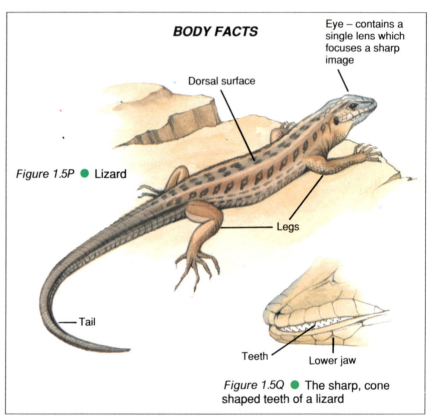

Eye – contains a single lens which focuses a sharp image

Dorsal surface

Figure 1.5P ● Lizard

Legs

Tail

Teeth

Lower jaw

Figure 1.5Q ● The sharp, cone shaped teeth of a lizard

FOOD FACTS

The cone-shaped teeth of a lizard grip and chew food. In different mammals teeth are different shapes adapted for cutting, piercing or grinding food.

Homo sapiens is a species of mammal. Us! Humans have mammalian features and characteristics. We care for our young, which the female feeds in early life with milk from her mammary glands (breasts). We have hair, which helps to conserve body heat, though we do not have as much as other mammals. We have a small tail bone called the coccyx at the base of our backbone. Most mammals move by using all four limbs, but we stand upright and our forelimbs are adapted as hands, able to carry out different tasks. Our large brains and our dexterous hands mean that we have more control over our environment than the members of any other species.

Animal Kingdom 6: Vertebrates

The frog is an amphibian. The adult frog lives on land but breeds in water-filled ditches and ponds. Fertilised eggs hatch into swimming tadpoles which grow and develop into adults. The change from tadpole to adult is called metamorphosis. Toads and newts are also amphibians. Toads, like frogs, divide their time between water and land, but newts rarely leave the water even as adults. The frog's skin is soft and wet and in dry air the body quickly loses water. So frogs, toads and most amphibians live in wet environments.

The carp is a species of bony fish. Its skeleton is made of bone. It lives on the bottom of muddy rivers and lakes. Different species of sea-living bony fish, like cod, plaice and haddock, are important sources of food. Bony fish have a swim-bladder, a gas-filled sac that helps the fish to control its depth in the water.

Fish without a swim bladder, for example sharks and skates, use their fins to control their depth. In such fish the skeleton is made of cartilage: they are cartilaginous fish. Stroking the body of a cartilaginous fish from tail to head feels like stroking sand paper, but scales covering the body of a bony fish feel smooth. The gill flap is another feature which distinguishes the two types of fish, bony fish have them but cartilaginous fish do not.

The pigeon is a domestic bird which is descended from the rock dove which makes its home on cliffs and craggy hills. In towns and cities, tall buildings substitute for cliffs (as far as the pigeons are concerned). Birds are covered with feathers which make flying possible and keep in body heat. Feathers are covered in oil which keeps out water and prevents them from getting waterlogged in the rain. The oil is produced from a special gland commonly called the 'parson's nose'. Instead of teeth, birds have a beak which is adapted differently in different species to deal with various types of food. Birds lay eggs protected by a hard shell, usually in a nest.

SUMMARY

Living things are organised into five kingdoms which represent different ways of life. Some living things are very small and only visible under the microscope, others are very large. All are adapted for their way of life.

CHECKPOINT

❶ Look at Figure 1.5A. With the help of named examples, summarise the different way of life represented by the five kingdoms.

❷ Compare the appearance of yeast, bread mould and the field mushroom.

❸ Why can fern plants live in a drier environment than mosses?

❹ Using the buttercup and beech tree as examples, explain the meaning of the words annual and perennial. State an important function of wood.

❺ (a) What kind of symmetry is shown by the body of (i) a mussel and (ii) a jellyfish?
   (b) Explain how the type of symmetry suits the way each animal spends its life.

❻ Snails and frogs live in water and on land. What prevents the animals losing too much water when on land?

❼ List the features that make humans mammals. Which ones help us to have more control over our lives and the environment than any other kind of living thing?

TOPIC 2

ECOLOGY

 FIRST THOUGHTS

2.1 **Ecosystems, habitats and communities**

Ecology is the study of how living things interact with each other and with their environment – the study of the balance of nature. The people involved in this study are called ecologists.

LOOK AT LINKS
for **species**
See Topic 1.3.

If, as ecologists, we are to carry out a scientific study we need to look at a localised part of the environment called an **ecosystem**. Ecosystems exist in great variety both on land and in water (see Figure 2.1A).

All ecosystems are made up of two parts: the **living** and the **non-living**. The non-living part of an ecosystem is called the **habitat**. It is the place where organisms live and therefore has the conditions which they need for their existence such as light, oxygen and a suitable temperature. The living part of an ecosystem is the plants and animals that inhabit the habitat and is known as the **community**. Each community is made up of many populations of plants and animals. A **population** is a group of individuals of the same **species**. Each population in the community is adapted to living in a particular habitat.

Figure 2.1A ● Ecosystems

Figure 2.1B ● Adapting to environmental conditions

(a) Periwinkles resist drying up by shutting their opercula (trap doors) when the tide goes out

(b) Limpets stick firmly to the rocks when the tide goes out and their tough shells protect them from extremes of temperature

(c) Gulls are major predators on rocky shores

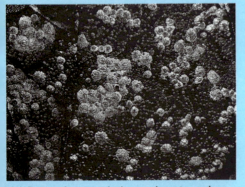

(e) Barnacles attach themselves to rocks and compete for space with each other and with seaweeds

In a sea-shore ecosystem there are many different animal and plant populations in the community. Each of these has been successful in adapting to a number of environmental conditions which occur on the seashore:

- **Tides.** Many animals and plants remain on the shore after the tide has gone out. They must be able to withstand the harsh conditions.
- **Evaporation.** Organisms have to conserve their body water otherwise they would dry up (see Figure 2.1B (a)).
- **Temperature.** Air temperature tends to fluctuate more quickly and to a greater extent than the temperature of the sea (see Figure 2.1B (b)).
- **Predators.** Shore animals have to contend with sea birds when the tide is out and with predatory fish when the tide is in (see Figure 2.1B (c) and (d)).
- **Food and space.** Seaweeds compete for space on the rocks and for light to carry out photosynthesis. Animals compete for food and for shelter (see Figure 2.1B (e)).
- **Wave action.** When the tide comes back in seaweeds and seashore animals have to withstand the battering of waves (see Figure 2.1B (f)).

(d) Predatory fish return to the shore with the tide

(f) Seaweeds are able to bend with the waves and are covered with slimy mucilage which protects them from being torn to pieces

Figure 2.1C ●

(a) Competition

(b) Predation

(c) Parasites

(d) Grazer

(e) Decomposers

Each ecosystem has a particular set of **environmental factors** to which animals and plants have to be adapted if they are to survive. These include:

- Light – which is needed by plants in order to carry out photosynthesis.
- Oxygen – for respiration.
- Carbon dioxide – for photosynthesis in plants.
- Water – as an essential constituent of cells.
- Temperature – which affects the rate at which reactions take place in cells.
- Nutrients – which are essential for growth, energy and reproduction.

The presence of other organisms within the community gives rise to the following factors:

- **Competition.** Individuals in a population of plants or animals have to compete not only with each other but with other species as well (see Figure 2.1C (a)).
- **Predation.** A predator is an animal that has to kill to feed. The prey must evade the predator if it is to survive (see Figure 2.1C (b)).
- **Parasites and disease.** A parasite usually lives on or within the host organism, using the host or the host's food supply as its own source of food (see Figure 2.1C (c)). Some diseases are caused by parasites.
- **Grazers** are animals that feed entirely upon plants (see Figure 2.1C (d)).
- **Decomposers** are animals such as **bacteria** and **fungi** which break down dead and decaying material (see Figure 2.1C (e)).
- **The effect of humans.** People affect ecosystems in a number of ways. Industry, farming, forestry and housing have all contributed to the loss or deterioration of many natural ecosystems.

Look at the two very different ecosystems in Figure 2.1D. One represents a pond the other a town or city ecosystem. In each case make a list of the environmental factors that will affect the organisms that live there.

We will see that although ecosystems may appear to be very different, their structures are based upon feeding and energy transfer and therefore follow a common pattern. All the ecosystems together make up the **biosphere**.

Figure 2.1D ● **(a) A pond ecosystem**

IT'S A FACT

The leaves of a tree differ depending on their position. Those at the top are in direct sunlight a lot of the time. These 'sun leaves' are smaller in area and are packed with chloroplasts. The 'shade leaves' found lower in the tree canopy, have a much larger area and far fewer chloroplasts. They contribute very little to photosynthesis.

LOOK AT LINKS
for **bacteria** and **fungi**
See Topic 1.5.

SUMMARY

Ecosystems are the basic unit of study in ecology. They are made up of the habitat (non-living surrounding) and the community (living plants and animals). A community consists of populations of plants and animals. Each population must be adapted to the conditions that exist in the ecosystem if it is to survive.

CHECKPOINT

❶ Match the list of terms below with the descriptions that follow.

habitat biosphere population competition community ecosystem

(a) All the ecosystems on the Earth.
(b) A group of individuals of the same species.
(c) Organisms that interact within the same ecosystem.
(d) Made up of the habitat and the community.

❷ Match the animals and plants in Column A below with their correct habitats in Column B:

Column A	Column B
Lichen	Wood
Trout	Path
Hawthorn	Rocky shore
Groundsel	Pond
Squirrel	Moorland
Heather	River
Frog	Hedge
Crab	Wall

❸ Environmental conditions differ in different habitats. In each case, choose two physical conditions, e.g. climate, soil, wave action and suggest the problems that they would pose for animals and plants living in the following habitats:
(a) Fast-flowing mountain stream.
(b) Coniferous plantation.
(c) Saltmarsh.
(d) Acid heathland.

❹ Distinguish, as clearly as possible, between the following:
(a) Ecosystem and habitat.
(b) Population and community.
(c) Competitors and predators.
(d) Grazers and decomposers.

(b) A town ecosystem

2.2 Community structure

Figure 2.2A ● Green plants trap the radiant energy of sunlight

LOOK AT LINKS
for **teeth**
See Topic 11.5.

Producers and consumers

In any community, the organisms can be classified by their method of feeding. Basically, either they can make their own food or they can't.

Autotrophs are able to make their own food from simple substances such as carbon dioxide and water. Green plants are autotrophs. They use light as a source of energy to make sugars. Many bacteria are also autotrophs. Some obtain their energy from light and others from simple chemical reactions. Since autotrophs ultimately provide food for all the other members of the community, we call them **producers**. (See Figure 2.2A).

Heterotrophs cannot make their own food so they have to eat it. For this reason they are called **consumers**:

- **Primary consumers** are herbivores. They eat producers (see Figure 2.1C(d)).
- **Secondary consumers** are carnivores. They eat herbivores (see Figure 2.1C(b)).
- **Tertiary consumers** are carnivores that eat secondary consumers.

Each category of feeding is known as a **trophic level** (from the Greek *trophos* which means 'a feeder'). *But how do we know what an animal feeds upon?* We can get a good idea by looking at **teeth** and other feeding structures. We can also investigate contents of the animal's gut or study its feeding behaviour.

Decomposers

Some consumers obtain their energy from dead and decaying material and are called decomposers. These are mainly bacteria and fungi and they perform an important role in the cycling of nutrients. As they break down dead material, they make it available as nutrients for the growth of new plants. (See Figure 2.2B.)

Figure 2.2B ● Decomposition

Food chains and food webs

Feeding methods are a useful way of showing the relationships that exist in a community. We can highlight these by means of **food chains**. These show the passage of food (and therefore energy) from one organism to another. Food chains always start with a producer. The food then passes

IT'S A FACT

Animal plankton contains the immature larvae of many sea-shore animals. The larvae of crabs look very different from the adults. They also feed upon different food and are carried by currents to eventually settle on new parts of the shore.

IT *Ecology Foodweb* (program)

Use the program to look at a woodland and a pond foodweb. You can then change the balance of the ecosystem. Finish off with the quiz.

to a primary consumer, next to a secondary consumer and so on. The number of links (producer and consumers) in a food chain may vary, but is seldom more than five. (See Figure 2.2C.)

Food chains can show the feeding in any community. The arrows represent the transfer of food between different trophic levels. So it is easy to see the particular role that each organism plays in the community. The role of an organism is known as its **niche**. The niche can be identified in other ways apart from feeding, for example it may depend upon where the organism makes its home or how it reproduces.

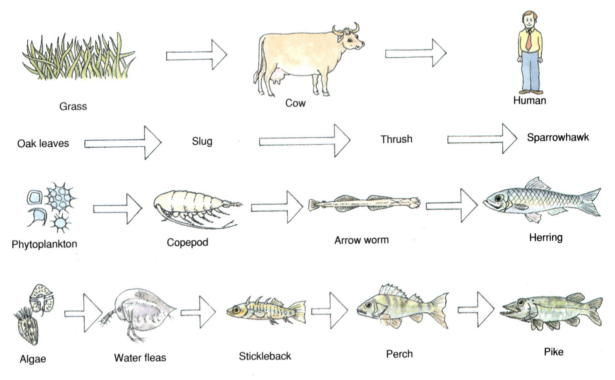

Figure 2.2C ● Food chains

Several food chains exist in a community and these will link up to form a **food web** (see Figure 2.2D). This is because the diet of an animal consists of a number of different species. A food web represents a more complete picture of the feeding relationships in a community.

SUMMARY

Producers are able to make their own food. They include green plants and some bacteria. Producers provide food for primary consumers (herbivores). These in turn provide food for secondary consumers (carnivores). Decomposers feed upon dead and decaying material. These feeding relationships can be shown by the use of food chains and food webs. The role that an animal of plant plays in a community is known as its niche.

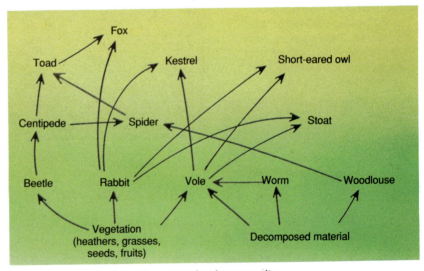

Figure 2.2D ● Food web for a moorland community

❶ Study the food web and answer the questions which follow.

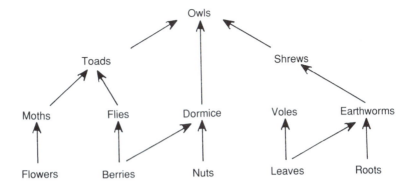

(a) From the diagram name (i) two primary consumers, (ii) one secondary consumer.
(b) Construct, using the food web above, two different food chains, each with four links. The arrows should point from eaten to eater.
(c) Explain why every food chain must begin with a green plant.
(d) If all the owls were killed, what would happen to the size of the populations of (i) dormice, (ii) earthworms?
(e) In the food web shown, the moths depend upon the flowers and the dormice depend upon the hazel tree for food. Give one example to show how (i) the flowers may depend on the moths, (ii) the hazel tree may depend upon the dormice.

(WJEC)

2.3 🌍 Ecological pyramids

Food chains and food webs can describe the feeding relationships that occur in a community. However they give us no information about the numbers of individuals involved. It takes many plants to support a few herbivores. Similarly there must be far more prey than there are predators. (See Figure 2.3A.)

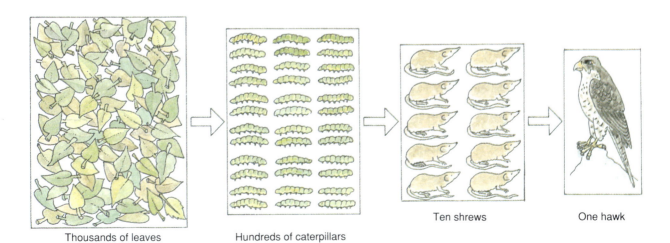

Thousands of leaves   Hundreds of caterpillars   Ten shrews   One hawk

Figure 2.3A ● More plants than herbivores – more prey than predators

39

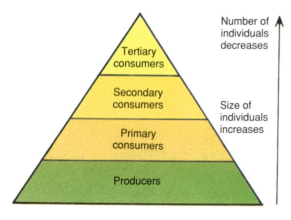

Figure 2.3B ● Pyramid of numbers

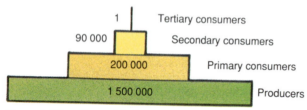

Figure 2.3C ● Pyramid of numbers for a grassland community in 0.1 hectare

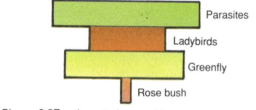

Figure 2.3D ● Oak tree pyramid

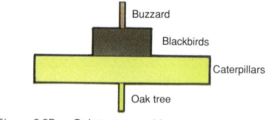

Figure 2.3E ● Inverted pyramid

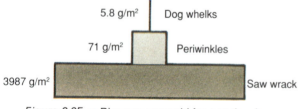

Figure 2.3F ● Biomass pyramid for a rocky shore

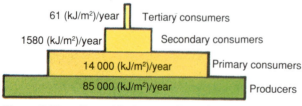

Figure 2.3G ● Pyramid of energy for a river

Pyramids of number

Pyramids of number give information about the numbers of individuals in each trophic level. They are drawn up by counting the number of individuals in a certain area, say one square metre. These numbers are then plotted like bar charts but horizontally (see Figure 2.3B). The producers are placed at the base of the pyramid. Above them are placed the primary consumers. The secondary consumers are placed on top of these and so on. Note that as we go up the pyramid:

• The numbers of individuals decrease.
• The size of each individual increases.

Figure 2.3C shows a pyramid of numbers drawn from some results taken from a grassland community.

A problem with this method is that it fails to take into account the size of the organisms at each trophic level. For instance one oak tree will support many more herbivores than one grass plant. Figure 2.3D shows a pyramid of this kind. Pyramids appear top heavy if we include parasites since many parasites will feed on the secondary consumers. Such a pyramid is said to be inverted as shown in Figure 2.3E.

Pyramids of biomass

One way of overcoming this problem of the size of organisms, is to chart the **biomass** at each trophic level. A representative sample of the organisms at each trophic level is weighed. The mass is then multiplied by the estimated number in the community. In practice the dry mass is used since fresh mass varies so much with water content. The sample is heated in an oven at 110 °C until there is no further change in mass (this is often neither practicable nor desirable if it involves destroying animals and plants in the process). Figure 2.3F shows a biomass pyramid for a rocky shore community.

Biomass pyramids also have their drawbacks:

• Some organisms grow at a much faster rate than others, for example grass does not have a large biomass, but it carries on growing at a very fast rate.
• The biomass of an individual can vary during the year – a beech tree will have a much greater biomass in June than it has in November. *Why do you think this is?*

Pyramids of energy

Pyramids of energy provide the most accurate representation of the feeding relationships in a community. They give information about the amount of new tissue at each trophic level over a certain period of time. They make it possible to compare different communities accurately. They can also be used to

Lake Web and Bio-wood
(program)

In an environment such as a lake or a wood, survival depends on making decisions about what to eat, what to ignore and what to escape from. Try surviving in *Lake Web* or *Bio-wood*.

compare the production of food by different methods of farming. Figure 2.3G shows an energy pyramid for a river.

Shortening the food chain

Look at the pyramid of numbers in Figure 2.3H. It shows the estimated numbers of individuals that could be supported on 1000 tonnes of grass in ten months.

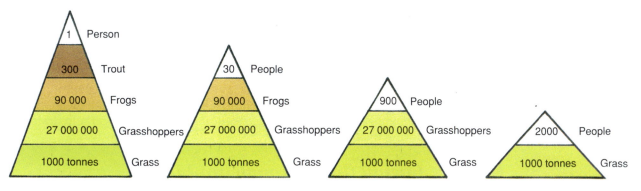

Figure 2.3H ● Shortening the food chain

SUMMARY

Pyramids of number tell us about the numbers of individuals at each trophic level. Biomass pyramids give information about the mass of material present. More accurate are energy pyramids that show the transfer of energy through a community. A short food chain can support far more people than one with many links in the chain.

If the food chain was shortened and people decided to eat frogs instead of trout, 30 people could be supported in this way (assuming that a person could get by on 10 frogs a day). *What would happen if the food chain were shortened still further and people ate grasshoppers?* Assuming that 100 grasshoppers a day would keep one person satisfied, then 900 people could be supported. If we now eliminate the grasshoppers from the food chain, then by feeding on grass alone, 2000 people could live on land that could only support one person when his or her diet was trout.

What message is there for us from this exercise? It tells us that a vegetarian diet can sustain far more people. By eliminating links in the food chain, more individuals at the end of the food chain can be fed. *Why do you think that the diets of people in underdeveloped countries tend to be made up mostly of plants?* In the Western developed countries people have a varied diet including large amounts of poultry, fish, lamb, beef and pork. *What does this tell you about the economies of these countries?*

With the huge increase in the world's population, what is likely to happen to the price of meat and what may be the inevitable change in human diets in the future?

CHECKPOINT

1 The diagram below shows a food web.

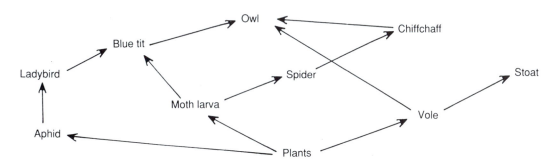

(a) Name the ultimate source of energy for all the organisms in this food web.
(b) From this food web name one example of a secondary consumer.
(c) Construct a food chain which includes examples of animals from four different feeding levels in this food web.
(d) If all the spiders were killed by disease, suggest the effects this could have on the food web shown.

2.4 Energy flow through ecosystems

We have talked about the passage of food between trophic levels, but it is far more accurate to describe the passage or flow of energy through an ecosystem. Energy can be converted. For example, the radiant energy of sunlight is converted into the chemical bond energy in carbohydrates during photosynthesis.

Energy flow through producers

LOOK AT LINKS
for **energy conversion** and **absorption of light** by chlorophyll
See Topic 11.1

The energy in all ecosystems has its origin in sunlight. Green plants (and some bacteria) are the only organisms that can use this energy to convert carbon dioxide and water into carbohydrates. Photosynthesis is far less efficient than we may imagine, because much of the sunlight that falls on to plants, is not absorbed. The light may not be absorbed because:
• much of it is reflected from the surface of the leaf,
• a large amount passes straight through the leaf,
• only light falling in the range of wavelengths from 380 nm (red) to 720 nm (blue) can be absorbed by chlorophyll.
The overall efficiency of energy conversion in photosynthesis is less than 8%.

What does the plant do with this energy? The plant needs energy to drive its own chemical reactions during respiration. The remainder of the energy is used in growth and repair, which increase the biomass of the plant. This represents the food and energy that is available to primary consumers and so can be transferred up to a higher trophic level. Some energy will be transferred from the plant when leaves are shed, when fruits and seeds are dispersed and eventually when the plant dies. Decomposers will benefit from this since they will gain energy from the dead plant tissues. Figure 2.4A summarises the energy flow through a green plant.

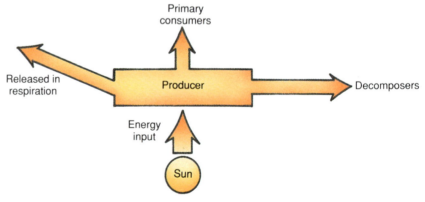

Figure 2.4A ● Energy flow through a green plant

Energy flow through consumers

Transfer of energy from producers to primary consumers, that is from plants to herbivores also involves a loss of energy. It is estimated that for every 100 g of plant material eaten, only 10 g ends up as herbivore biomass. This represents an effective loss of energy between trophic levels of approximately 90%. The reasons for the inefficiency of this transfer of energy are:

- Some of the plant material passes out of the body of the herbivore as faeces without being digested.
- A lot of energy is used in respiration by the herbivore.
- Some energy passes to decomposers in dead remains.

Similar losses of energy from the ecosystem occur between each subsequent trophic level. Carnivores are able to achieve a 20% efficiency. That is to say, 20% of the herbivore biomass ends up as carnivore biomass. This is possible because proteins are more efficiently digested than are carbohydrates.

● **The energy budget of a primary consumer**

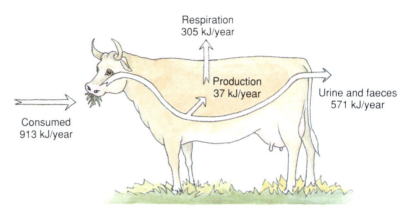

Respiration
305 kJ/year

Production
37 kJ/year

Urine and faeces
571 kJ/year

Consumed
913 kJ/year

Figure 2.4B ● The energy intake and output of a cow

Figure 2.4B shows the energy intake and output of a cow. Of the energy in the grass which a cow eats, over half is passed out of the body in faeces. Some goes to increase cow biomass (production). A lot of energy is lost from the body as heat produced in respiration and in the urine. The energy budget of the cow can be summarised in the following equation:

$$\text{Energy intake} = \text{Energy transfer in respiration} + \text{Energy transfer in production} + \text{Energy in urine} + \text{Energy in faeces}$$

SUMMARY

Energy enters ecosystems as sunlight and leaves it as heat. Only a fraction of the light falling on a leaf will end up as new plant biomass. There is great loss of energy from the ecosystem as it flows between trophic levels.

CHECKPOINT

❶ Look at Figure 2.4B.
  (a) Would you say that the cow is efficient in converting grass into biomass (production)? Explain your answer.
  (b) What percentage of the energy intake is present in faeces and urine? Faeces provide a rich food source for a number of organisms. Name some of them.
  (c) What percentage of the energy intake is used up in respiration?
  (d) Cows spend a great deal of their time grazing. In view of your other answers, why do you think this is?
  (e) How do you think the energy budget of a secondary consumer would compare with the example here?

❷ Look at the figure below and answer the following questions (the size of the arrows represents the relative amounts of energy passed on at each stage).

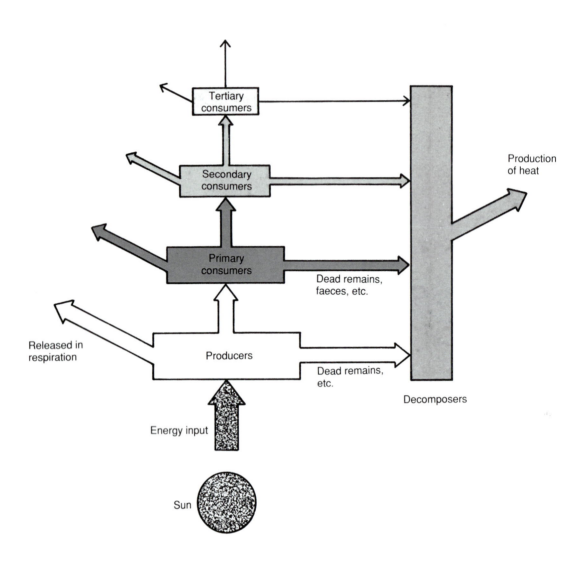

(a) What is the original source of energy for the ecosystem?
(b) How is energy eventually lost from the ecosystem?
(c) Would you say that the transfer of energy through the ecosystem was a cyclical process where the energy is reused or a flow in one direction? Give reasons for your choice.
(d) In what ways is energy lost from (i) the producers and (ii) the consumers?
(e) Predict what might happen to the biomass of the primary consumers when the rate of photosynthesis increases.
(f) How might an increase in the number of primary consumers affect
    (i) the number of producers,
    (ii) the number of secondary consumers?

❸ From every one square metre of grass it eats, a cow obtains 3000 kJ of energy. It uses 100 kJ for growth, 1000 kJ are lost as heat and 1900 kJ are lost in faeces.
(a) What percentage of the energy in one square metre of grass
    (i) is used in growth,
    (ii) passes through the gut and is not absorbed?
(b) If beef has an energy value of 12 kJ per gram, how many square metres of grass are needed to produce 100 g of beef?

2.5 🌍 Decomposition and cycles

Decomposition

Figure 2.5A ● Animal decomposition

Eventually all plants and animals die. You may have seen the remains of a blackbird caught by the family cat, a hedgehog run over and left at the side of the road or a dead gull washed up on the shore (see Figure 2.5A). In autumn dead leaves form a carpet on the ground (see Figure 2.5B). All this dead material eventually disappears. *Where does it go?*

Figure 2.5B ● Plant decomposition

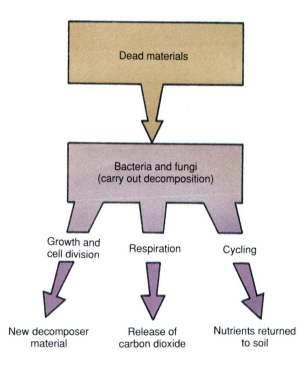

Figure 2.5C ● Activities of decomposers

IT'S A FACT

In the UK decomposition is quicker in the summer than in the winter. Maggots are able to decompose up to 80% of small mammal and bird carcasses in the summer. In the winter, there are few maggots around so decomposition is mainly carried out by microorganisms.

There are many animals that feed upon dead remains, we call them **scavengers**. Crows, maggots, woodlice and vultures all make a living by feeding upon dead material. However scavengers cannot account for the removal of all dead material and even without them it would slowly disappear anyway due to the action of decomposers. These are microorganisms, mainly fungi and bacteria, that break down dead material and also make food go rotten. They release enzymes which digest the dead remains in just the same way as food in your gut is digested. The simple products of digestion are absorbed by the decomposers and used for their own growth. (See Figure 2.5C.)

Cycles

LOOK AT LINKS
for **the carbon cycle**
See Topic 2.7.

LOOK AT LINKS
for **the nitrogen cycle**
See Topic 2.6.

IT
Cycles in Nature
(program)

Investigate the three cycles, water, nitrogen and carbon, using the program.

SUMMARY

Decomposers break down dead remains and release nutrients into the ecosystem. These nutrients are absorbed by plants and passed along food chains. Nutrients such as carbon, nitrogen and sulphur are continuously being cycled through ecosystems.

Not all of the products of decomposition are used to make new fungi and bacteria: most are returned to the soil as nutrients. These nutrients are then made available to plants and in turn to animals for growth. This cycling of nutrients is vital if life is to continue. Most living matter (95%) is made up of just six elements: carbon, hydrogen, nitrogen, oxygen, phosphorus and sulphur.

A constant supply of nutrients containing these elements is essential if living organisms are to continue to make important compounds like proteins, fats and carbohydrates. Figure 2.5D shows the pattern of cycling nutrients which takes place in all ecosystems, on land, in freshwater and in the sea.

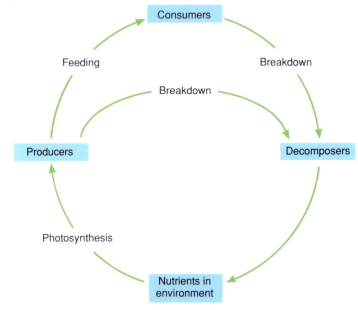

Figure 2.5D ● Cycling of nutrients in an ecosystem

2.6 The nitrogen cycle

LOOK AT LINKS
The importance of proteins as food is described in Topic 11.3.

Figure 2.6A ● Nodules containing nitrogen fixing bacteria on the roots of a legume

Nitrogen is an essential element in **proteins**. Some plants have nodules on their roots which contain nitrogen-fixing bacteria. These bacteria **fix** gaseous nitrogen, that is, convert it into nitrogen compounds. From these nitrogen compounds, the plants can synthesise proteins. Members of the legume family, such as peas, beans and clover, have nitrogen-fixing bacteria.

Plants other than legumes synthesise proteins from nitrates. *How do nitrates get into the soil?* Nitrogen and oxygen combine in the atmosphere during lightning storms and in the engines of motor vehicles during combustion. They form nitrogen oxides (compounds of nitrogen and oxygen). These gases react with water to form nitric acid. Rain showers bring nitric acid out of the atmosphere and wash it into the ground, where it reacts with minerals to form nitrates. Plants take in these nitrates through their roots. They use them to synthesise proteins. Animals obtain the proteins they need by eating plants or by eating the flesh of other animals.

In the excreta of animals and the decay products of animals and plants, **ammonium salts** are present. Nitrifying bacteria in the soil convert ammonium salts into nitrates. Both nitrates and ammonium salts can be removed from the soil by **denitrifying bacteria**, which convert the compounds into nitrogen. To make the soil more fertile, farmers add both nitrates and ammonium salts as fertilisers. The balance of processes which put nitrogen into the air and processes which remove nitrogen from the air is called the **nitrogen cycle** (see Figure 2.6B).

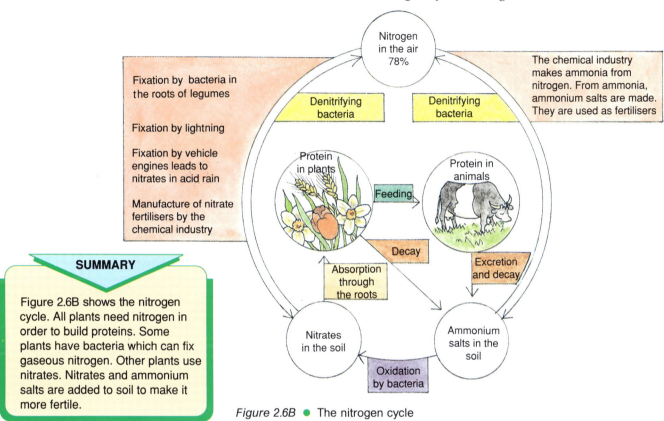

SUMMARY

Figure 2.6B shows the nitrogen cycle. All plants need nitrogen in order to build proteins. Some plants have bacteria which can fix gaseous nitrogen. Other plants use nitrates. Nitrates and ammonium salts are added to soil to make it more fertile.

Figure 2.6B ● The nitrogen cycle

2.7 **The carbon cycle**

FIRST THOUGHTS

The percentage by volume of carbon dioxide in clean, dry air is only 0.03%. Perhaps you think this makes carbon dioxide sound rather unimportant. Through studying this section, you may change your mind about the importance of carbon dioxide!

Plants need carbon dioxide, and animals, including ourselves, need plants. Plants take in carbon dioxide through their leaves and water through their roots. They use these compounds to **synthesise** (build) sugars. The reaction is called **photosynthesis** (photo means light). It takes place in green leaves in the presence of sunlight. Oxygen is formed in photosynthesis.

Photosynthesis (in plants)

catalysed by chlorophyll in green leaves

Sunlight + Carbon dioxide + Water ➔ Glucose + Oxygen
(a sugar)

The energy of sunlight is converted into the energy of the chemical bonds in glucose.

Animals eat foods which contain starches and sugars. They inhale (breathe in) air. Inhaled air dissolves in the blood supply to the lungs. In the cells, some of the oxygen in the dissolved air oxidises sugars to carbon dioxide and water and energy is released. This process is called **respiration**. Plants also respire to obtain energy.

LOOK AT LINKS
For a fuller account of photosynthesis and respiration.
See Topics 11.1–12.1.

LOOK AT LINKS

A reaction like photosynthesis, in which energy is taken in, is called an endothermic reaction. A reaction like respiration, in which energy is given out, is called an exothermic reaction.
See Topics 11.1 and 12.2.

Respiration (in plants and animals)

Glucose + Oxygen → Carbon dioxide + Water + Energy

The processes which take carbon dioxide from the air and those which put carbon dioxide into the air are balanced so that the percentage of carbon dioxide in the air stays at 0.03%. This balance is called the **carbon cycle** (see Figure 2.7A).

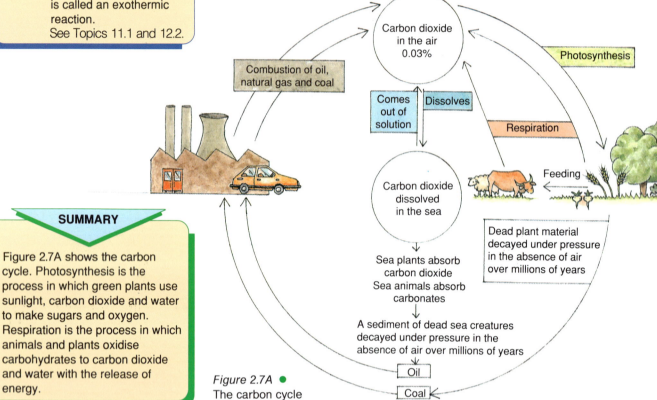

Figure 2.7A ●
The carbon cycle

SUMMARY

Figure 2.7A shows the carbon cycle. Photosynthesis is the process in which green plants use sunlight, carbon dioxide and water to make sugars and oxygen. Respiration is the process in which animals and plants oxidise carbohydrates to carbon dioxide and water with the release of energy.

CHECKPOINT

❶ *Process A* Nitrogen-fixing bacteria convert nitrogen gas into nitrates.

   Process B Nitrifying bacteria use oxygen to convert ammonium compounds (from decaying plant and animal matter) into nitrates.

   Process C Denitrifying bacteria turn nitrates into nitrogen.

   (a) Say where *Process A* takes place.
   (b) Say what effect the presence of air in the soil will have on *Process B*.
   (c) Say what effect waterlogged soil which lacks air will have on *Process C*.
   (d) Explain why plants grow well in well-drained, aerated soil.
   (e) A farmer wants to grow a good crop of wheat without using a fertiliser. What could he plant in the field the previous year to ensure a good crop?
   (f) Explain why garden manure and compost fertilise the soil.

❷ Explain why nitrogen is used in (a) food packaging (b) oil tankers (c) hospitals and (d) food storage.

❸ Name two processes which add carbon dioxide to the atmosphere.

❹ Suggest a place where you would expect the percentage of carbon dioxide in the atmosphere to be lower than average.

❺ Suggest two places where you would expect the percentage of carbon dioxide in the atmosphere to be higher than average.

TOPIC 3 POPULATIONS

3.1 Population growth

Figure 3.1A ● Populations come in all shapes and sizes

A **population** is a group of individuals of the same species living in a particular habitat (see Figure 3.1A). The following are all populations:
- aphids on a sycamore tree,
- barnacles on a rocky shore,
- bacteria growing on a food medium in a petri dish,
- duckweed on the surface of a pond,
- a shoal of herring in the sea.

One of the reasons ecologists study populations is to gain information about the rate at which their numbers change. Such studies are of practical use if the species is a pest or if it causes disease. For instance, population studies of the locust are important since a swarm of locusts can devour a harvest. Studies of mosquito populations can help scientists to control the spread of malaria. Investigations into the human population explosion are very important for the future of mankind.

Why do individuals live in populations? There may be a number of reasons:
- The habitat provides food, shelter and other vital factors.
- Individuals come together to breed.
- Individuals may gain more protection as a population, for example a shoal of fish or a colony of gulls.

However, overcrowding may result in competition within the population. Individuals may compete for food, space, light and other resources. Inevitably some will not survive this competition and numbers may decline.

Think about a plant or animal species colonising a new area. A few individuals enter the new habitat. Provided there are no predators and that there is no shortage of food, the individuals will thrive and reproduce. At first their numbers will increase slowly. Later, as the population grows, so does the rate of increase (2 becomes 4 then 8, 16, 32, 64, 128, 256, 512 and so on). This sort of growth where each generation is double the size of the previous one is the maximum rate of growth and is called exponential growth. Exponential growth can only take place under ideal environmental conditions. It cannot go on for long. Eventually limiting factors such as lack of food and overcrowding cause the rate of increase to slow and eventually to level off. After this the population remains constant as long as the rate at which individuals are born is equal to the rate at which individuals die. (See Figure 3.1B.)

A population growth curve like the one in Figure 3.1B can quite easily be produced by growing some yeast cells in sugar solution. Every half hour take up a small sample with a pipette and look at a drop of it under the microscope. By counting the number of yeast cells in the microscope's field of view, you can follow the development of the yeast population. The exercise is more accurate if a counting slide is used. The slide is divided into squares and the number of yeast cells in each square is counted at the same magnification each time.

Lag phase	Rapid growth	Growth slowing	Population constant

Number of individuals

Figure 3.1B ● Population growth curve

3.2 Checks on population increase

So what prevents populations increasing? It could be lack of food, overcrowding, or perhaps the build up of poisonous waste products produced by the individuals of the population itself. A number of environmental factors limit the growth of populations. These environmental factors act as natural checks which prevent the population from growing too large. Among them are:

- **Food, water and oxygen**. These affect the growth of most populations. Individuals in the population compete for them and if they are in short supply then some individuals will not survive and the population will level off or decline.
- **Light**. Light is especially necessary for green plants which carry out photosynthesis. Plants compete for light in areas of dense vegetation. (See Figure 3.2A.)

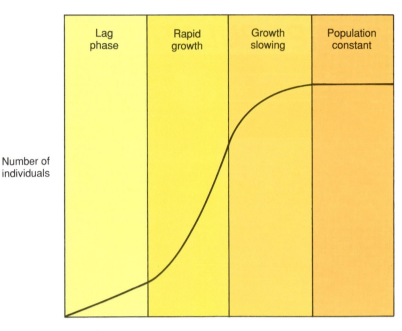

Figure 3.2A ● Competing for light

Old Park Farm (program)

What is happening to the bat population? Use the program to investigate bats' roosting and feeding areas and to find out about the ecological problems which threaten the bat population.

- **Lack of shelter**. Animals compete for shelter in order to avoid predators or harsh climate (see **Climate** below).
- **Overcrowding**. Overcrowding often leads to unhygienic conditions which favour the spread of disease. It also results in stress.
- **Predators**. These are natural checks on population increase. The size of a predator population depends on the numbers of prey. An increase in the number of predators means that more prey will be caught and the number of prey will fall. The predators' food supply is reduced and this leads in turn to a drop in the number of predators.
- **Accumulation of toxic wastes**. Natural waste products such as carbon dioxide and nitrogenous waste such as ammonia and urea become toxic if they are allowed to build up. Obviously, as the population increases, more waste is produced and this can restrict further growth.
- **Disease**. Disease can spread very quickly through large populations. Monocultures of arable crops are susceptible to disease-causing organisms.
- **Climate**. Harsh weather conditions can reduce populations. Animals and plants can die if the temperature is very high or very low. Drought and flood, storms and gales all take their toll on populations.

Sometimes populations decrease dramatically in a short space of time. They do not level off but crash dramatically. This may occur because the population runs out of food or if it has been overcome by disease. The danger is often a temporary or a seasonal affair. For example, the numbers of aphids on a rose bush decline very quickly if ladybirds start to prey on them.

3.3 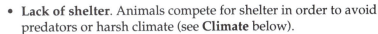 Population size

Golden Eagle (program)

As chief warden of a Scottish Wildlife reserve, you have been asked to double the Golden Eagle population. Use the program to try this. You will find it helpful to make notes of things you try.

Birth rates as well as death rates will determine the size of a population. Some species are capable of phenomenal increases in population. Many insects lay hundreds of eggs, frogs lay thousands and fish produce millions.

Few animals stay in one place, they tend to move into and out of the habitat in which the population lives. We call the movement of individuals into a population **immigration** and the movement of individuals out of a population **emigration** (see Figure 3.3A). In a stable population

$$\text{Birth rate} + \text{Immigration rate} = \text{Death rate} + \text{Emigration rate}$$
$$(B) \qquad\qquad (I) \qquad\qquad (D) \qquad\qquad (E)$$

If the population is increasing then

$$B + I \text{ will be more than } D + E$$

If a population is declining then

$$D + E \text{ will be more than } B + I$$

Estimating the size of a population

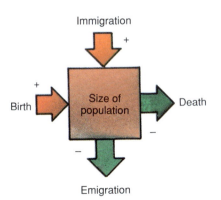

Ecologists often need to estimate the size of a particular population when they are carrying out a scientific study. In the case of plants there is usually little difficulty. A quadrat is laid down on the vegetation (a quadrat is a sampling device which consists of a metal or wooden

Figure 3.3A ● Factors affecting the size of a population

Flat stone

Stick support

Yoghurt carton sunk into soil

Holes for drainage

Ground slopes away for drainage

Figure 3.3B ● A pitfall trap is one way of trapping small animals prior to counting

**Ecosoft /
Junior Ecosoft**
(program)

If you have collected data from fieldwork or laboratory activities, you can put them into the programs *Ecosoft* or *Junior Ecosoft*. These programs help you to sort out and present the data attractively and clearly.

rectangle, usually one metre square). The number of individuals of a particular plant occurring within the quadrat is then counted. The number of quadrats needed to cover the habitat is estimated and multiplied by the number of plants per quadrat to give the total population.

Sampling animals is more difficult. Animals tend to move about – they certainly do not stay in a quadrat. Also many animals only come out at night so they have to be trapped before they can be counted (see Figure 3.3B). Ecologists use a method called mark, release, recapture to estimate many animal populations. It works like this:

- Collect a number of animals from the habitat.
- Mark the captured animals. Ecologists try to make the mark in some way that is harmless and does not make the animals conspicuous to predators, for example, a small dot in a dull colour on a beetle's carapace.
- Release the marked animals into the same habitat they came from and leave them for a day or two to mix freely back into the rest of the population.
- Make a second collection of a similar number of animals from the same habitat. Note the number of marked individuals in this second sample.
- The total population of each species can be estimated by using the following

$$\text{Total population} = \frac{\text{Number in first catch} \times \text{Number in second catch}}{\text{Number of marked individuals in second catch}}$$

Worked example 20 ground beetles were caught in a pitfall. They were each marked with a small dot of paint and when it was dry, they were released. After two days a second sample of 18 beetles was caught. Of these, six were marked individuals from the first occasion. Estimate the total population of beetles.

Solution
Number in first catch = 20
Number in second catch = 18
Number of marked individuals in second catch = 6
$$\text{Total population} = \frac{20 \times 18}{6} = 60$$

═══════════════════════════ **CHECKPOINT** ═══════════════════════════

❶ By using a net to sweep long grass, 90 froghoppers (small insects) were collected. Each was marked and released back into the grass. The next day a second sweep produced 80 froghoppers and of these six had a mark on them. Estimate the total population of froghoppers in the field.

❷ It is important that the handling and marking of animals causes them no harm. Explain why it is also important that:
(a) the mark does not rub off too soon,
(b) the mark does not make the animal conspicuous.

❸ On the seashore, periwinkles are excellent animals for this sort of exercise. They are easy to find under rocks and seaweed and are most active when the tide is in. A sample of 18 periwinkles was marked and released and after a period of four days, 26 were collected. Of these, 12 were marked individuals. Estimate the periwinkle population for the stretch of shoreline.

3.4 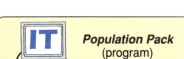 Human populations

Population Pack
(program)

The *Population Pack* lets you investigate human population growth, using various information and models.

The human population is presently growing at an alarming rate, but it has not always been so. For thousands of years the rate of population increase was fairly slow. The huge increase that we see today has really only occurred over the last 300 years (see Figure 3.4A). We can see from Figure 3.4B that this rate of increase is far greater in so-called under-developed countries than in Europe and North America. *Why is this?*

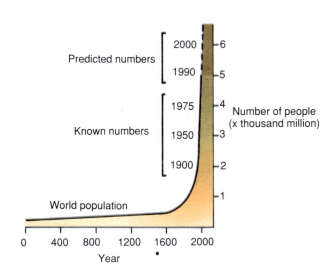

Figure 3.4A ● World population growth based on present

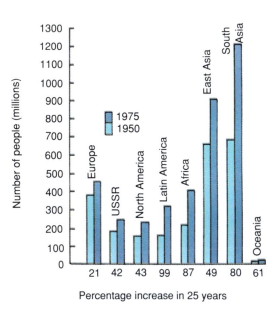

Figure 3.4B ● Where the population is growing

There are many factors which contribute to this population increase:
- Improved agriculture has meant greater food production.
- Public health has improved, for instance there are cleaner water supplies and improved sanitation.
- Medical provision such as vaccination has improved and new drugs have been introduced to combat disease.
- Advances have been made in the control of disease-causing organisms.

All these factors contribute to lower infant mortality and increased life expectancy. The average life expectancy has increased, in Europe and North America, to 68 for men and 73 for women. In India, the figures have for a long time been less than half these ages because of the high death rate among infants. However, with improved health and living conditions, the average life expectancy in India has risen to 56.

There are unforeseeable population checks, such as famine, floods, war, earthquakes and other natural and man-made disasters. In Bangladesh between 1970 and 1975, famine, floods and fighting resulted in a yearly death rate of 30 per 1000. The rapid spread of Acquired Immune Deficiency Syndrome (AIDS) particularly in developing countries, could not have been predicted a few years ago. However the overall trend is one of rapid increase in the world's population. Estimates suggest that at the current rate, it could double in the next 30 years. *But the Earth has limited resources and limited space, so what are we to do?* Clearly the biggest objective must be to reduce the birth rate. This will involve education in under-developed countries and may also challenge the religious and moral principles of many.

CHECKPOINT

The rate of population growth is influenced by the proportion of young people in the population, in particular by women of child-bearing age. The figure below shows population pyramids for India and the USA. The base of the pyramid shows the percentage of children below the age of five years in each population. The oldest people appear at the top of the pyramids.

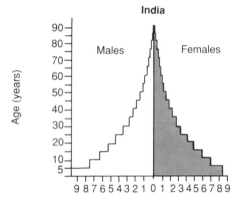

India

Percentage of the population in each age group
(For India 1% = approximately 4 million people)

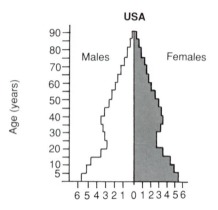

USA

Percentage of the population in each age group
(For USA 1% = approximately 2 million people)

❶ (a) (i) Which country has the higher number of children below the age of 5 years?
    (ii) Which country has the higher number of old people over 80 years?
(b) Which country has (i) the higher birth rate (ii) the lower death rate?
(c) Explain your answers to (a) and (b) above.

❷ (a) What forecast can you make about the numbers of child-bearing women in each country in 15 years time?
(b) What will happen to the birth rate in each country as a result?

❸ (a) Which of the population pyramids of these two countries most resembles the pattern found in Britain?
(b) What is likely to happen to the number of adult workers in relation to elderly dependents as the average age of the population increases?
(c) What implications does this have for (i) health and social services and (ii) employment in this country?

3.5 Competition

We all have an idea of what is meant by competition. In a race, all the competitors strive as hard as they can to win, but at the end there is only one winner. In nature, the same thing happens. Individuals compete for resources that are in short supply. These may include food, space, light and other things that are vital for life. Clearly only those individuals able to compete successfully will survive so competition restricts the size to which a population will grow. Competition between individuals is of two types:

• Competition between individuals of the same species.
• Competition between individuals of different species.

● Competition between individuals of the same species

This sort of competition arises from the fact that plants and animals tend to produce far more offspring than the habitat can support. They do this to ensure the survival of the species and to enable the population to colonise new areas. As a result, there will be competition between individuals and only those that are the best adapted will survive to breed. They will be the ones who pass on their genes to the next generation. This competition is called 'survival of the fittest'. By this we mean that only as the result of a 'struggle for existence' do the best adapted individuals survive to breed and pass on their characteristics to the next generation. Competition will therefore contribute to natural selection. (See Figure 3.5A.)

LOOK AT LINKS
for **natural selection**
See Topic 21.

Figure 3.5A ● Competition between individuals of the same species

● Competition between individuals of different species

There will be competition for scarce resources between the different species in a community. Weeds compete successfully with crops. They germinate rapidly and grow quickly to establish themselves before the crop matures. They occupy space and take light, water and minerals from the soil, which could have been used by the crop. By the time the crop plants have grown to their full height, the weeds will already have flowered and set seed, so ensuring their survival.

Animals of different species quite often compete for food. For instance predators like owls and weasels compete for the same prey, for example shrews. Different species of fly larvae will compete for the limited food available within a cow pat. Animals also compete for territory, for shelter and for nesting sites. Many different species are able to co-exist in a particular habitat because they occupy different **niches**.

Competition is most intense when different species attempt to fill the same niche. For example, *Elminius*, the Australian barnacle arrived on the UK shores during the Second World War when it hitched a ride on the

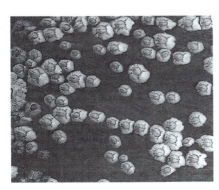

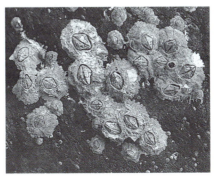

Figure 3.5B ● (a) *Elminius*   (b) *Semibalanus*   (c) *Chthamalus*

LOOK AT LINKS
for **niches**
See Topic 2.2.

hulls of ships. There are two common native species of barnacle, *Semibalanus* and *Chthamalus* (see Figure 3.5B). In many situations, *Elminius* is able to 'out compete' both *Semibalanus* and *Chthamalus*, the reasons for this include:

- *Elminius* is more tolerant to low salinity.
- *Elminius* can withstand lower temperatures than *Chthamalus*.
- *Elminius* can withstand higher temperatures than *Semibalanus*.
- The feeding rate and rate of growth of *Elminius* are faster than those of *Semibalanus* and *Chthamalus*.

The ability of *Elminius* to fill its niche at the expense of the two native species of barnacle, demonstrates **competitive exclusion**. Competitive exclusion means that two species cannot co-exist if they both have the same niche.

Succession

What was it like when plants first colonised a habitat? We can get a good idea if we look at an area of ground that has been disturbed, perhaps by developers, by farming or by fire. Within a few weeks some weed and grass species will have grown. These species are known as **pioneers**, since they are the first to colonise. With the passage of time the pioneer plants may be replaced by a wider range of taller herb species. These may in turn be replaced by other more competitive scrub vegetation. This gradual change in the composition of the species is known as **succession**.

We can see succession taking place if arable land is neglected or on a smaller scale, if we do not mow the lawn. Other plant species soon appear: in the lawn, daisies and dandelions, on arable land, poppies and thistles. In most lowland areas of Britain the process of succession would eventually end as deciduous woodland if we were to let it, but it would take over 150 years. By then, the numbers of different species would have increased and the community as a whole would have become more complex and more stable. This sort of community changes very little and is known as a **climax community** (see Figure 3.5C). Other examples of climax communities include tropical rain forests and coral reefs. They contain a huge variety of different species so it is vital that we conserve them by protecting them from destruction.

TRY THIS

Why not persuade your teacher or your parents to allow a small area of lawn at school or at home to remain uncut? You will be surprised how many different plants appear. These will flower and attract insects and soon you will have provided a refuge for wildlife right on your doorstep!

SUMMARY

Competition exists both within a single species and between different species. Only those animals and plants that are the best adapted will survive this competition to fill a niche, others will be excluded. The sequence of changes in plants, which ends at a climax community is called succession.

Figure 3.5C ● A climax community - deciduous woodland

3.6 Predators and prey

Predators are animals that catch and kill other animals called prey. Thus predation has a major effect upon the number of the prey population. It is also true to say that the prey will have an effect upon the predator population. If the prey is scarce then some of the predators will starve.

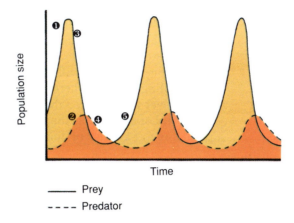

Figure 3.6A ● Predator–prey relationship

Figure 3.6A shows the relationship between the numbers of predators and prey. The relationship is self-regulating as we shall see. Notice that the since predators tend to reproduce more slowly, the predator cycle is smaller than the prey cycle and also lags behind it.

The events taking place in the predator–prey cycle are as follows:

❶ If conditions such as food supply are good, then the prey will breed and increase in number.

❷ Since there is now more food available, the predator will thrive and its numbers will increase.

❸ The rate of predation will now increase and as a result the number of prey will decline.

❹ Since there is now less food available to the predator, many will starve.

❺ With fewer predators, the prey numbers will increase again and so it goes on.

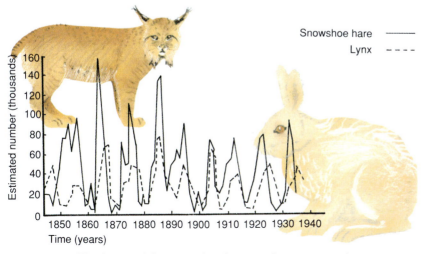

Figure 3.6B ● The lynx and the snowshoe hare predator–prey cycle

Other factors will affect the size of both populations. The incidence of disease, harsh climate and lack of shelter will all play their part. The most widely documented example of this predator–prey cycle is that of the snowshoe hare and its predator, the lynx. Records of their numbers were kept by the Hudson Bay Fur Company in Canada between 1845 and 1935. The numbers of hare and lynx pelts brought in by trappers were recorded. The fluctuations in the numbers of predator and prey are shown in Figure 3.6B.

What makes a successful predator?

A predator must be able to hunt and kill efficiently. *But how can a predator make life easier for itself?*

- By not being dependent on one particular species for prey there is less risk of starvation. If the numbers of one species of prey decline the predator can switch to another.
- Catching young, old and sick prey. The predator needs to expend less energy if its prey can be caught easily. A result is the weeding out of weaker individuals in the prey population. Those that remain are the best adapted individuals, who pass on their genes to the next generation.
- Catching large prey will provide more food for the predator per kill.
- Migrating to areas where prey is more plentiful.

  However well adapted the predator becomes, it should not be too successful or else it risks the possibility of wiping out its food source. The predators most at risk of doing this are humans.

The prey's guide to survival

In order to avoid being captured, many prey species have a great variety of adaptations. These deter the predator, though they are never completely successful.

- Many try to out-run, out-swim or out-fly the predator.
- Staying in large groups, for example herds of antelope and shoals of fish, is an attempt to distract the predator from concentrating on one particularly individual (see Figure 3.6C).
- Some animals, such as bees and wasps, will sting to avoid capture. Others just taste horrible, for instance ladybirds, so the predator thinks twice before eating one again.
- Some prey species possess warning colours that tell the predator to 'keep clear'. The hoverfly has striking yellow and black stripes that remind the predator of a wasp or bee (see Figure 3.6D). A hungry frog that has been stung previously by a wasp will think again before it attempts to eat a hoverfly. In fact, the hoverfly has no sting, so it is quite harmless.
- Camouflage is often used by prey to escape capture. The problem is that it is also used by the predator to avoid being seen when stalking the prey. Many animals try to blend in with their surroundings. Others try to look like something else, twigs or leaves for instance (see Figure 3.6E).
- Some prey species try shock tactics. They try to startle their predators into abandoning their attack. The eyed hawkmoth can quickly open its wings to reveal a pair of huge, coloured eye-spots. This often has the effect of scaring off a hungry bird (see Figure 3.6F).

Figure 3.6C ● Hiding in the crowd

Figure 3.6D ● A hoverfly displaying the 'danger' colours of a wasp

Figure 3.6E ● Two camouflaged moths

Figure 3.6F ● The eyed hawkmoth

Man the predator

Can you think of a modern day example of how man hunts and kills his food?
Most of our food, animal and vegetable, is produced by farming, but one
activity in which man remains the primitive hunter is fishing (see Figure
3.6G).

Over the past 50 years, the intensity of fishing in many areas has
increased to such an extent that fish stocks are threatened. In many cases
the size of a particular stock has been so reduced that it is no longer
profitable to fish. In more severe cases the entire stock has been brought
to the verge of extinction.

Figure 3.6G ● A trawler in action

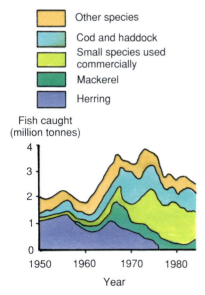

Other species
Cod and haddock
Small species used
commercially
Mackerel
Herring

Fish caught
(million tonnes)

Year

Figure 3.6H ● Total catches of fish in
the North Sea

This situation of overfishing has been brought about by improved technology, which has made it possible to increase the amount of fish caught dramatically. The power of fishing vessels has increased, making it possible to use bigger nets. The nets are now plastic. They are stronger, lighter and almost transparent in the water. Sonar has made it possible to detect fish shoals exactly, so the net can be positioned in such a way that an entire shoal is trapped. Refrigeration enables a catch to be preserved so factory ships can stay at sea for longer periods of time and roam further afield.

The rise in world fish landings after the Second World War was truly remarkable (see Figure 3.6H). Between 1940 and 1970, landings increased from 20 million tonnes per year to 70 million tonnes per year. Since then landings have not increased significantly. *Why is this?* If we look at the catch effort curve (see Figure 3.6I), we can see that an increase in fishing effort will, for a time, bring about an increase in the catch. The maximum which is reached is termed the **maximum sustainable yield**. If this level of fishing effort is exceeded, then landings will be reduced. This overfished situation can be solved only if the intensity of fishing is reduced. By reducing catches, the fisherman would, in the long term, have bigger landings of better quality fish. But this is not what happens – faced with declining catches the fisherman tries to catch more fish. So a situation of extreme overfishing is reached where both the existence of the stock and the fisherman's livelihood are threatened.

Another example of man's greed and short-sightedness is the over-exploitation of the whale for its meat and oil. Predation by man has been the main cause of the decline in a number of whale species (see Figure 3.6J). In the 1940s the blue whale had been hunted almost to the point of extinction. Having exhausted the stock of blue whales, whaling fleets turned instead to another species, the fin whale. Over the 1960s hunting reduced their numbers to danger level. The sei whale was the next to be over-exploited and its numbers fell dramatically …

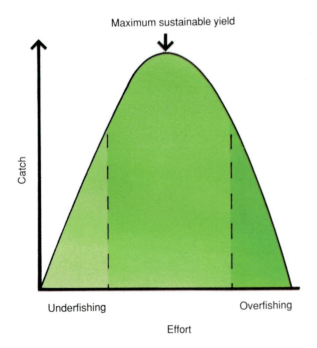

IT'S A FACT

News Report – May 1994
Conservation in action
Nearly all of the member countries of the International Whaling Commission have voted to make 18 million square kilometres of ocean surrounding Antarctica a safe haven for whales.
What effect do you think this action will have on the numbers of different species of whale?

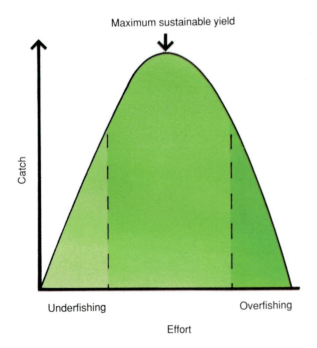

Figure 3.6I ● The catch–effort curve for fishing

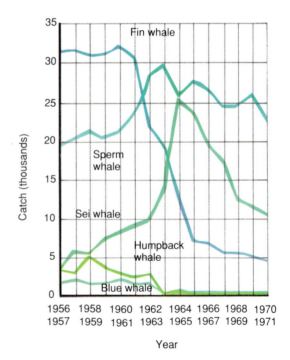

Figure 3.6J ● The decline in numbers of whale species due to overfishing

3.7 Parasitism and other associations

Parasites often cause a sense of revulsion in people, because of the harm that they can cause. In fact, probably every living animal has some of these uninvited guests. Some parasites cause disease and death and therefore reduce the size of a population (see Figure 3.7A).

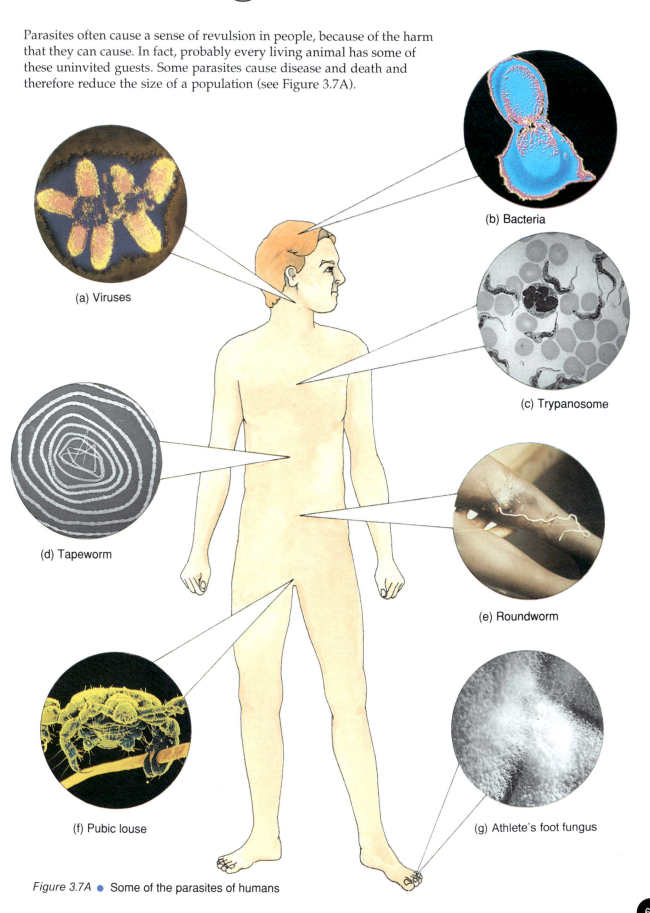

(a) Viruses

(b) Bacteria

(c) Trypanosome

(d) Tapeworm

(e) Roundworm

(f) Pubic louse

(g) Athlete's foot fungus

Figure 3.7A ● Some of the parasites of humans

Parasitism is one of a number of different associations that can exist between two organisms. **Symbiosis** is a general term for close associations in which either one of the partners benefits to the detriment of the other or both benefit or neither appears to benefit. Benefits can be food, shelter or protection.

Table 3.1 ● Some associations

Association	How each participant is affected
Competition	Neither population benefits, they inhibit each other
Predation	The predator benefits at the expense of the prey
Parasitism	The parasite benefits to the detriment of the host
Commensalism	One participant, the commensal, benefits whilst the other, the host, is unaffected
Mutualism	The association benefits both partners

Parasitism

Parasitism is a one-sided relationship between two organisms. The parasite obtains food at the host's expense, either by consuming the host itself or by consuming the host's food supply. Parasites of humans include fungi, bacteria, viruses, protozoa, worms and arthropods. Microorganisms that cause disease are termed **pathogens**. When an animal is healthy, it has little difficulty supporting all its parasites with food. However, if the animal is weakened by starvation or drought for instance, then parasites can have a fatal effect upon it.

There is a great variety of parasites. The degree to which they are tied to the host varies. **Ectoparasites** attach themselves to the outside of the host and many only form temporary associations, for example, a tick will attach itself to a cow long enough to take a blood meal and then drop off. An **endoparasite** lives inside the host's body which becomes the parasite's habitat, giving it food, shelter and protection. The malarial parasite only leaves the human body when it is carried to a new host by a mosquito.

● *Parasitic adaptions*

Parasites have very specialised life styles. This is particularly true of endoparasites which have to adapt to living inside a host that may well react adversely to them. To cope with their mode of existence, parasites have evolved a number of adaptations.

* Endoparasites often have poorly developed organs of locomotion and digestion. They move very little and many absorb the pre-digested food of the host.
* They have poorly developed senses of sight, hearing and smell as there are few stimuli they are likely to encounter. Parasitic flowering plants have no roots and may lack chloroplasts. *Why is this?*
* Many endoparasites have well developed organs for attachment to the host, for example the hooks and suckers of the tapeworm.
* Parasites such as the flea have specialised mouthparts for penetrating the host's tissue.
* If the endoparasite is to live inside the host, then it must be able to resist the host's **immune reactions**. For this reason parasitic roundworms have a thick resistant cuticle.
* The host tissue may lack oxygen. This is true of the human intestine where the hookworm lives.

IT'S A FACT

The lousiest animal in the world is a small mammal about the size of a rabbit called the hyrax. It harbours 25 different species of lice.

IT'S A FACT

Some wasps have parasitic larvae. The adult wasp lays its eggs inside the body of another insect such as a caterpillar. The larvae hatch out and entirely consume the caterpillar from the inside. After forming pupae, the adult wasps eventually emerge.

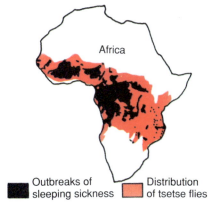

Figure 3.7B ● Tsetse fly and sleeping sickness

Outbreaks of sleeping sickness

Distribution of tsetse flies

● How do parasites find new hosts?

If a parasite cannot ensure that some of its offspring are able to infect a new host, then the species will die out. Parasites have numerous adaptations that have enabled them to meet this aim.

* Many are capable of producing hundreds of thousands of eggs every day in the hope that some will find their way to a new host.
* Some are hermaphrodite, that is they have male and female reproductive organs. Consequently when one mates with a partner, both donate sperm and twice as many fertilised eggs are produced.
* In some parasites the male and female get very attached to each other, literally. In the case of the fluke that causes bilharzia, the male carries the female in a groove in his body to ensure that they will mate.
* Many parasites use another organism to transmit their offspring to the host. Many insects acts as **vectors** in this way. The tsetse fly transmits the protozoan that causes sleeping sickness from one person to another. The flies have piercing mouthparts that penetrate the human skin like a hypodermic needle and then inject the protozoan into a new host (see Figure 3.7B).
* In some cases, the parasite may use a secondary host to enable it to complete its life cycle. The pork tapeworm has two hosts in its life cycle. A human is the primary host and a pig the secondary host. The tapeworm lives in the human intestine. Its millions of eggs are passed out in the faeces and some may be eaten by a pig. These hatch out and the larvae find their way into the muscle of the pig. If this is eaten as under-cooked pork, then the adult tapeworm will develop in the new primary host (see Figure 3.7C)

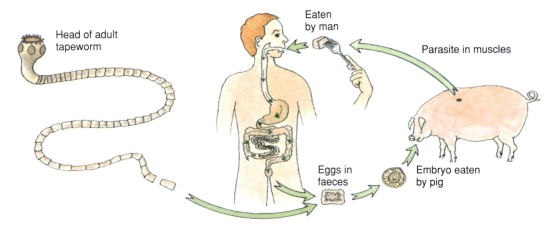

Head of adult tapeworm

Eaten by man

Parasite in muscles

Eggs in faeces

Embryo eaten by pig

Figure 3.7C ● Life cycle of the pork tapeworm

IT'S A FACT

Pilot fish are commensals that accompany large sharks wherever they swim. Since sharks are messy feeders the pilot fish are able to pick up scraps of food. No one knows why the shark tolerates the pilot fish. If a shark is caught its companions will reluctantly swim off to find another protector.

Mutualism

Mutualism is an association between organisms of different species in which each partner benefits. Sometimes the partners become so dependent upon each other that they are unable to exist separately.

The roots of legumes such as peas, beans and clover, have lumps on them called **root nodules**. These contain **nitrogen-fixing bacteria**. The bacteria can convert nitrogen from the air into compounds which the plant can use to make protein. In return, the bacteria obtain carbohydrates from the plant's roots. Many bacteria and protozoa live in the large intestines of mammals. They break down the cellulose in plant material to sugars. They are especially important to herbivores in helping them to digest the cellulose which forms a large part of their diet. In return, the microorganisms get some of the food and a warm environment.

LOOK AT LINKS
for **root nodules** and **nitrogen fixing bacteria**
See Topic 2.6.

A looser example of mutualism is that between oxpecker birds and animals such as buffaloes and antelopes. The oxpecker removes ticks and other ectoparasites from the skin of the game animal and often alerts it if predators approach. Cleaner fish do a similar job for larger fish, removing lice and other ectoparasites.

Commensalism

This is an association between two organisms of different species, in which one benefits without harming the other. The sea anemone is able to hitch a ride on the crab. The crab is a messy feeder and scraps of food float towards the sea anemone which picks them up in its tentacles (see Figure 3.7D). In this example, the sea anemone is the commensal, its presence is of no advantage or disadvantage to the crab.

Figure 3.7D ● Crab and sea anenome

SUMMARY

Parasites live in or on their hosts to which they cause harm. In order to be successful, parasites have evolved many adaptations. Some parasites use vectors and secondary hosts to ensure the spread of their offspring.

CHECKPOINT

❶ Why are under-nourished people more likely to be affected by parasites than people who are well-fed?

❷ Look at the figure below. It shows the life cycle of the malarial mosquito. The female mosquito is the vector of the protozoan, *Plasmodium*, that causes malaria.
  (a) Suggest ways in which the eggs, larvae and pupae of the mosquito are well adapted for living in water.
  (b) Attempts to control the parasite aim to break the life cycle of the vector. Suggest ways in which each of the stages in the life cycle could be eradicated.
  (c) How can the adult mosquito be prevented from biting people and so spreading the disease?

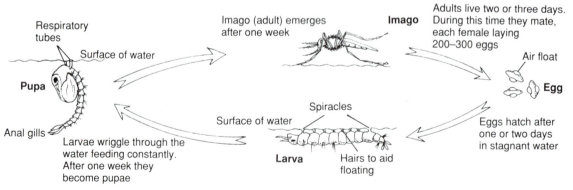

TOPIC 4 INSECTS AND HUMANS

4.1 Beneficial and harmful insects

FIRST THOUGHTS

There are more insect species and more insects living than all other animal species put together. Many insects are beneficial; some are harmful. This section introduces you to them.

IT'S A FACT

Houseflies, tsetse flies, screw-worm flies and mosquitoes are examples of true flies. They belong to the group called *Diptera* which literally means 'two-winged'. Most insects have four wings.

LOOK AT LINKS
for more about **sieve tubes** and **aphids feeding**
See Topic 14.1.

LOOK AT LINKS
for more about **asexual reproduction in aphids**
See Topic 15.1.

SUMMARY

Different predators reduce the numbers of aphids. However, insecticides are needed to minimise damage by aphids to agricultural crops.

LOOK AT LINKS
for **greenhouses**
See Topic 11.1;
for more about **phloem** and the **transport of food in plants**
See Topic 14.1;
for **parasites**
See Topic 3.7.

Blackfly and greenfly are everyday names for **aphids** although they are not flies at all. They feed on the sugar-rich **sap** flowing through the **sieve** tubes of plants. The hollow needle-like mouthparts (called stylets – see p. 233) are tapped into a stem. Sap, under pressure in the sieve tubes, is forced through the mouth parts and into the aphid gut.

Aphids spend a lot of time feeding. Large colonies can quickly build up and cover a plant, which wilts and may die through loss of food and water.

Aphids also transmit **virus diseases**. While aphids are feeding, their mouth parts inject viruses into the sieve tubes. The sap stream transports the viruses around the plant, which sickens and dies. Virus-free seed potatoes are grown in Scotland where aphids are generally scarce. The risk of aphid-borne viruses is therefore less than elsewhere in the United Kingdom. Potato growers can reduce the risk of devastation by raising their crops from Scottish seed potatoes.

Controlling aphids

Crowded colonies of aphids are a juicy target for insect predators. Figure 4.1A (overleaf) shows them in action. Despite the attentions of predators, however, aphids are major pests of agricultural crops. As well as potatoes, broad beans, barley, wheat and sugar beet are just some of the crops which suffer serious damage.

Varying the sowing date, the number of plants sown (crowded plants are less vulnerable to aphid attack than plants spread apart) and the use of varieties of plant which are resistant to attack help to control the impact of aphids. However, insecticides may have to be used if damage is to be kept to acceptable levels.

Systemics – so called because the chemicals pass into the soil in solution to be absorbed by the roots of the plant – are the type of insecticide most often used to control aphids. While feeding, the aphid takes in a large dose of insecticide and is killed. *Why do you think systemics are an efficient way of controlling pests that feed on plant sap? What advantage does their use have for wildlife?*

Controlling whitefly

Whitefly is a common pest of greenhouse crops. The moth-like insect covers the undersides of leaves, feeding on the sugary liquid transported in the phloem tissue of the plants. The sticky juice which the whiteflies excrete slows crop growth and makes fruit dirty and unattractive.

Encarsia formosa is a small parasitic wasp. It lays its eggs in whitefly. The eggs hatch and the wasp larvae feed on the tissues of the whitefly and kill them. Adult wasps emerge from the dead whitefly. The wasp gives reliable **biological control** of whitefly, providing the control of temperature in the greenhouse is accurate. If the temperature is too low, the parasite is sluggish and the whitefly gains the upper hand. If the temperature is too high, the parasite overwhelms the whitefly leaving itself without hosts for the next generation of wasps. *What would then happen to the wasp population?*

(a) Earwig

(b) Ground beetle

(c) Blue tit

(d) *Aphidoletes* larva

(e) *Aphidoletes* adult

(f) Lacewing larva

(g) Lacewing adult

(h) Hover-fly larva

(i) Hover-fly adult

(j) *Anthocoris* nymph

(k) *Anthocoris* adult

(l) Ladybird larva

(m) Ladybird adult

(n) Mealy cabbage aphid
(winged adult)

Figure 4.1A ●
Aphid
predators

Table 4.1 ● Advantages and disadvantages of biological control

Advantages	Disadvantages
Harmless – programmes of biological control do not harm people, wildlife or the environment	**Slow** – biological control acts slowly
Low cost – biological control is relatively cheap	**Unpredictable** – changes in the environment may cause sudden outbreaks of pest problems
Specific – control agents (predators and parasites) usually attack only the pest species and do not harm other wildlife	**Pests survive** – a pest must persist even if at a low level for the control agent to survive long-term
No resistance – pests are unlikely to develop resistance to controlling agents	**No insecticides** – if a crop has several pests then biological control against one means insecticides cannot be used against the others. *Why? Briefly summarise your opinions*

LOOK AT LINKS

for more about **X-rays**
See *KS: Physics*, Topic 14.2.

SCIENCE AT WORK

Screw worm flies are bred for sterile male campaigns at a factory in Mexico. Female flies feed and lay their eggs on blood and meat. Larvae are collected as they begin to pupate. Pupae are sterilised with 5000 rads of gamma radiation from a cobalt source.

LOOK AT LINKS

for more about **gamma radiation** and **radioactivity**
See *KS: Physics*, Topic 10.1.

Controlling screw worm fly

Screw worm flies are a serious pest of cattle. An insect bite, even a small scratch, attracts the female fly to lay her eggs around the edge of the wound (called **strike**). After a few days the larvae (**maggots**) hatch and eat into the tissue increasing the size of the wound. Victims of attacks quickly lose condition, and unless they are treated with insecticides to kill the maggots they eventually die.

Treatment of livestock for strike is difficult where animals are scattered over large areas. In the 1950s scientists in North America wondered if releasing sterile males into the environment would control screw worm fly. They knew that:

- female screw worm flies only mate once
- exposing male flies to X-rays causes lethal mutations in their sperm, sterilising them
- sterile males are as effective as wild fertile males in finding mates
- eggs fertilised by sperm from sterile males fail to hatch.

Success with trial experiments on the island of Curaçao off the coast of Venezuela encouraged scientists to repeat their efforts in North America. Beginning in Florida in 1957, the campaign gathered pace, and in 1958 50 million flies were sterilised and released every week. Invasion by wild flies across the Mexican border was countered by moving the campaign westward. A factory was set up producing millions of sterilised pupae which were boxed for airlifting and dropping by parachute into the countryside.

By the mid-1960s, screw worm fly ceased to be a problem. Later outbreaks, particularly in Texas, were soon brought under control, but gave a warning that in the long term, screw worm fly is still a threat.

4.2 Malaria and mosquitoes

Female mosquitoes (genus: *Anopheles*) transmit the protist parasites (genus: *Plasmodium*) that cause the disease malaria in humans. Think about these facts:

• 2400 million people are at risk from malaria
• 280 million people are infected with malaria parasites
• 110 million people develop the disease each year
• 2 million people die from malaria each year.

How can malaria be treated or prevented? What is the role of mosquitoes in its transmission? Read this section and find out.

LOOK AT LINKS
for more about **parasites**
See Topic 3.7.

LOOK AT LINKS
for more about **vertebrates**
See Topic 21.5.

LOOK AT LINKS
for more about **tsetse flies**
See Topic 3.7.

IT'S A FACT

The mosquitoes *Culex*, *Aedes* and *Mansonia* are vectors for different diseases but not malaria. They transmit the tiny worms that cause **filariasis**. The worms block lymph vessels and lymph nodes (see p. 248). As a result limbs may swell to alarming proportions.

The parasite and the disease

There are four species of the protist *Plasmodium* responsible for human malaria. The parasite infects liver cells and red blood cells.

Once the parasite is in the blood stream, flu-like symptoms are followed up to 30 days later by high fevers, teeth-chattering chills and heavy sweating. The timing of the attacks depends on the species of *Plasmodium*. *Plasmodium vivax* is the most common species. It causes a mild form of the disease. *Plasmodium falciparum* is the most dangerous. It may cause red blood cells to become sticky, blocking blood vessels to the kidneys, brain, intestines and lungs. Unless treated the victim dies within a few days.

The geography of malaria

Plasmodium needs warmth to reproduce inside the mosquito. Malaria therefore flourishes in the warmer tropical and equatorial regions of the world. Poorer countries are particularly vulnerable. They are short of expert help and also find it difficult to finance measures for treatment and prevention.

Mosquitoes

Mosquitoes suck sugary liquids. Nectar and honey are favourite foods. Females need a blood meal as well before their eggs can develop. Most vertebrates are hosts for female mosquitoes.

Only female *Anopheles* mosquitoes transmit the malaria parasite to humans. About 30 different species world-wide are responsible. As in the tsetse fly, the female's pointed mouthparts pierce capillary blood vessels beneath the surface of the skin and take up blood like a hypodermic needle (Figure 4.2A). If the person is infected with *Plasmodium*, then the parasite is sucked up as well and then passed on when she feeds on another person.

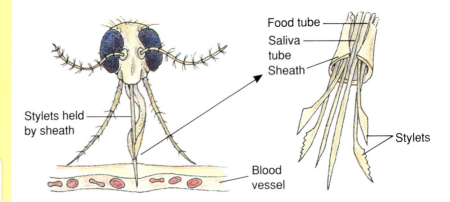

Figure 4.2A ● Female mosquitoes feed on blood. Their mouthparts are long and sharply pointed forming stylets. They are surrounded by a sheath. Inside is a tube through which saliva passes and a food tube. When feeding, the sheath bends back and holds the stylets in place as they pierce the victim's skin until a blood vessel is reached. The food tube is then pushed in and saliva pumped down through the salivary tube into the wound.

The malaria life-cycle

Since *Plasmodium* infects blood and female mosquitoes feed on blood, female mosquitoes are efficient vectors of the parasite. Female mosquitoes are unaffected by it. As we have seen, people are affected by it. Figure 4.2B shows the roles of mosquitoes and people in the malaria life cycle.

LOOK AT LINKS
for more about **vectors** and **hosts**
See Topic 3.7.

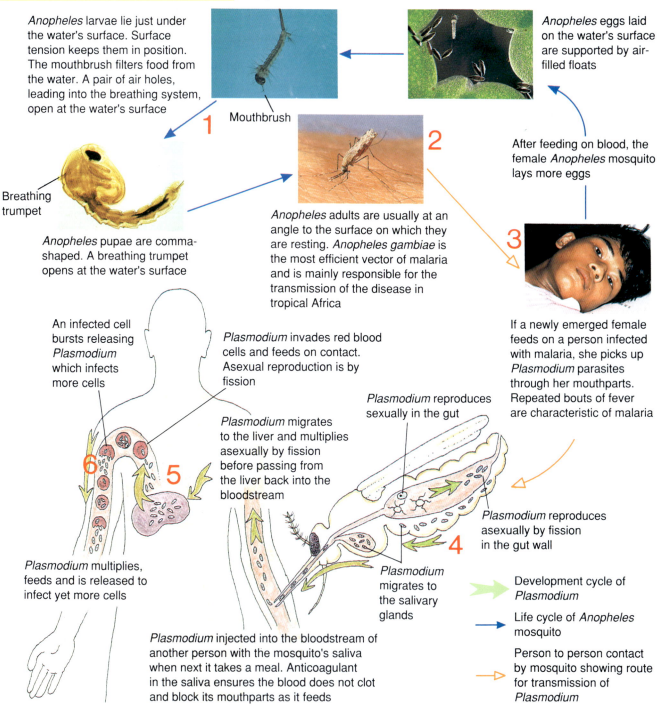

Anopheles larvae lie just under the water's surface. Surface tension keeps them in position. The mouthbrush filters food from the water. A pair of air holes, leading into the breathing system, open at the water's surface

1 Mouthbrush

Breathing trumpet

Anopheles pupae are comma-shaped. A breathing trumpet opens at the water's surface

2

Anopheles adults are usually at an angle to the surface on which they are resting. *Anopheles gambiae* is the most efficient vector of malaria and is mainly responsible for the transmission of the disease in tropical Africa

Anopheles eggs laid on the water's surface are supported by air-filled floats

After feeding on blood, the female *Anopheles* mosquito lays more eggs

3

If a newly emerged female feeds on a person infected with malaria, she picks up *Plasmodium* parasites through her mouthparts. Repeated bouts of fever are characteristic of malaria

An infected cell bursts releasing *Plasmodium* which infects more cells

Plasmodium invades red blood cells and feeds on contact. Asexual reproduction is by fission

Plasmodium migrates to the liver and multiplies asexually by fission before passing from the liver back into the bloodstream

Plasmodium reproduces sexually in the gut

6 **5**

Plasmodium multiplies, feeds and is released to infect yet more cells

Plasmodium migrates to the salivary glands

Plasmodium reproduces asexually by fission in the gut wall

4

Plasmodium injected into the bloodstream of another person with the mosquito's saliva when next it takes a meal. Anticoagulant in the saliva ensures the blood does not clot and block its mouthparts as it feeds

➤ Development cycle of *Plasmodium*

➤ Life cycle of *Anopheles* mosquito

⇨ Person to person contact by mosquito showing route for transmission of *Plasmodium*

Figure 4.2B ● The malaria life cycle

LOOK AT LINKS
for more about **fission**
See Topic 15.1.

Repeated invasions of and release from red blood cells of *Plasmodium* are tracked by a cycle of chills and fevers. The victim's body temperature soars when the red cells burst releasing parasites into the blood stream. Eventually enormous numbers build up, causing serious and sometimes fatal illness.

TRY THIS

Match up the numbers in Figure 4.2B with different control measures and devise your own anti-malaria campaign. Explain fully the reasoning behind your plan.

LOOK AT LINKS

for more about **biological control**
See Topic 4.1;
for more about **persistent insecticides**
See Topic 6.8;
for more about **drug resistance**
See Topic 21.4;
for more about **antigens, vaccines** and the **immune response**
See Topic 14.2.

LOOK AT LINKS

for more about the **exchange of materials across the placenta**
See Topic 15.5;
find out about **passive** and **acquired immunity** on pp. 237–9.

FIRST THOUGHTS

Houseflies are a health hazard. They feed on faeces and on our food. This section makes the connection between the housefly's habits and disease.

IT'S A FACT

Up to the 1950s, sticky fly-papers hung from the ceiling were used to catch flies in the home. In summer a character nicknamed 'Flypaper Joe' called each week selling fly-papers from a band worn around his top hat.

Treatment and prevention of malaria

Fighting malaria depends on destroying the mosquito vector, the parasite or preventing contact between the vector and people. Find the numbers on Figure 4.2B. They indicate where control measures are most effective. Control measures include:

- **draining** marshes, ponds and ditches, thus preventing the female mosquito from laying eggs, and the eggs from developing into larvae.
- **biological control** in the form of fish which eat mosquito larvae.
- **insecticides** sprayed on the water's surface killing mosquito larvae and pupae.
- **drugs** like quinine, chloroquine and quinacrine, which prevent fission in the red blood cells of the host.
- **vaccines** which inhibit different stages in the parasite's life cycle.
- **bed nets** soaked with insecticide to protect the resting person not only from mosquitoes but from other biting insects as well.
- **chemical repellents** sprayed on to the skin and clothes to deter mosquitoes from landing on the body.

Natural immunity

In many parts of Africa people are partly immune to malaria. The body produces antibodies in response to *Plasmodium*. The antibodies prevent infection in the liver cells and red blood cells, or destroy red blood cells which are already infected. Other immune responses have similar effects.

The children of an infected mother are given some protection by her antibodies, which cross the placenta before birth. This **passive** immunity disappears after 4–6 months. However if the child survives to reach 5–6 years, then repeated infections result in **acquired** immunity which gives a high level of protection.

4.3 Houseflies spread disease

Houseflies feed on most organic matter. Look at the fly's trail shown in Figure 4.3A. How many sources of food for flies can you find in Figure 4.3B? Using both figures as evidence, explain why flies are a health hazard.

Figure 4.3A ● The petri dish is covered with a thin layer of blood agar (a jelly-like substance on which bacteria grow). A housefly has walked over the agar surface. The petri dish was sealed and then incubated at 37°C for 4 days. *Why was the petri dish sealed and then incubated?*

Figure 4.3B ● Faeces and rotting organic matter are homes for microorganisms which cause disease. Houseflies pick up and transfer microorganisms as they move from one food source to another.

The problem

Figure 4.3C shows the mouthparts of the housefly. Digestive juices are released through the fine tubes on to the food on which the fly has landed. The digested food is sucked up. In the process, microorganisms on the mouth-parts and in the digestive juice are transferred from the fly to the food. If the fly's food is also human food, then people are in danger from the disease-causing microorganisms the fly leaves behind (see Figure 4.3D).

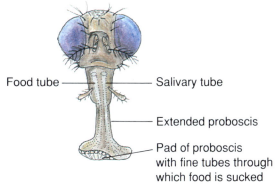

Food tube ———   ——— Salivary tube

——— Extended proboscis

——— Pad of proboscis
with fine tubes through
which food is sucked

Figure 4.3C ● The proboscis of the housefly has a two-lobed pad at its tip. In the pad are many thin tubes. During feeding, digestive juices are pumped down these tubes on to the food. The juices make the food liquid. The liquid food is then sucked up the tubes and into the mouth.

LOOK AT LINKS

for more about **absorption of food**, the **movement of food through the gut** and **faeces**

See Topic 11.6.

IT'S A FACT

Giving a solution of table salt and sugar to victims of diarrhoea saves thousands of lives daily. The treatment is called **oral rehydration**. The salt and sugar help the body to absorb water from the gut, replacing the water lost in the faeces. A sachet of the mixture costs only a few pence. Patients rapidly improve after the treatment.

LOOK AT LINKS

for more about **Kingdom Bacteria**

See Topic 1.5.

LOOK AT LINKS

for more about **antibiotic drugs**

See Topic 20.5.

SUMMARY

Houseflies transfer disease-causing microorganisms from organic matter to human food as they move from one food source to another. Strict hygiene controls flies and drugs help to treat the diseases they spread.

Fly-borne diseases

The bacteria (*Shigella spp*) or protista (*Entamoeba histolytica*) that cause **dysentery** are transferred to human food by houseflies. Dysentery results in diarrhoea. Contractions of the gut (peristalsis) are more frequent than normal, and food and water are not absorbed through the gut wall. The faeces are liquid. The patient loses water and feels very weak. Children are particularly vulnerable and may die through dehydration (loss of water).

Dysentery appears wherever people are crowded together in unhygienic conditions. During the Second World War, the British Army in North Africa kept outbreaks of dysentery to a minimum by strict attention to hygiene and the control of flies. The picture was different in the Crimean War of 1854 when filth provided ideal breeding conditions for flies, and 8000 British soldiers suffered from dysentery. More than 2000 died of the disease.

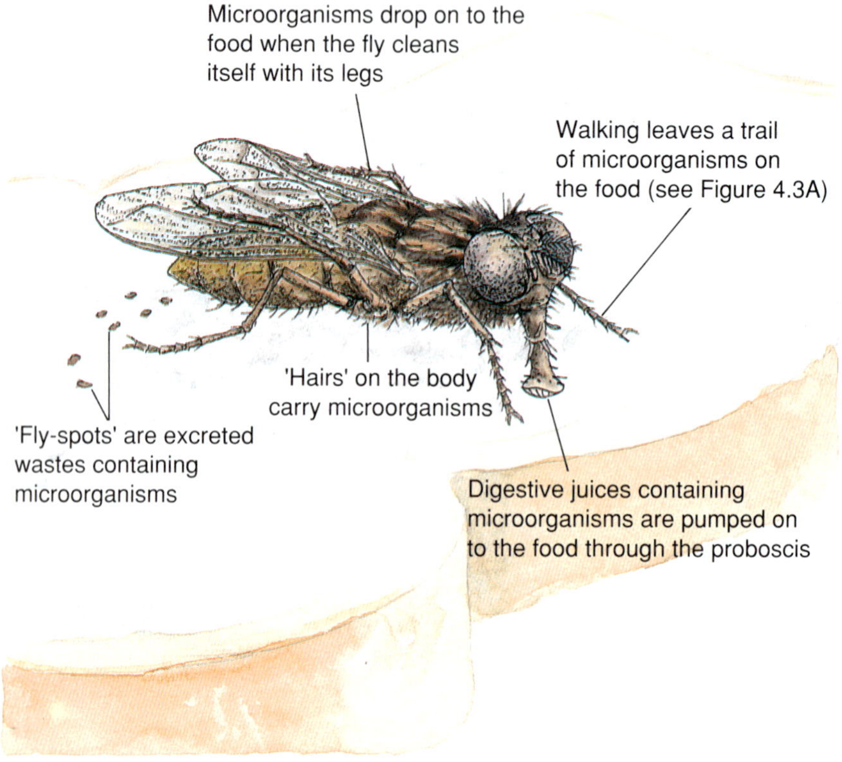

Microorganisms drop on to the food when the fly cleans itself with its legs

Walking leaves a trail of microorganisms on the food (see Figure 4.3A)

'Hairs' on the body carry microorganisms

'Fly-spots' are excreted wastes containing microorganisms

Digestive juices containing microorganisms are pumped on to the food through the proboscis

Figure 4.3D ● The housefly is a vector for microorganisms that cause disease.

Typhoid fever is caused by the bacterium *Salmonella typhosum* carried in contaminated water. Houseflies also transfer the bacterium to human food. After about fourteen days, the infected person develops a fever which grows steadily worse. Delirium and uncontrollable shakes appear about two weeks later, and the victim may die unless treated with antibiotic drugs.

Cholera bacteria (*Vibrio cholerae*) are also carried in contaminated water and transferred to food by houseflies. Symptoms of cholera include vomiting and passing large amounts of liquid faeces. Loss of salts from the body causes severe cramps in the limbs and, unless treated with antibiotics, the patient may die from the effects of dehydration.

Improvements in treatment, health, personal hygiene and sanitation have helped to reduce the extent and effects of fly-borne diseases. However, remember that flies carry microorganisms. Infections are reduced by covering food to prevent flies from landing and feeding.

'Busy as a bee' is an old saying, but how busy are bees? What are they busy doing and why? Reading this section will help you to answer these questions.

Figure 4.4A shows honeybees collecting nectar from a flower. The bees pass from flower to flower and then fly away. If you were in the field following them, the bees would probably lead you to their hive. Landing near the entrance, 'your' bees would join hundreds of others coming and going in seemingly ceaseless activity.

Figure 4.4A ● Honeybees on a dandelion

Observation of honeybees has shown that:
- foraging bees usually visit the same type of flower at the same time, day after day
- they have a sense of direction
- they are social animals, each one living as a member of a large colony
- guards at the hive entrance check returning bees by touching them with their antennae (feelers). If one of the bees is from another hive, the guards drive it away.

Communicating and sharing jobs seem to be important aspects of honeybee biology.

Bees produce honey and other valuable products, as well as pollinating flowers. Understanding honeybees, therefore, will help you to understand their importance.

LOOK AT LINKS
for more about **pollination**
See Topic 15.2.

Sharing the jobs

All the members of a colony of honeybees are the offspring of one mother. She is called the **queen**. Her only job is to lay eggs. Most of the offspring are female. Their jobs are rearing larvae, cleaning and repairing the nest and foraging for food. They are called **workers** and are **sterile** because their ovaries are undeveloped. About 200 individuals are males whose only job is to mate with a queen. They are called **drones**. A large honeybee nest may contain up to 80 000 individuals. Its survival depends on its members carrying out their particular duties. The queen survives for several years, but workers and drones are short-lived. Summer-hatched workers live for about six or seven weeks. Those hatched in the autumn live with the queen through the winter. They carry on with their jobs in early spring. In this way the colony continues from year to year.

The hive

The hive provides protection for the nest of combs that honeybees build inside. Combs hang vertically in parallel rows. Each one is made from wax produced by the worker bees from wax glands on the underside of their abdomens. The workers mould the wax in their mouthparts into six-sided cells which lie horizontally back to back. Some cells are used as **brood chambers** for rearing larvae, others as pots for storing honey and pollen. Figure 4.4B shows the arrangement. Notice that the cells slope gently backwards so that larvae or honey do not spill out.

Why do bees build hexagonal cells? Six sides mean that there is no wasted space between cells. The wall of each cell also forms the wall of the cell next to it, so wax for building is not wasted. The hexagon shape is also best suited for the roundish shape of developing larvae

A frame of honey comb – bees build combs vertically. When a super is full, the wax caps covering the cells of the comb on each frame are cut away. The frames are placed in a centrifuge and the honey spun down. The honey is then run into containers

Comb honey is cut out with a stainless steel comb cutter and boxed for sale.

Queen (royal) cells are larger than other brood cells. They hang downwards from the surface of the comb like acorns

More supers can be added if needed

Frame slots into super

Super

Queen excluder

Brood box

Entrance to hive

Support

Brood cells form the brood area. Workers are reared in cells about 5 mm in diameter. Drones in cells about 6 mm in diameter (black area on combs)

Pollen is a source of protein for adults and developing larvae. A little honey is added to pollen stored for winter. The mixture keeps fresh for a long time and is called 'bee-bread' (shaded area on combs)

Queen excluders fit over the brood box, allowing workers into the honey supers but confining the queen to the brood box. If the queen were to enter the supers, she would lay eggs up the centre frames so that honey, larvae and eggs would be mixed up. The honey would be unsaleable.

Figure 4.4B ● Three-dimensional section through a bee hive. The queen excluder separates the brood box from the supers. Each super holds the frames on which the bees build their combs. During spring and early summer, further supers are added as the bees make more honey and therefore need more space for storage. The hive rests on blocks which help ventilation and prevent the floor from becoming damp

The workers' time-table

Worker bees change their jobs as they grow older. There are three phases:
- young workers look after developing larvae in the brood cells
- older workers build honeycombs and keep the nest clean. They also ventilate the hive by fanning their wings at the entrance. Fanning also helps to lower the temperature of the hive in hot weather
- the oldest workers collect nectar and pollen from flowers. They are not 'house' bees like their younger sisters. Their job is outside the hive.

The time-table however is not rigid. The timing of each phase may alter to suit circumstances.

Food

Bees visit flowers to collect nectar from which they make honey. Nectar is a dilute solution of various natural sugars. Figure 4.4C shows that the nectar is sucked through the mouthparts into the honey sac. Here chemical reactions which change the nectar into honey begin.

The bee uses some of the mixture as food for herself. The rest she brings up and transfers to a 'house' bee on her return to the hive. Further processing occurs as the 'house' bee passes the liquid on to other bees. Water evaporates from the mixture and the end result is honey which is stored in the honey pot cells of the comb. Further evaporation reduces the water content of the honey to less than 20%. The bees then seal the cells with cappings of wax.

TRY THIS

'Busy as a bee'
Assume that a foraging bee visits one thousand flowers to fill her honey sac with nectar once. Also assume that she empties and refills her honey sac fifty times to make 5 g of honey. In good weather a hive of bees may produce 1 kg of honey daily. *How many flower visits are needed to make this amount of honey?*

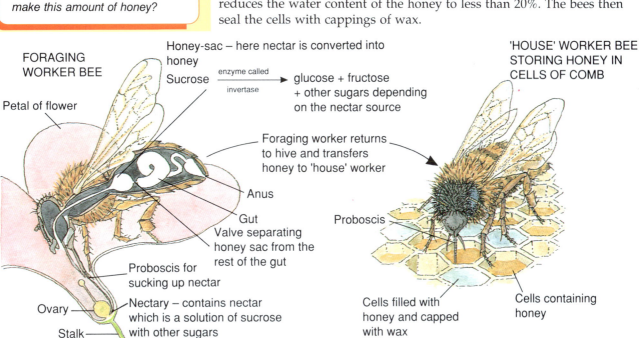

FORAGING WORKER BEE

Honey-sac – here nectar is converted into honey

Sucrose → (enzyme called invertase) → glucose + fructose + other sugars depending on the nectar source

Foraging worker returns to hive and transfers honey to 'house' worker

Anus

Gut

Valve separating honey sac from the rest of the gut

Petal of flower

Proboscis for sucking up nectar

Ovary

Stalk

Nectary – contains nectar which is a solution of sucrose with other sugars

'HOUSE' WORKER BEE STORING HONEY IN CELLS OF COMB

Proboscis

Cells filled with honey and capped with wax

Cells containing honey

Figure 4.4C ● Worker bees making and storing honey

Honeybees collect pollen as well as nectar. Pollen is a source of protein. Grains of pollen stick to the 'hairs' of the bees' bodies when they visit flowers. The pollen grains are combed from the body with a row of bristles (**pollen comb**) on each of the front legs. The pollen is mixed with a little nectar to make two balls of pollen paste which are then passed to the **pollen basket** on each back leg (see Figure 15.2E on p. 264). When bees return to the nest they push the pollen paste into food-storage cells in the comb. 'House' bees pack down the pollen and when the cells are full seal them with caps of wax. Pollen and honey are eaten in the winter when no other food is available.

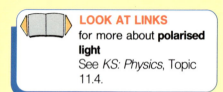

LOOK AT LINKS
for more about **polarised light**
See *KS: Physics*, Topic 11.4.

Communication

Foraging bees find their way to and from the hive by using the position of the Sun. If the Sun is covered by cloud, they can still find their way because their eyes 'see' **polarised** light which gives them the direction of the Sun. When they return to the hive foraging bees tell other bees where the food is by dancing, either at the hive entrance or on the honeycombs (see Figure 4.4D).

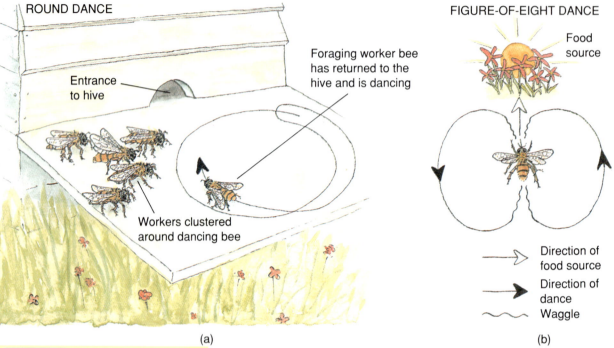

ROUND DANCE

Entrance to hive

Foraging worker bee has returned to the hive and is dancing

Workers clustered around dancing bee

FIGURE-OF-EIGHT DANCE

Food source

→ Direction of food source

➤ Direction of dance

〜 Waggle

(a)   (b)

LOOK AT LINKS
for more about **pheromones**
See Topic 16.5.

Figure 4.4D ● (a) The round dance tells other bees that food is up to 90 m away. The dancer runs around in circles, first one way then the other. The bees nearest to her leave the nest and look for the food on which the dancing worker has left her scent. (b) The figure-of-eight dance tells other workers that the food is further away. This dance also gives information about the direction in which workers must fly when they leave the nest to find food:

● if the dancing worker moves straight up the face of the comb during the straight section of the dance, then the food is in the direction of the Sun

● if the dancer moves straight down the comb then the food is in the direction away from the Sun

● if the dance is at an angle then it shows the angle to the Sun at which the workers must fly to find the food.
 Dancing bees also waggle their bodies on the straight section of the dance. These movements tell other bees how far they must fly to the food.

SCIENCE AT WORK

Pheromones are used to help farmers decide the best time for spraying crops with insecticide to protect the crops against insect pests. For example in apple orchards, traps baited with pheromone are set up for apple codling moths. There is one trap per hectare of trees. If in any one week five or more codling moths are attracted by the pheromone and caught in a trap, then the farmer knows that the orchard should be sprayed with insecticide to prevent damage by the codling moth. This method prevents the farmer wasting money on unnecessary insecticide, because spraying can be carried out when it will be most effective and cause least damage to beneficial insects and other forms of wildlife.

Substances with powerful smells, called **pheromones**, are also important for communication. For example, the queen produces a pheromone called queen substance which is transferred to the workers when they touch her with their mouthparts. The queen substance:

● tells all the other bees of the hive that the queen is present and in good health

● makes sure that the ovaries of workers remain undeveloped

● prevents workers from building royal cells in the comb

● acts as a sex attractant to drones during the mating flight.

Swarming and mating

When the number of bees is too large for the nest, the queen leaves with a swarm of worker bees (called the **prime** swarm) to begin a new nest elsewhere. The original nest has a new queen who emerges from one of the pupae in the brood area of the nest (see Figure 4.4B).

She leaves the nest with a flight of drones who are attracted to her by the smell of the queen substance she gives off. They take part in a nuptial flight during which the fastest flying male mates with the queen. He dies in the act! The newly mated queen and remaining drones return to the nest. At the end of the summer, workers chase the drones out of the nest to die.

The next generation

Before she begins to lay eggs, the new queen and her workers destroy all the royal larvae and newly emerged queens. A nest has only one queen! She then settles to the task of laying. Figure 4.4E shows what happens.

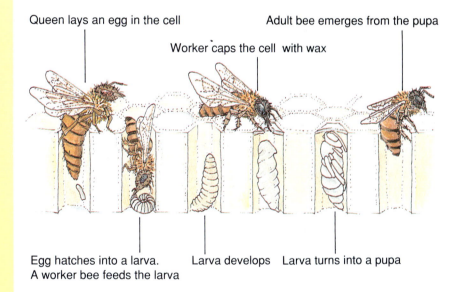

Queen lays an egg in the cell

Adult bee emerges from the pupa

Worker caps the cell with wax

Egg hatches into a larva. A worker bee feeds the larva

Larva develops

Larva turns into a pupa

Figure 4.4E ● Section through the brood chambers of a comb showing the next generation from egg to emerging adult

IT'S A FACT

A queen honeybee lays up to 2000 eggs each day during the summer.

LOOK AT LINKS

The production of new individuals from unfertilised eggs is called **parthenogenesis** See Topic 15.1.

The production of drones, workers or a queen depends on:
- whether or not the bees develop from fertilised eggs
- the diet of the developing larvae.

Unfertilised eggs develop into drones, fertilised eggs into females. The type of female depends on the diet fed to the developing larvae by the workers. A new queen is produced if the larvae are fed only on royal jelly. This is a rich food solution secreted by the workers from a **nurse** gland in their mouths. New workers are produced if the larvae have a diet of honey and pre-digested pollen.

Why feed bees? When honey is harvested, alternative food such as sugar syrup must be provided to see bees through the long winter months.

RESOURCE ACTIVITY PACK

The value of honeybees

Honey was used as a sweetener long before sugar became popular. Its flavour depends on the source of nectar, which varies from month to month. Of all the sources, rape is the most important. Much farmland in the UK is covered with its bright yellow flowers in late spring and early summer. Bees love rape flowers and eagerly visit them to forage on the nectar and pollen. The honey produced is distinctive and strong in flavour.

Propolis is a dark brown substance which bees obtain from plants like the horse chestnut tree. Its sticky resin is used to seal and repair cracks in the honey combs and to block up holes. Bees also smear the frames and wax surfaces of the cells of the honeycombs with propolis. This prevents the growth of bacteria and moulds which would otherwise flourish in the warm humid environment of the hive. The value of honey, propolis and other honeybee products is illustrated in Figure 4.4F.

Honey is sugar-rich and a high-energy nutrient (see p.176). Nectar, collected from different flowers is converted into honey with distinctive taste

Royal jelly is used in health foods and beauty products. It is high in protein and vitamins

Beeswax comes mostly from the cappings cut from combs when honey is extracted. It is used to make church candles, ointments, beauty products, high quality furniture polish and to coat tablets

Propolis is scraped from frames, queen excluders and from sheets of plastic placed specially over frames. After a few weeks, the sheets are removed, frozen and the propolis which has been deposited flaked off. Propolis has antiseptic qualities and is used to treat sore throats, bed sores and other infections. Violin makers varnish instruments with propolis

Figure 4.4F ● Honeybee products

SUMMARY

Honeybees are valuable to us because of their biology. They produce honey, propolis and other products. They are also important pollinators of food crops, wild plants and garden flowers.

Although we usually think of honeybees as producers of honey, their value as pollinators of food crops, wild plants and garden flowers outstrips their value as a source of food. Farmers pay bee-keepers to take their hives into fields and orchards so that the bees can pollinate the crop in flower.

? THEME QUESTIONS

● *Topic 1*

1 Read the descriptions of the following animal or plant groups and try to decide which group each refers to.
   (a) Lays eggs with a shell on land. Have wings and feathers but no teeth.
   (b) Body divided into three parts. Have antennae (feelers), six legs and wings.
   (c) Have chlorophyll but only primitive roots, stems and leaves. Reproduce by means of spores.
   (d) Not segmented. Body often surrounded by a shell. Move by means of a muscular 'foot'.
   (e) No chlorophyll. Feed either by parasitic or saprophytic means. Reproduce by means of spores.
   (f) Warm blooded. Have hair and produce milk from special glands.

2 Use the key to identify the seashore animals **A**, **B**, **C**, **D**, **E** and **F**.

❶	Animal divided into segments	Go to ❷
	Not divided into segments	Go to ❸
❷	Animal has large pincers	Lobster
	No large pincers	Centipede
❸	Animal has a shell	Go to ❹
	No shell	Go to ❺
❹	Shell with two pieces	Cockle
	Shell with one piece	Winkle
❺	Animal has five arms	Starfish
	Animal has many tentacles	Sea anenome

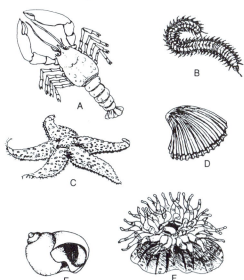

● *Topic 2*

3 The diagram near the top of the next column shows some of the feeding relationships in a British oak wood.
   (a) (i) Name the source of energy for **all** the organisms in this food web.
   (ii) Name **three** substances which the oak tree could obtain from the environment and which would contribute to its structure.

(b) From the diagram, select
   (i) one carnivore,
   (ii) one producer,
   (iii) one food chain which includes **four** different organisms.

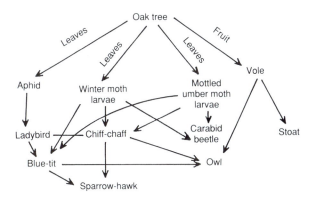

(c) Draw a simple diagram to compare the likely numbers of each of the species in the food chain which you have selected in (b) (iii).
(d) Explain how energy is lost to the environment from the food web.
(e) In some years there are exceptionally large numbers of winter moth and mottled umber moth larvae. Describe **two** probable effects of this on other organisms in the food web.
(f) Each autumn the oak trees shed their leaves. Explain how the elements contained in the cellulose in the leaves are made available for the growth of trees in subsequent years. (JMB)

4 When sewage effluent was discharged into a river, the oxygen concentration at the point of discharge rapidly dropped.
   (a) Suggest why this happened.
   (b) The numbers and types of organisms in a part of a non-polluted river are shown below.

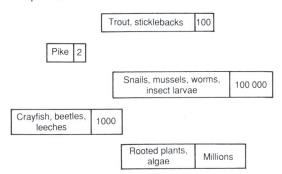

Construct a pyramid of numbers, (not to scale) using all this data.
   (c) (i) Name the organism you would remove if you wanted to increase the number of trout in the river.
   (ii) State one further effect of removing this organism from the river. (WJEC)

● *Topic 3*

5 The graph shows the average weight/day of haddock caught by trawlers in the North Sea from 1905–1960.

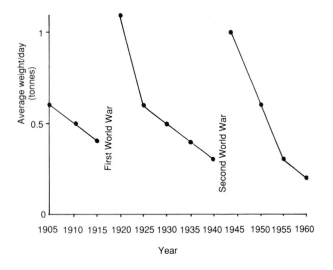

During the World Wars, fishing trawlers were very much reduced. During peace time, the number of trawlers remained the same.

(a) Explain the decrease in the weight of haddock caught between 1920 and 1940.

(b) Explain how the lack of fishing during the war years helped to conserve the haddock.

(c) In 1960 the captains of two trawlers were prosecuted for using nets with a mesh smaller than is permitted.
(i) Explain why a large mesh helps to conserve fish stocks.
(ii) Suggest **three** other ways of conserving haddock in the North Sea.
(WJEC)

● *Topic 4*

6 (a) Using only the following information answer questions (i) and (ii).

> About 500 million people are at risk from malaria, some 200 million have severe attacks and there is an annual death rate of 2 million people. Malaria is caused by the protozoan pathogen *Plasmodium*. The pathogen is transmitted by the *Anopheles* mosquito. *Anopheles* is called a vector and *Plasmodium* is called a parasite.

(i) How many people die from malaria each year?
(ii) Name the pathogen and vector involved with malaria.

(b) Using only the information in the table at the top of the next column, answer questions (i), (ii) and (iii).
(i) Which species of *Plasmodium* causes benign tertiary malaria?
(ii) Name the disease caused by *Plasmodium falciparum*.
(iii) What is the difference between the severity of the diseases caused by *Plasmodium falciparum* and *Plasmodium vivax*?

Cause	Disease	Severity of disease	Frequency of attacks
Plasmodium vivax	Benign tertiary malaria	Rarely fatal	Every 48 hours
Plasmodium falciparum	Malignant tertiary malaria	Often fatal	Every 48 hours
Plasmodium malariae	Quaternary malaria	Mild compared with others	Every 72 hours

(c) Using only the following information answer questions (i) and (ii).

> **The adult mosquito can be killed by insecticides such as DDT, gamma BHC or Dieldrin. These are sprayed in and around buildings and may remain effective for weeks. Mosquito larvae are destroyed by spraying oil on the water surface of ponds, lakes and swamps. The oil reduces the surface tension of the water and the larvae sink below the surface where they cannot obtain oxygen. The adult insect can be prevented from getting to the human host. Fly gauze over windows and doors will keep the insect out. Sleeping with mosquito nets over beds at night helps to reduce the risk of being bitten.**

(i) Name **two** insecticides which can be used in controlling the spread of malaria.
(ii) Some people live near lakes. Suggest how spraying oil on the surface of a lake might reduce their chance of getting malaria.
(MEG)

7 Oral rehydration treatment is a low cost way of preventing death from diarrhoea caused by micro-organisms carried by flies.
This is a graph of deaths from diarrhoea in Egypt.

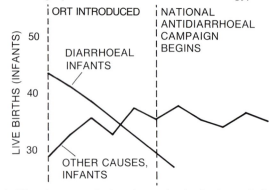

(a) What happened after the oral rehydration solutions were introduced?

(b) Apart from telling people about oral rehydration solutions, what else would you expect the National Antidiarrhoeal campaign to tell people?

(c) Suggest how you would carry out an investigation to look at the ways in which different concentrations of oral rehydration solutions affect living cells.

Human Impact on the Environment

Land is the environment of millions of kinds of plants and animals, including humans. We use part of the land for agriculture, to grow crops and to raise livestock. Modern methods of agriculture are intensive: they are mechanised and require extensive use of fertilisers and pesticides. Conflicts arise between utilising the land for the maximum benefit of humans and allowing plants and animals to flourish in their natural habitats. In this theme you will be able to weigh up some of the pros and cons of different farming methods.

A modern industrial society makes plenty of goods to keep us free from squalor and starvation. Plenty has led to pollution. The air is polluted by objectionable gases, dust and smoke from our cars and factories. Rivers and lakes are polluted by waste from factories, fields and our own bodies. The land is polluted by rubbish tips. Urgent measures are needed to combat pollution. If we take the necessary steps, we can avoid disaster. We can reduce the pollutants in our atmosphere; we can make our rivers and lakes clean again; by making use of much of our rubbish, we can conserve the Earth's resources.

TOPIC
5

THE PHYSICAL ENVIRONMENT

5.1 Soil

Rocks are acted upon by rain, snow and frost. They break up into small particles. The process is called **weathering**. The small particles become part of the soil (see Figure 5.1A). Soil also contains **organic** matter: matter that was once part of plants or animals. This is the part of the soil called **humus** (see Figure 5.1C). Water is another ingredient of soil. Soils differ in the size of the particles, in the humus content, the water content, the pH and the chemicals which are present.

The difference in particle size means that different soils have different properties (see Figure 5.1B and Table 5.1).

LOOK AT LINKS
for **weathering**
See *KS: Chemistry*,
Topic 2.3.

Figure 5.1A ● Clay and sandy soils

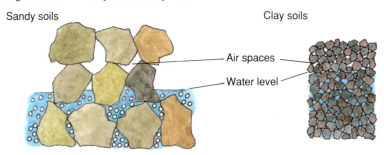

Sandy soils

Clay soils

Air spaces

Water level

Table 5.1 ● The properties of sandy soils and clay soils

	Sandy soils	*Clay soils*
Particle size	Large, > 0.2 mm	Small, < 0.002 mm
Air spaces	Large	Small
Drainage	Rapid, leaving a dry soil	Slow, leaving a wet soil
Temperature	Fluctuates, tending to be higher	More consistent, tending to be lower
Cultivation	Easy to dig and plough because they are dry and loose	Difficult to dig and plough because they are wet and sticky
For plant growth …	Plants may suffer from lack of water. Minerals may be leached (washed) from the soil by rain	Plant roots may lack oxygen if the soil becomes waterlogged. The mineral content is high since minerals tend to stick to clay particles.

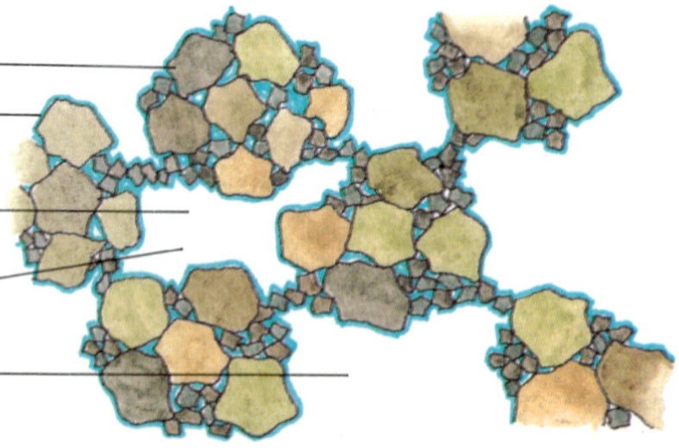

Soil crumbs form when sand and clay particles stick together
A film of water binds to the soil crumbs
Excess water drains out of the air spaces so that the soil does not become waterlogged
Air spaces provide oxygen for the growth of roots and for useful microbes
Air spaces make the loam less dense and easier to cultivate

Most soils are a mixture of particles of different sizes. They therefore combine the properties listed in Table 5.1, and the advantages and disadvantages are balanced. Such a mixture is called **loam** and is ideal for growing plants.

The 'feel' of a soil depends on the proportion of large and small

Figure 5.1B ● The crumb structure of loam

particles in it. This 'feel' is called **texture**. The texture of a soil decides the amount of air in the soil and the speed at which water drains from the soil. These factors affect the plants which grow in the soil and also the animals and micro-organisms which keep the soil in good condition.

Humus

Part of the soil is called **humus**. It has been formed from the decayed remains of dead organisms. A variety of organisms take part in the formation of humus. Figure 5.1C shows some of them at work in the dead wood of a fallen tree.

❶ Detritus feeders break up the tree into crumbly pieces. Woodlice, wood-boring beetles and millipedes are detritus feeders (that is they feed on small particles of dead material). Their activity increases the surface area of the wood which fungi and bacteria can attack.

❷ Fungi feed on the dead wood making it decompose and decay. They are saprophytes.

❸ The shredded material and the wastes which the detritus feeders produce is called humus. It is a dark coloured, fibrous material. The original tree is by this time unrecognisable.

❺ The nutrients in the soil help new plants to grow.

❹ Decomposers (fungi and bacteria) break down humus. Minerals and organic compounds are released into the soil.

Figure 5.1C ● The formation of humus

Humus improves soils by the following means:
- It improves the texture of clay soils by helping the particles to stick together to form crumbs.
- It improves the texture of sandy soils by increasing the soil's ability to hold water.
- Humus reduces the leaching of minerals.
- By absorbing water, humus makes soil more fertile. This is especially important in sandy soils.
- Humus provides food for detritus feeders, e.g. woodlice and earthworms, which in turn fertilise the soil.
- Its high water content enables humus to absorb heat and warm the soil.

Water table

The roots of a plant anchor it in the ground and extract nutrients from the soil. Many plants have enough room for their roots in a layer of soil 15 cm deep: some plants need a layer of soil 4 m deep.

The natural level of water is called the **water table** (see Figure 5.1D) Plants need to have their roots above this, so that they can draw water from the water table without becoming wet. Roots should not be immersed in water the whole time. By means of a good drainage system, it is possible to lower the water table. Farms are careful to ensure good drainage in their fields. Sometimes, they install underground pipes. Good drainage helps air to enter the soil; it helps seeds to start growing, and it helps to control plant diseases.

SUMMARY

Sandy soil: easy to cultivate; contains air, which assists growth of plant roots; needs frequent rain; may lack minerals as minerals are leached from the soil.
Clay soil: sticky and difficult to cultivate in wet weather; can sustain crop growth in dry periods; may become waterlogged and lack air; rich in minerals.
Loam: a mixture of sandy soil and clay soil; easily worked; contains air; holds water without becoming waterlogged.
Humus: improves all soils; binds sand particles together; increases water retention.

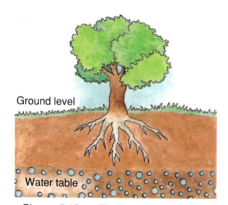

Ground level

Water table

Figure 5.1D ● The water table

Soil as a habitat for animals

Soil is teeming with living organisms. One hectare (10 000 m²) of soil contains a million earthworms and millions of mites, springtails, millipedes, centipedes, woodlice, fly larvae, beetles, ants and microorganisms. Some soil organisms are pests, for example wireworms burrow into root crops such as potatoes and ruin them. Others play a vital part in maintaining the fertility of the soil. Some bacteria feed on the dead remains of plants and animals to make humus. Other bacteria **fix** nitrogen; that is, convert nitrogen in the air into nitrogen compounds, on which plants can feed. Earthworms improve the texture of the soil. Moving through it, they open channels for roots to spread through. As well as mixing the soil, they help air to enter and help water to drain from the soil.

Water

Plants obtain water from the soil. Plants lose water constantly as water vapour passes out through openings in the leaves of plants into the atmosphere. This process is called **transpiration**. Plants take in water through their roots to replace that lost by transpiration. Dissolved in the water are many nutrients the plants need. During a dry spell, plants need to be watered (see Figure 5.1E).

Figure 5.1E ● Many plants need to be watered during dry spells

Air

Soil must be aerated (contain air) so that plant roots can obtain air. Many of the microorganisms in the soil also need air. If soil is packed down by tramping feet as hikers walk across the land, or by heavy vehicles, such as tractors, it will contain little air.

Nutrients

Plants need certain chemical elements. In photosynthesis, plants obtain carbon and oxygen from air, and hydrogen and oxygen from water. Land plants obtain the rest of the elements which they need from the soil. Aquatic plants obtain their nutrients from water. Table 5.2 shows the elements which plants obtain from soil. The major elements are needed in large amounts (on a kilogram per hectare scale), and the trace elements are needed in small amounts (on a gram per hectare scale).

LOOK AT LINKS
for **nitrogen fixation**
See Topic 2.6.

LOOK AT LINKS
for **plant nutrition**
See Topic 11.1.

LOOK AT LINKS
for **transpiration**
See Topic 14.1.

Table 5.2 ● Elements obtained by plants from soil

Major elements	Importance	Trace elements
Nitrogen	Needed for protein synthesis	Manganese
Phosphorus	Needed for development of roots, energy-transfer reactions, nucleic acids	Copper
Iron
Zinc
Potassium	Needed in photosynthesis	Chlorine
Magnesium	Present in chlorophyll	Boron
Sulphur	Needed for synthesis of some proteins	Molybdenum
Calcium	Needed for transport	

Soil Analysis
(program)

Have you experimented with soil in a laboratory? You can try some computer experiments as well using this software.

The natural cycle is for plants to die and decay, returning chemicals to the soil. When crops are harvested, the chemicals they contain are not returned to the soil. Fertilisers are needed to replenish the soil with the nutrients which the crops have taken out of it.

pH

A soil may be acidic or alkaline or neutral. For each type of plant, there is a range of soil pH over which the plant can be grown successfully.

Table 5.3 ● Crop growth and pH

Crop	Potatoes	Swedes	Oats	Wheat	Sugar beet	Kale	Barley
pH	2	4.7–5.6	4.8–6.3	6.0–7.5	7.0–7.5	9–11	12

Farmers have a choice. They can measure the pH of their soil and then choose a crop which will grow well on it. Alternatively, they can alter the pH of the soil to suit the crop which they want to grow. Often soils are too acidic. Rainwater, being weakly acidic, neutralises bases which are present in topsoil. When soils are too acidic, farmers add lime (calcium oxide). This neutralises the excess of acid. Occasionally, soils are too alkaline. Then iron(II) sulphate is added. This salt is weakly acidic in solution and neutralises the excess of alkali in the soil.

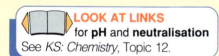

LOOK AT LINKS
for **pH** and **neutralisation**
See *KS: Chemistry*, Topic 12.

Erosion

The top few centimetres of soil contain the most nutrients. This **topsoil** can be **eroded**: that is, washed away by high rainfall or blown away by strong winds. When farmers convert grassland into arable fields, erosion may follow (see Figure 5.1G). A field of winter wheat can lose $2.4\,kg/m^2$ of soil or more to erosion by wind and rain each year. Loss of soil leads to weaker plants which are more likely to become diseased.

There are many methods of preventing soil erosion.
- Grass prevents erosion: its roots keep the soil in place.
- Hedges reduce erosion by reducing the speed of the wind.
- Forests reduce erosion as the roots of the trees hold soil in place.
- Farmers plough land so that furrows lie across the natural slope of the land; rainwater will then not run down the furrows and cause erosion.
- A terraced hillside is less likely to be eroded than a natural hillside. By slowing down the flow of rain down a hillside, terraces give the water time to soak in and nourish the crop (see Figure 5.1F).

The flooding which Pakistan experienced in 1988 was made much worse by deforestation. The acute shortage of fuel had made people cut down trees for burning. As a result, when the rains came, the flood waters were not held back by the forests which had previously helped to avoid a disaster.

Figure 5.1F ● Terracing to prevent soil erosion

Figure 5.1G ● Soil erosion

SUMMARY

Soil contains small particles of weathered rock and humus – decayed organic matter. Plants need enough soil for their roots to grow in. They also need water, air and certain chemical elements as nutrients. The soil must have a suitable pH. Topsoil can be eroded by rain and by wind. Soil needs to be cultivated if it is to grow crops.

Cultivation

Soil needs to be cultivated, that is, broken up. A gardener digging the garden and a farmer ploughing the land are both cultivating the soil. The benefits of cultivation are:
- uprooting weeds,
- mixing fertilisers and decaying plants with the soil,
- introducing air into the soil,
- giving plants enough depth for their roots to grow.

CHECKPOINT

❶ Christine measured 50.0 cm³ of soil in a large measuring cylinder. She added 50.0 cm³ of water from a second measuring cylinder. After she had stirred the soil to dislodge all the air bubbles, the final volume was 90 cm³.
 (a) Why did Christine take care to tap the soil down in the measuring cylinder?
 (b) What volume of air was driven out of the soil?
 (c) What is the percentage by volume of air in this sample of soil?

❷ Christopher weighed 20.0 g of soil and put it into an oven at 100 °C. After an hour, he took the sample from the oven, let it cool and reweighed it. The mass was 15.0 g.

(a) What is the mass of the water lost on heating?
(b) What is the percentage by mass of water in the soil?
(c) What should Christopher have done to make sure that *all* the water has been driven out of the soil sample?

3 Caroline took 10.0 g of the dry soil from Christopher's experiment. She heated it strongly in a crucible for 20 minutes. The humus burnt away to form carbon dioxide and water. The mass of the soil was 9.0 g.
(a) What mass of humus was present in the soil sample?
(b) What is the percentage by mass of humus in this soil sample (that is, Caroline's dry soil)?
(percentage by mass of humus = (mass of humus/total mass of soil) x 100)
(c) Why did Caroline not use a fresh sample of soil for her experiment?
(d) What should she have done to make sure that all the humus had been burnt away?

4 A friend of yours is given an indoor potted plant. She asks your advice about looking after it. What advice do you give her about:
(a) watering (how often),
(b) the size of pot (what to do when the plant grows),
(c) forking the soil in the pot occasionally?

5 A propagator is a heated tray covered with a plastic lid. It contains potting compost. Its purpose is to help seeds to propagate. Why do seeds sown in a propagator have a better chance than seeds sown in a field?

6 Miss Sue N Dunn decides that there ought to be an easier way of gardening. She does not like digging so she pulls out the weeds and rakes over the topsoil. Then she plants her seeds. What are the disadvantages of Sue's method?

7 Give two reasons in each case why the following are important features of soil: good drainage, cultivation, microorganisms, the level of the water table.

5.2 Air and water

FIRST THOUGHTS

As you study this section think about the vital importance of air as a source of:
- oxygen – which supports the life of plants and animals
- nitrogen – the basis of natural and synthetic fertilisers

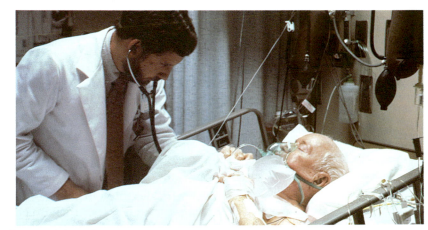

Figure 5.2A ● Oxygen in hospitals

Hospitals need oxygen for patients who have difficulty in breathing (see Figure 5.2A). Tiny premature babies often need oxygen. They may be put into an oxygen high pressure chamber. This chamber contains oxygen at 3–4 atmospheres pressure, which makes 15–20 times the normal concentration of oxygen dissolve in the blood. People who are recovering from heart attacks and strokes also benefit from being given oxygen. Patients who are having operations are given an anaesthetic mixed with oxygen.

In normal circumstances, we can obtain all the oxygen we need from the air. Figure 5.2B shows the percentages of oxygen, nitrogen and other gases in pure, dry air. Water vapour and pollutants may also be present in air.

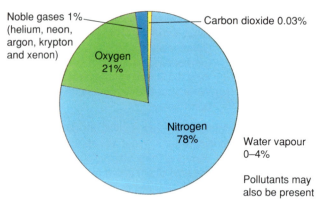

Noble gases 1% (helium, neon, argon, krypton and xenon)

Carbon dioxide 0.03%

Oxygen 21%

Nitrogen 78%

Water vapour 0–4%

Pollutants may also be present

Figure 5.2B ● Composition of clean, dry air in percentage by volume

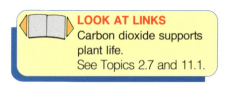

LOOK AT LINKS
Carbon dioxide supports plant life.
See Topics 2.7 and 11.1.

Nitrogen is a colourless, odourless gas which is slightly soluble in water. It does not readily take part in chemical reactions. Many uses of nitrogen depend on its unreactive nature. Liquid nitrogen (below –196 °C) is used when an inert (chemically unreactive) refrigerant is needed. The food industry uses it for the fast freezing of foods.

Vets use the technique of artificial insemination to enable a prize bull to fertilise a large number of dairy cows. They carry the semen of the bull in a type of vacuum flask filled with liquid nitrogen.

Many foods are packed in an atmosphere of nitrogen. This prevents the oils and fats in the foods from reacting with oxygen to form rancid products. As a precaution against fire, nitrogen is used to purge oil tankers and road tankers. The silos where grain is stored are flushed out with nitrogen because dry grain is easily ignited.

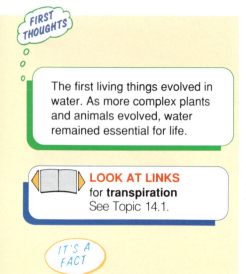

FIRST THOUGHTS

The first living things evolved in water. As more complex plants and animals evolved, water remained essential for life.

LOOK AT LINKS
for **transpiration**
See Topic 14.1.

IT'S A FACT

A large tree can lose 300 litres of water vapour in an hour by transpiration.

The water cycle

Where does all the rain come from? Why does the atmosphere never run out of water? Four-fifths of the world's surface is covered by water. From oceans, rivers and lakes, water evaporates into the atmosphere. Plants give out water vapour in **transpiration**. As it rises into a cooler part of the atmosphere, water vapour condenses to form clouds of tiny droplets. If the clouds are blown upward and cooled further, larger drops of water form and fall to the ground as rain (or snow). *Where does the rain go?* Rain water trickles through soil, where some is taken up by plants. The rest passes through porous rocks to become part of rivers, lakes, ground water and the sea. This chain of events is called the **water cycle** (see Figure 5.2C).

As rain water falls through the air, it dissolves oxygen, nitrogen and carbon dioxide. The dissolved carbon dioxide forms a solution of the weak acid, carbonic acid. Natural rain water is therefore weakly acidic. In regions where the air is polluted, rain water may dissolve sulphur dioxide and oxides of nitrogen, which make it strongly acidic; it is then called **acid rain**. As rain water trickles through porous rocks, it dissolves salts from the rocks. The dissolved salts are carried into the sea. When sea water evaporates, the salts remain behind.

LOOK AT LINKS
for **clouds**
See *KS: Physics*, Topics 3.4 and 3.5.

LOOK AT LINKS
for **acid rain**
See Topic 7.5.

SUMMARY

The water cycle:
• Evaporation, transpiration and respiration send water vapour into the atmosphere.
• Condensation forms clouds which return water to the Earth as rain, hail or snow.

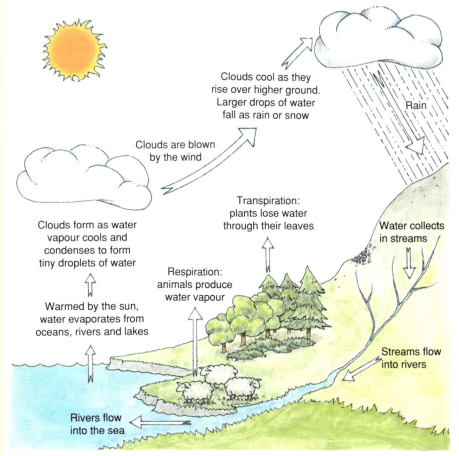

Clouds cool as they rise over higher ground. Larger drops of water fall as rain or snow

Clouds are blown by the wind

Rain

Transpiration: plants lose water through their leaves

Water collects in streams

Clouds form as water vapour cools and condenses to form tiny droplets of water

Respiration: animals produce water vapour

Warmed by the sun, water evaporates from oceans, rivers and lakes

Streams flow into rivers

Rivers flow into the sea

Figure 5.2C ● The water cycle

FIRST THOUGHTS

Oxygen sensors connected to a computer are used to monitor the concentration of dissolved oxygen in commercial fish tanks. Fish farmers and breeders are warned when the concentration starts dropping and can take preventive action before any damage is done.

SUMMARY

Air dissolves in water. The dissolved oxygen in water keeps fish alive. If too much organic matter, e.g. sewage, is discharged into a river, the dissolved oxygen is used up in oxidising the organic matter, and the fish die.

Dissolved oxygen

The fact that oxygen dissolves in water is vitally important. The solubility is low: water can dissolve no more than 10 g oxygen per tonne of water, that is 10 p.p.m. (parts per million). This is high enough to sustain fish and other water-living animals and plants. When the level of dissolved oxygen falls below 5 p.p.m. aquatic plants and animals start to suffer.

Water is able to purify itself of many of the pollutants which we pour into it. Bacteria which are present in water feed on plant and animal debris. These bacteria are **aerobic** (they need oxygen). They use dissolved oxygen to oxidise organic material (material from plants and animals) to harmless products, such as carbon dioxide and water. This is how the bacteria obtain the energy which they need to sustain life. If a lot of untreated sewage is discharged into a river, the dissolved oxygen is used up more rapidly than it is replaced, and the aerobic bacteria die. Then **anaerobic** bacteria (which do not need oxygen) attack the organic matter. They produce unpleasant-smelling decay products.

Some synthetic (manufactured) materials, e.g. plastics, cannot be oxidised by bacteria. These materials are nonbiodegradable, and they last for a very long time in water.

The earliest human settlements were always beside rivers. The settlers needed water to drink and used the river to carry away their sewage and other waste. Obtaining clean water is more difficult now.

SCIENCE AT WORK

The water industry makes use of computers. Sensors detect the pH, oxygen concentration and other qualities of the water and relay the measurements to a computer. This constant monitoring enables the industry to control the quality of the water it provides.

IT

Water Treatment
(program)

This program allows you to control and maintain a town's domestic water supply.

Water treatment

The water that we use is taken mainly from lakes and rivers. Water treatment plants purify the water to make it safe to drink. They do this by:
- filtration to remove solid matter followed by...
- bacterial oxidation to get rid of organic matter and...
- treatment with chlorine to kill germs.

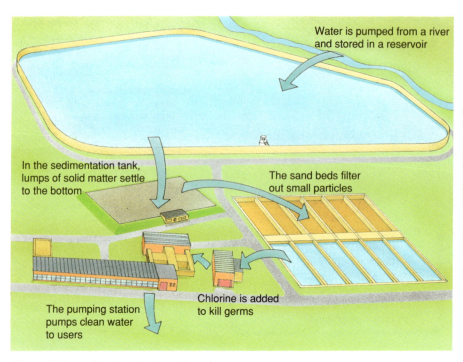

Figure 5.2D ● A water treatment works

CHECKPOINT

❶ The table shows the world consumption of water over the past 30 years.

Year	World consumption of water (millions of tonnes per day)
1960	10.0
1970	11.5
1975	13.0
1980	15.0
1985	17.0
1990	20.0

(a) On graph paper, plot the consumption (on the vertical axis) against the year (on the horizontal axis).
(b) Say what has happened to the demand for water over the past 30 years.
(c) Suggest three reasons for the change.
(d) From your graph, predict what the consumption of water will be in the year 2000.

❷ Explain why oxygen is used in hospitals and what advantage it has over air.

5.3 The world problem

Our effect upon ecosystems is highlighted daily in the newspapers and on television (see Figure 5.3A). Not a day goes by without our hearing of a new environmental crisis: the pollution of our air and water, soil erosion and the creation of deserts, the population explosion and the shortages of food and shelter, destruction of natural habitats such as tropical rain forests and vanishing wildlife.

Figure 5.3A ● Environmental disaster - deforestation in the Amazon

The depletion of the ozone layer, acid rain, the increase in the greenhouse effect have all brought home to us the fact that our planet is fragile. Disasters such as oil spills and chemical leaks bring environmental problems to our attention, but less serious incidents happen everyday. *What is our response?* It is easy to think that these disasters are far too vast and are impossible to cope with. *But why?* Over the past fifty years we have made great strides in reducing poverty. Improvements have come in health: infant mortality has been reduced and many diseases can now be controlled. Advances have also been made in combatting food shortages, improving welfare and education. In 1990 some nations began to reduce their stocks of arms. If all nations did this, there would be more money to spend on undoing the damage which we have done to the Earth. One thing is sure, we cannot continue to use the Earth as a rubbish tip. This would only be storing up unsolvable problems for future generations. We have only one Earth.

Agricultural chemicals

Between 1945 and 1975, the world's food output doubled. This 'Green Revolution' was brought about by improved agricultural technology, advances in crop breeding and the use of agricultural chemicals. The rapid rise in production now seems to be slowing and a number of unexpected side-effects are surfacing. The structure and properties of the soil seem to have been changed. Over-use of fertilisers, the addition of lime to counter acidity and constant ploughing have taken their toll. The Soviet Union in the 1950s and 60s made huge efforts to produce more grain. It is now suffering the after-effects. There are areas where the soil is exhausted and the land has to remain fallow.

● Fertilisers

Fertilisers improve crop growth by giving the plants nutrients that they need, such as nitrates and phosphates (see Figure 5.3B). Some of these chemicals are not taken up by crops and find their way into our waterways when they are washed through the soil by rain. Once this agricultural 'run-off' gets into the surface waters it can have adverse effects on the environment.

Figure 5.3B ● Applying artificial fertiliser

Figure 5.3C ● Crops are often sprayed with herbicides and insecticides

● Herbicides

LOOK AT LINKS
for **dioxin**
See Topic 6.7.

Herbicides are chemicals used to eliminate weeds. Weeds compete with the crop plant for light, space, water and soil nutrients. They can seriously reduce crop yield. Many herbicides are made from the chemical dinitrophenol. During the manufacture of dinitrophenol, a harmful impurity is formed called **dioxin**.

● Insecticides

LOOK AT LINKS
for **insecticides**
See Topic 6.8.

Insecticides are chemicals that kill pest insects. The use of insecticides has saved millions of lives. Malaria, typhus and yellow fever are all carried by insects. It is estimated that without pesticides, crop losses of 45% would occur in the tropics. However insecticides have proved to have dangerous side effects.

The cottony-cushion scale insect became a pest in the orange groves of California when accidentally introduced from Australia. It had no natural predators in California so scientists went to Australia and found a species of ladybird. They introduced this species to California where it quickly reduced the scale insect numbers so that it was no longer a pest. This is an example of biological control.

Alternative methods of controlling pests are being explored. **Biological control** involves the introduction of the pest's natural predator in order to reduce its numbers. A classic example of biological control comes from Australia, where farmers planted the prickly-pear cactus in an attempt to keep wallabies from eating their crops. The prickly-pear grew so well that it covered millions of acres of grassland and became a greater nuisance than the wallabies had been. The answer to the problem was found by introducing a moth from Argentina. The moth's caterpillars fed on the cactus. The spread of the cactus was halted and much of the land reclaimed for agricultural use. Parasitic wasps have also been used successfully to control many crop pests throughout the world.

Radiation

Radiation probably represents the most feared type of pollutant. Sources of non-natural radiation come from the testing and use of nuclear weapons and from leakages from nuclear reactors (see Figure 5.3D). Radioactive waste from the latter is a mixture of different isotopes. These all **decay** (give off radiation) at different rates. The half-life of a radioactive isotope is the time taken for its activity to decay to half of its original level.

Figure 5.3D ● A nuclear reactor

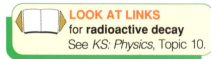

LOOK AT LINKS
for **radioactive decay**
See *KS: Physics*, Topic 10.

LOOK AT LINKS
for **the greenhouse effect**
See Topic 7.10;
for **acid rain** and **sulphur dioxide**
See Topics 7.4 and 7.5;
for **carbon monoxide**
See Topic 7.6;
for **CFCs**
See Topic 7.9.

Polluting gases

- **Carbon dioxide** is not only produced when we breathe, it is released into the atmosphere whenever fossil fuels, such as coal, oil and natural gas are burnt. Carbon dioxide enhances the **greenhouse effect**.
- **Sulphur dioxide** is released into the atmosphere when fossil fuels are burnt. Together with oxides of nitrogen in car exhaust fumes, sulphur dioxide eventually falls as **acid rain**.
- **Carbon monoxide** is produced by car exhausts. Haemoglobin in red blood cells takes up carbon monoxide in preference to oxygen.
- **CFCs** (chlorofluorohydrocarbons) are used in refrigerators and as propellants in aerosol sprays. It has been found that the release of CFCs into the atmosphere is destroying the ozone layer.

Pollution of lakes and rivers

Our waterways are often the sink of domestic, industrial and agricultural wastes.

● *Sewage*

Sewage is often discharged into our rivers and is the biggest single pollutant (see Figure 5.3E). It effects resemble those of **eutrophication**. It encourages rapid growth of bacteria which use up oxygen from the water during respiration. The result is that the river will be depleted of fish and many invertebrates because of the lack of oxygen. It takes time for the sewage to be broken down by the bacteria, so the amount of dissolved oxygen in the river only reaches its original, unpolluted level some way downstream.

● *Heat*

Water is used to cool many industrial processes. The largest user is the electricity industry in its power stations (see Figure 5.3F). Water is returned to the river at a higher temperature than the river temperature. Oxygen becomes less soluble in water as the temperature increases.

LOOK AT LINKS
for **pollution by sewage**
See Topic 8.3.

LOOK AT LINKS
for **eutrophication**
See Topic 8.4.

LOOK AT LINKS
for **thermal pollution**
See Topic 8.2.

Pollution of the seas

We are at long last beginning to realise that the dumping of waste into our seas is storing up long-term problems for the future. Our rivers empty toxic chemicals into the sea as they drain from the land and are discharged as waste from industries. They include:

- **Fertilisers and sewage** increase the growth of algae and threaten the balance of natural ecosystems such as the Great Barrier Reef in Northern Australia. The rapid growth of the algae is killing the coral, which grows best in low nutrient concentrations.

- **Insecticides** such as DDT, aldrin and dieldrin, used in the protection of crops, also drain off the land into rivers. Shellfish are able to concentrate pesticides in their tissues.

- **Radionuclides** can also reach high levels in tissues. They can be found in high concentrations near coastal nuclear power stations.

- **Toxic metals** such as mercury, cadmium, copper and lead can also be concentrated up food chains. **Minamata disease** is one of the most striking examples of the adverse effects of heavy metal pollution. Cadmium has been found in high concentrations in the bodies of fish in the Bristol Channel. Lead concentrations in the North Sea have increased three-fold and copper dumped off the Dutch coast resulted in the deaths of thousands of fish.

LOOK AT LINKS
for **Minamata disease**
See Topic 8.1.

Figure 5.3E ● Sewage – the biggest water pollutant

Figure 5.3F ● Power station cooling towers

● *Oil pollution*

Modern oil tankers carry millions of tonnes of unrefined, crude oil. The pollution from spillages that we see in the media is made worse by the practice of washing out oil tanks at sea. The advent of large scale offshore drilling brings further risk. Names like the *Exxon Valdez*, *Torrey Canyon* and *Amoco Cadiz* conjure up familiar pictures of oil covered beaches. The sufferers include fish-eating birds, whose feathers become clogged, so they are unable to fly and soon die of exposure, drowning or starvation (see Figure 5.3G). Marine mammals such as seals and sea otters can be affected by the oil penetrating their fur. The detergents sprayed on the shorelines of Britain to disperse the oil were toxic to much intertidal life.

Figure 5.3G ● Oiled sea bird

LOOK AT LINKS
for **oil pollution**
See Topic 8.6.

SUMMARY

Much threatens the delicate balance of life in our ecosystems due to the intervention of man. We have a duty of care to redress this balance and protect it for future generations.

5.4 Conservation

Habitat destruction and exploitation have meant that thousands of species of plants and animals have been brought to the verge of extinction. In the face of economic development, the growth of human populations and the increase in agriculture to feed them, habitats are vanishing quickly and their species with them. The destruction of tropical rain forests means that 500 to 1000 species of plants are becoming extinct every year. They are disappearing even before they can be discovered by man and their potential value, for instance in providing new medicines, is wasted. Animal species are now disappearing at the rate of one species per day. Humans, through greed and ignorance, are felling forests that not only harbour a huge diversity of species, but provide the oxygen we breathe and use up carbon dioxide.

IT'S A FACT

The Amazon rain forest is thought to supply us with one third of the world's oxygen. Its destruction means a build up of carbon dioxide, enhancing the greenhouse effect.

Figure 5.4A ● Species are disappearing from the rain forest at a rate of 500 to 1000 a year

Figure 5.4B ● The Kenyan authorities burned hundreds of tonnes of poached ivory to prevent it reaching the market-places of the world

Ecology and Conservation
(program)

Slapton Ley is a nature reserve in Devon. It needs good management. *Can you do the job?* The videodisc gives you the background information. You can control the reserve by computer. Beware – other people might have different ideas about how to manage the reserve.

Ecodisc
(videodisc)

Use the program to look at a woodland and a pond foodweb. You can then change the balance of the ecosystem. Finish off with the quiz.

We can benefit future generations by preserving our environment. We have a duty of care to manage wild species which can be of direct use to us in terms of food and other materials. We must realise that over-exploitation of non-renewable resources will leave us poorer in the future. It is also in our interests to safeguard the existence of species that are a part of the delicate balance of nature in our world. There are many animals that we shall never see. The giant otter, wood bison, Parma wallaby, spectacled bear and Atlantic walrus are thought to be extinct or close to extinction. The destruction of habitats to enable agriculture to feed ever more human mouths is one cause. The over-exploitation of species for commercial use is another. In a world where species are disappearing so fast, perhaps we should look again at the sort of legacy we wish our children to inherit.

Many of the species that are approaching a premature end, could in fact become food for us if they were managed. The Saiga antelope in the USSR has been saved from the brink of extinction and has now increased to numbers that allow some animals to be killed for food. The same is true of the American bison and the eland.

Plant and animal breeders combine desirable characteristics to produce improved strains, e.g. crops with high yields and farm animals with more meat. To produce the new varieties, breeders often use the original wild ancestors of modern stock. If we fail to preserve rare breeds their gene banks will not be available for use at some future date.

We have to take action if fish and whale stocks are to be conserved. There has been a serious decline in the numbers of many marine molluscs from tropical waters, such as cowries, scallops, cone shells and clams. This is because too many are harvested and exported to the USA and Europe, leaving too few to breed.

In the 1970s there was a large increase in the amount of illegal ivory poaching. During the decade half of the elephants in Kenya were lost to poaching and Uganda lost 90% (see Figure 5.4B). Approximately 50 tonnes of ivory is used in Europe each year, equivalent to 10 000 elephants. Illegal poaching also concentrates the profits in the hands of a few. Proper management with the sale of ivory from animals that have died naturally and from those killed to control numbers, would help to improve the economies of some African countries.

Sometimes the conservation of wildlife can be a personal responsibility of each of us. The trade in furs and other animal skins is an obvious example (see Figure 5.4C). If we stopped buying furs, then the slaughter

97

Figure 5.4C ● The fur trade

Figure 5.4D ● Birdcages and birds for sale. Many of the birds sold in such markets are endangered species.

SUMMARY

Conservation of species and protection of their habitats has had a low priority in the eyes of the nations of the world in the past. It must constantly be brought to their attention that the world's resources are limited. If we do not act we will pay a high price in terms of the loss of beautiful, fascinating and essential wildlife.

would stop, and after all there are plenty of imitation furs available. Prosperity increased in the 1960s and more women were able to afford furs. In 1968 the USA imported 10 000 leopard skins, 1000 cheetah skins, 13 000 jaguar skins and 130 000 ocelot skins. Similar numbers were imported into Europe. The shooting and trapping took their toll on the wild species. At last the Convention on International Trade in Endangered Species (CITES) asked governments to pass laws to protect threatened species. But some governments have yet to recognise CITES and widespread poaching and illegal trading still occur.

The trade in exotic birds takes 10 million birds from the wild every year; about half of these die even before they reach their destination. High prices are paid for parrots, waxbills, lovebirds and other exotic species. Also birds of prey, such as the peregrine falcon are exported to be used in falconry. Despite the protection of CITES illegal trafficking continues, with the loss of many birds in transit (see Figure 5.4D).

People have introduced new species of animals into many countries. Some of these have been beneficial; others have upset the native animal species. Cattle have been introduced into the Americas, Australia and Africa, where there has been a large increase in meat production. On the other hand, rabbits were introduced into Australia only to become a pest and to threaten native species. Rats have spread all round the world on ships and have carried their diseases with them.

The news has not all been bleak. In the early 1970s, the tiger was on the verge of extinction, with just 1800 left. Hunting, trapping and the destruction of its natural habitat were the main causes. In 1973, the World Wildlife Fund, with the co-operation of the Indian government, launched 'Operation Tiger'. They attempted not only to save the tiger but also to conserve a whole ecosystem. Laws were passed to protect the species and tiger reserves were set up. Poachers were given very stiff sentences. The operation has resulted in a dramatic increase in the tiger population.

Captive breeding programmes have also seen some species saved from extinction. The Arabian oryx has been bred in captivity in Phoenix Zoo, Arizona. Herds have since been set up in other parts of the world and some even released into the wild in Oman. Other species saved by captive breeding include Przewalski's horse, the European bison and Pere David's deer.

Some years ago, a number of international organisations combined forces to put forward the World Conservation Strategy (WCS). This plan was to persuade nations to pass laws which would protect plant and animal species from destruction and preserve their habitats. The WCS aim is to conserve the natural processes and cycles of the biosphere. If this aim is achieved, then habitats will be preserved and ecosystems will be saved. The WCS plans to preserve the Earth's variety of species by controlling the killing of animals and the destruction of plants. Some governments agree with the WCS aims, but have so far failed to put the programme into action. Other governments are indifferent. At present, all hope of progress in conservation lies with the independent organisations.

CHECKPOINT

❶ Give examples of the relationship between economic development and damage to the environment.

❷ How do rain forests affect the world's climate?

❸ Summarise the programme of the World Conservation Strategy for the 1990s.

TOPIC 6 FARMING

6.1 Intensive farming

In 1990 there were twice as many people in the world as there were in 1950. Farmers have to supply more food to meet the needs of the growing world population.

TRY THIS

The population of the world is increasing.

Year	World population (millions)
1800	907
1900	1610
1950	2509
1970	3650
1983	4670
2000	—

On graph paper, plot the figures in the table. Extrapolate your graph to the year 2000. What population do you forecast?

When the human population began to rise sharply in the nineteenth century, the need for food increased. Farming methods had to become more **intensive**, that is, to produce as much food as possible from the land available for raising crops (arable farming) and animals (livestock farming).

Figure 6.1A ● Intensive cereal production

The field of wheat in Figure 6.1A has been grown by intensive methods. Chemical fertilisers have been applied to produce stronger, larger plants in a shorter time. Chemical pesticides have been used: insecticides have been sprayed to kill insect pests and herbicides have been applied to kill weeds. The use of all these chemicals costs the farmer money. The farmer has to balance this expense against the increased yield of wheat grain. Most of our bread is made from cereals which have been grown by intensive methods.

Most of the beef and pork we eat is produced by intensive methods. Cattle and pigs will find their own food if free to graze. The pigs shown in Figure 6.1B have been kept in pens and given fodder which has been bought in, not produced on the farm. As a result of the lack of exercise, cattle and pigs gain weight more quickly and so are ready for slaughter earlier. The farmer has to balance the cost of feed against the shorter time for which he has to feed the cattle.

Scientists and engineers have developed new technologies to help farmers to produce food in great quantities. In the following sections we shall look at the various factors that contribute to intensive farming. Sometimes the new methods have an unwanted effect on the environment. We have to balance the advantages of the new methods against the impact which they have on the environment.

Figure 6.1B ● Intensive pork production

CHECKPOINT

❶ (a) Why are cereals so important for the diet of the human population? (See Topic 11.)

(b) Use the bar chart opposite to work out the ratio:
Cereal production in 1985/
Cereal production in 1965.

(c) What changes in farming methods have made the increase possible?

(d) Which block shows the greatest percentage increase over the preceding figure?

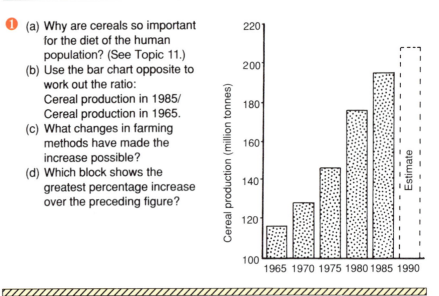

6.2  Irrigation

In some parts of the world, the rainfall is too low for crops to grow well. Yields can be increased by **irrigation**, that is, by supplying water. The extra water is led in along canals or through overhead pipes. Much of the world's food is grown on irrigated land. However, it is very costly for developing countries to buy, install and maintain the necessary equipment.

Figure 6.2A ● An irrigation pipe

● The Aswan High Dam, Lake Nasser and the Nile

Lake Nasser in Egypt is 500 km long. From it flows the River Nile. When there is heavy rainfall, the level of water in Lake Nasser rises. The Aswan High Dam was built to prevent Lake Nasser overflowing into the Nile. Instead, the lake serves as a reservoir of water in dry years. In 1985 when other African countries, such as Ethiopia, suffered from drought, Egypt had the water from Lake Nasser. There have, however, been some drawbacks.

• Before the dam was built, the Nile used to flood. The silt which it carried to vast areas on its banks was rich in minerals. Now that the Nile no longer carries this fertiliser to the farms on its banks, farmers must buy synthetic fertilisers.

LOOK AT LINKS
for **flukes**
See Topic 1.5.

- The sodium chloride content of the banks of the river has increased because it is no longer washed out by flood waters. Many crops do not grow well in soil with a high salt content.
- Irrigation canals carry water from the Nile to the farmlands on its banks. They also carry snails from Lake Nasser. Some of the snails act as hosts to a tiny worm called the **bilharzia fluke**. These worms can infect anyone wading in the canal. When they do, they cause bilharzia, which is a serious disease, attacking the liver, lungs and heart. There has been a big increase in bilharzia since the dam was built because the slow-flowing water in the irrigation canals is ideal for the snails.
- Fishermen used to make good catches in the Nile Delta (see Figure 29.3B). Now that the Nile silt no longer reaches the delta, natural food chains are broken. There is little plankton for fish to live on, and catches are poor.

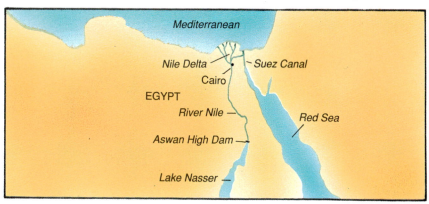

Figure 6.2B ● Egypt and the Nile

SUMMARY

Irrigation increases soil fertility. There may be some drawbacks, such as accumulation of salt and the spread of water-borne diseases.

CHECKPOINT

① (a) Name a benefit which the Aswan Dam has brought to Egypt.
(b) What drawbacks have there been to damming Lake Nasser?
(c) On balance, do you think the Aswan Dam has been a good investment for Egypt? Give reasons for your answer.

6.3 ● Mechanisation and its effect on wildlife

FIRST THOUGHTS

Since 1945, 200 000 km of hedges have been removed in England and Wales.

The change from traditional farming practices to intensive farming methods happened in the UK after 1945. An important part of the change was the invention of large machines for cultivation and harvesting. These modern machines work best in large fields where they do not have to stop for hedges and fences (see Figure 6.3A). Small fields were joined to form large fields by chopping down copses and uprooting hedges. The animals and plants which lived in the woods and hedgerows lost their habitats. Table 6.1 shows how some wildlife habitats have decreased. Table 6.2 shows how some animals have been affected.

Table 6.1 ● Losses in some habitats between 1945 and 1990

Habitat	Percentage lost
Unimproved grassland, including hay meadows rich in flowers	95
Lowland heaths	40
Lowland woods	40
Upland heaths	30

Figure 6.3A ● Large machinery means large fields

Table 6.2 ● Numbers of species in traditional and modernised farms

Group	Number of species of each group in the two habitats	
	Traditional farm habitat	Corresponding habitat in modernised farm
	(Hedges and natural grass verges)	(Wire fences and sown grass)
Mammals	20	5
Birds	37	6
Butterflies	17	0
	(Permanent ponds and ditches)	(Temporary ditches and piped water)
Amphibians	5	2
Fish	9	0
Dragonflies	11	0
Snails	25	3

Large fields mean that large machinery can be used more effectively. Crops can be harvested more quickly, with a resulting decrease in labour costs. Land drainage can be improved and boundary hedges can be dispensed with. However, the creation of large fields has some bad effects on the environment.

- Hedges and pockets of woodland provide habitats for our native wildlife. These habitats are lost when large fields are created.
- Heavy machinery can compact the land, which may then become waterlogged.
- Large machines use a lot of fuel.
- Hedges are natural windbreaks. When they are removed, the wind can cause more erosion of topsoil and more loss of water from the soil by evaporation.

SUMMARY

Combining small fields to form large fields makes agriculture more efficient. However, plants and animals lose their habitats when woods and hedges are destroyed, and soil erosion is increased.

CHECKPOINT

❶ (a) Describe a dragonfly to a child who has never seen one.
 (b) Say what harm it would do if there were no more dragonflies.

❷ Farmer Brown's daughter Jemima has married the boy next door, Farmer White's son Jake. The families have decided to combine the two farms and are planning to cut down hedgerows to make larger fields.
 (a) Write a letter to Jemima and Jake explaining why you want them to retain hedgerows.
 (b) Write a reply from the Whites, explaining what they can gain from having larger fields.

❸ (a) Plot the following figures as a bar graph.

Year	1947	1969	1980	1985
Length of hedgerows (x1000 km)	750	650	610	590

(b) Who benefits from the removal of hedgerows? Give two benefits which the destruction of hedgerows makes possible.

(c) Who loses by the removal of hedgerows? Give two disadvantages of destroying hedgerows.

6.4 Monoculture

FIRST THOUGHTS

Mechanisation, large fields and monoculture go together.

Intensive arable farms specialise in growing a small number of cash crops. Growing only one crop over a large area is called **monoculture** (see Figure 6.4A). It enables a farmer to obtain high yields which he can sell in bulk for a better price. Monoculture makes efficient use of expensive machinery and also reduces labour costs.

Monoculture also has drawbacks.

- It encourages an organism that feeds on the crop to increase to such large numbers that it becomes a pest. The farmer then has to pay the cost of treatment with pesticide.
- Growing a single crop exhausts the soil of the nutrients on which the crop depends. The farmer has to buy fertilisers to replace these nutrients.

Figure 6.4A ● Monoculture

6.5 Natural fertilisers: manure and slurry

FIRST THOUGHTS

Crops take nutrients from the soil, and these need to be replaced before the next crop is grown.

A substance which adds nutrients to the soil is called a **fertiliser**. In a mixed arable and livestock farm, the dung of farm animals is collected, left for a few months to decay and then used as **manure**, a natural fertiliser. When a herd of animals is reared indoors on an intensive farm, the dung has to be disposed of. Most intensive farms specialise, and a livestock farm may have no arable land to spread manure on. Figure 6.5A shows how the dung is collected.

LOOK AT LINKS
for **dissolved oxygen**
See Topic 20.6.

SUMMARY

Manure (rotted dung) and slurry (water and rotted dung) are natural fertilisers.

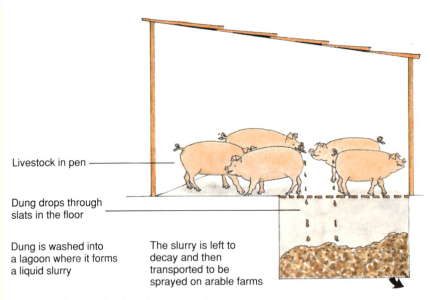

Livestock in pen

Dung drops through slats in the floor

Dung is washed into a lagoon where it forms a liquid slurry

The slurry is left to decay and then transported to be sprayed on arable farms

Figure 6.5A ● Collecting dung on an intensive farm

A serious hazard to the environment arises if slurry leaks from the lagoon. Then slurry may seep into rivers and lakes, where it will be broken down by bacteria. As the bacteria multiply, they use up dissolved oxygen in the water. The result is that fish and other forms of aquatic life die. The slurry may even find its way into ground water and contaminate drinking water.

6.6 Synthetic fertilisers

FIRST THOUGHTS

Most modern farms do not produce enough manure for their needs and use synthetic (man-made) fertilisers.

In the present century, we have seen the dramatic effects of synthetic fertilisers in the **green revolution**. This is the way fertilisers have turned barren areas in many parts of the world into green, fertile agricultural land. In India, the rice yield has increased by 50%; In Indonesia, the maize crop has doubled (see Figure 6.6A).

Figure 6.6A ● Compare the fertilised plot and the control plot

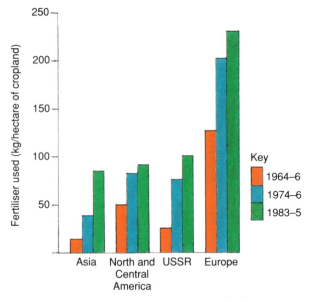

Figure 6.6B ● The worldwide use of fertilisers

How much fertiliser?

Most fertilisers concentrate on supplying the necessary nitrogen (N), phosphorus (P) and potassium (K). **NPK fertilisers** are consumed in huge quantities: in 1990, the world consumption was over 20 million tonnes (see Figure 6.6B). In the UK, the consumption of NPK fertilisers is about seven million tonnes a year. The fertilisers cost about £60 a tonne. Farmers and market gardeners need to invest a lot of money in fertilisers. They want good crops, but they do not want to spend more than they need on fertilisers. They can obtain expert advice from the Ministry of Agriculture, Fisheries and Food. Agricultural chemists at the Ministry will advise them on the type and quantity of fertiliser to apply and the best season of the year for applying it. Every farmer and grower has a different problem. The agricultural chemists must weight up the type of crop and the type of soil before they can recommend the most suitable treatment.

Drawbacks of using fertilisers

Using too much fertiliser is a waste of money (see Figure 6.6C). There is also another drawback. Rain leaches (washes out) soluble substances from the soil. The application of too much fertiliser results in the leaching of nitrates and phosphates into rivers, lakes and ground water. Phosphates and nitrates have caused **eutrophication** of lakes and rivers. Nitrates in the ground water find their way into the drinking water. Many people think this is a health hazard.

What can be done about the problem of leaching? There is less leaching from grassland because grass covers the topsoil all the year round. From arable land, about one third of the fertiliser applied is leached out. Matching the supply of fertiliser to the crop's ability to use it helps to reduce leaching. Winter crops, such as winter barley and winter wheat, reduce the leaching of fertiliser by winter rains.

Obviously, farmers cannot maintain their very high productivity without fertilisers. They are so efficient that the EC produces far more food than it needs. The excess of food is kept in vast stores. We pay the cost of storing all this food; it amounts to £150 per person per year in the UK. There is an argument for cutting back on production. To persuade farmers to use less fertiliser, the Government could either ration fertiliser or put a tax on it so that the farmers who use the most fertiliser and gain the biggest harvests would have the most fertiliser tax to pay.

IT'S A FACT

With 2–3 million tonnes of nitrogen (in the form of nitrates) going on to the soil of the UK each year, 600 000 tonnes of nitrate ion leach out every year.

LOOK AT LINKS
for the **eutrophication** (unintentional enrichment with nutrients) of lakes and rivers and the pollution of ground water by fertilisers. See Topic 20.6.

Fertiliser (program)

Investigate the chemistry of hydrogen, nitrogen and ammonium nitrate. You can also develop and operate a production system which is commercially viable.

IT'S A FACT

There are islands off the coast of South America which are inhabited by large flocks of sea birds. Mounds of their droppings, called guano, accumulate. Since sea birds eat a fish diet, which is high in protein, guano is rich in nitrogen compounds. For a century, European farmers imported guano from South America to use as a fertiliser.

Figure 6.6C ● Applying synthetic fertiliser to winter wheat

SUMMARY

Synthetic fertilisers enable farmers to obtain large yields per hectare. NPK fertilisers contain nitrogen (for protein synthesis), phosphorus (for healthy roots), potassium (to help photosynthesis) and some calcium and magnesium. There are disadvantages to the use of synthetic fertilisers.

Advantages of synthetic fertilisers

- Synthetic fertilisers are easy for the farmer to store and to apply.
- The application of fertiliser can be adjusted to the needs of the soil. Both the type and the amount of nutrient can be controlled.
- They replace bulky manure which cannot be applied evenly and is difficult to store. The farmer does not have to keep animals to produce manure.
- Land does not need a rest period between crops. The efficiency of a farm can be increased by growing one or two crops a year.

Disadvantages

- Synthetic fertilisers do not add humus to the soil. The structure of the soil deteriorates. Soil erosion may occur.
- Fertilisers which are not absorbed by the crop may be leached into rivers and lakes. The resulting eutrophication encourages the growth of algae.
- The manufacture of synthetic fertilisers uses a lot of fuel. This eats into the Earth's fuel reserves.

CHECKPOINT

1 The figure below shows how the yields of winter wheat and grass increase as more nitrogenous fertiliser is used. After studying the graphs, say what mass of nitrogen you would apply to a 100 hectare field to avoid waste and to give a maximum yield of (a) winter wheat and (b) grass.

Grass response to nitrogen fertiliser

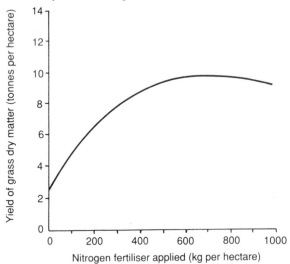

Winter wheat response to nitrogen fertiliser

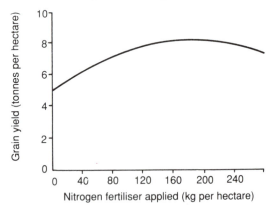

2 Refer to Figure 6.6B.
  (a) Compare the use of fertiliser in Europe and Asia in 1983–5 and in 1964–6. How many times more fertiliser was used per hectare in Europe in 1983–5 than was used in 1964–6?
  (b) By what factor did the use of fertiliser in Asia increase in 1983–5 compared with 1964–6?
  (c) Which area shows
   (i) the biggest percentage increase in the use of fertilisers per hectare from 1964–6 to 1983–5,
   (ii) the smallest percentage increase over the same period?

3 A dairy farmer spreads 240 tonnes of nitrochalk on his pasture. It costs £50 per tonne. If his milk cheque is £6000 a month, how long will it take him to recoup the cost of the fertiliser?

6.7 Herbicides

Farmers and gardeners have to cope with pests. These include weeds which compete with crops for nutrients, animals which eat crops, and disease-causing organisms like harmful fungi which attack plants and livestock. Substances which are used to kill pests are pesticides. There are three main types: herbicides, which kill plants, insecticides, which kill insects (Topic 6.8) and fungicides, which kill fungi (Topic 6.9).

Fertile soil is a good growing ground for weeds as well as for the crops that farmers sow. Weeds are plants that compete with crops for space, light and nutrients. At one time, farmers employed many farmhands, and one of the jobs they did was pulling out weeds by hand. Now, farmers employ fewer people and save labour by using chemical weedkillers – herbicides. By using a **selective herbicide**, they can kill weeds and leave the crop untouched (see Figure 6.7A). The most common selective herbicides are 2,4-D and 2,4,5-T. They kill broad-leaved plants, (dandelions, thistles and nettles).

Figure 6.7A ● A crop treated with herbicide and a control

Figure 6.7B ● Broad-leaved and narrow-leaved plants

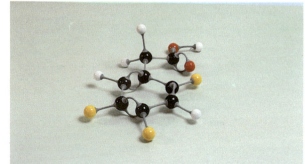

Figure 6.7C ● Models of 2,4-D and 2,4,5-T molecules

They do not harm plants with narrow leaves (grasses and cereals – see Figure 6.7B). These selective herbicides are organic compounds which contain chlorine (see Figure 6.7C). Other weedkillers, such as paraquat and sodium chlorate(V), are **non-selective**: they kill all plants.

Chemical weedkillers are of great benefit to us. It is difficult to see how scientific farming could continue without them. There have also been some tragic results of using herbicides.

SUMMARY

Selective weedkillers are of enormous benefit to agriculture. Nevertheless, one of them, 2,4,5-T, is now banned because it contains an impurity called dioxin, which has teratogenic effects and genetic effects.

Agent Orange

From 1961–71, there was a war in Vietnam between the USA and the communist Vietcong forces. The US planes found it difficult to carry out bomb attacks on the Vietcong because the Vietcong positions were hidden by dense jungle vegetation. US planes dropped a herbicide called Agent Orange, which killed all the trees and other vegetation where it landed. It destroyed the forests which sheltered Vietcong fighters and their bases and killed their crops. Agent Orange contained 2,4,5-T and an impurity, **dioxin**. At the time, no-one knew that dioxin has teratogenic effects (effects on unborn babies). Terribly deformed babies were born to Vietnamese mothers. Dioxin also has genetic effects. When US servicemen returned home and started families, some of them found to their dismay that they had become the fathers of deformed babies. Agent Orange had affected the fathers' genes. As a result of these tragedies, 2,4,5-T has been banned in many countries. The method of manufacture always produces some dioxin as impurity.

CHECKPOINT

❶ Explain what is meant by (a) herbicide and (b) selective herbicide.

❷ What are the advantages of selective herbicides over non-selective herbicides?

6.8 Insecticides

FIRST THOUGHTS

The previous section dealt with weeds – plant pests. Insects are the major animal pests, although many insects are beneficial. The pesticides which are used to kill insects are called insecticides.

Insects, weeds and plant diseases destroy about one third of the world's crops. Aphids attack the flowers of many fruits and vegetables. Cutworms chew off plants at soil level. Wireworms bore into potatoes and other root vegetables. Grain weevils eat stored grain. Unchecked, insects would consume most of the food we grow, and humans would starve (see Figure 6.8A). On the other hand, many insects are helpful. Some pollinate flowers, and some eat harmful insects.

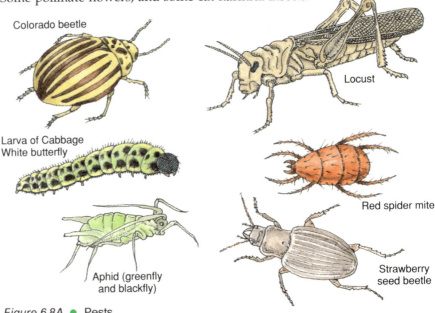

Figure 6.8A ● Pests

SUMMARY

Insecticides are used to protect crops against being eaten by insects. Insecticides have played a big part in increasing food supplies worldwide.

The first chemicals which were used as insecticides were general poisons, such as compounds of arsenic, lead and mercury. These compounds are poisonous to all animals which swallow them. Children and family pets were sometimes accidentally poisoned. These poisons are stable: they remain in the soil for many years.

The DDT story

In 1939, a scientist called Paul Mueller discovered that a substance called DDT was very poisonous to houseflies. His experiments soon showed that DDT was a very useful substance indeed, killing many insects such as lice and mosquitoes and agricultural pests such as potato-blight and grape-blight. The exciting aspect of DDT was that it left species other than insects unharmed. It is a **broad spectrum insecticide**. That is, it kills many types of insect. DDT is a **petrochemical**. Together with similar compounds, it is often referred to as an 'organochlorine' (see Figure 6.8B).

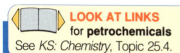

LOOK AT LINKS
for **petrochemicals**
See *KS: Chemistry*, Topic 25.4.

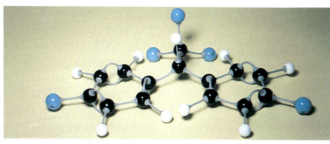

Figure 6.8B ● A model of a DDT molecule (dichlorodiphenyltrichloroethane)

DDT has saved more lives than any other single substance. It has saved million of lives and prevented billions of illnesses by killing insects which carry disease. It has saved people from hunger and starvation by killing insects which eat crops. Thanks to the use of DDT, 550 million people in tropical countries now no longer live in constant fear of malaria. The World Health Organisation says that DDT has saved five million lives. Cholera is another disease which is carried by insects and which can now be fought with DDT.

SCIENCE AT WORK

DDT came into production in time to be used in the Second World War. In wartime, the breakdown of normal hygienic living conditions means that both troops and civilians may suffer from lice, typhus and other afflictions. In the Italian city of Naples in 1943, the spread of lice caused a massive epidemic of typhus. The whole population of Naples was dusted with DDT, which killed the lice, and the epidemic was halted.

IT'S A FACT

Ten million tonnes of DDT have been applied to the Earth in the last 25 years.

Figure 6.8C ● Crop spraying today. In countries where DDT is banned, safer, often biodegradable chemicals are used.

IT'S A FACT

According to Oxfam, 375 000 people in the Third World are poisoned – 10 000 of them fatally – by pesticides each year. Pesticide controls are lax in some Third World countries. Manufacturers recommend that agricultural workers should wear protective boots, gloves, face masks and hoods while spraying pesticides. But workers often cannot afford this protective clothing.

SUMMARY

Insecticides fight disease by killing disease-carrying insects, e.g. lice and mosquitoes. The insecticide DDT has saved thousands of human lives. Worry over the build up of DDT in animals has led to strict limits on its use.

IT'S A FACT

Central Florida is the fruit basket of the USA. Without insecticides to kill mosquitoes, the area would be uninhabitable. To combat fruit fly, boxes of oranges and lemons are sprayed with an insecticide before they are sold. In 1984, scientists discovered that the insecticide used, EDB (1, 2-dibromoethane), is a carcinogen. Fruit-growers now use a different insecticide. Alternatives to spraying are refrigeration and irradiation. Both of these cost twelve times as much as spraying.

DDT has also contributed to agriculture. It was used successfully to control tea parasites in Sri Lanka and cotton pests in the USA. In the desert regions of North Africa, crops are often consumed by swarms of locusts. The result is famine, and a whole population will face starvation. DDT has reduced the harm done by locusts.

People predicted that DDT would exterminate all insect pests. In 1948, Paul Mueller was awarded the Nobel Prize for his discovery. No sooner had he received the prize than people began to discover some worrying effects of DDT. In 1962 Rachel Carson wrote a book called *Silent Spring*. She suggested that killing insects would deprive birds of their natural food, and birds would die out, giving a spring without birdsong. Many of her fears have been borne out. Since DDT does not break down easily, an application of DDT will remain in soil and water for many years.

In Topic 8, Figure 8.4D shows what happened when Clear Lake in California was sprayed with DDT to combat mosquitoes. People were surprised when aquatic birds, such as grebes, died. The concentration of DDT in the lake water was only 0.02 p.p.m. However, DDT is soluble in fat and therefore difficult to excrete. Along the food chain, the level of DDT built up: microorganisms 5 p.p.m., fish 250 p.p.m. and at the top of the food chain, grebes 1600 p.p.m., a lethal dose.

DDT has been carried through food chains to the remotest parts of Earth, far from the places where it was used. Figure 6.8D shows how DDT can build up along a food chain on land. A problem is that many insects are becoming immune to DDT. As a result, higher concentrations must be used in spraying.

❶ An elm tree is treated with 2,4,5,-T to stop Dutch elm disease

❷ A worm eats dead leaves which contain the pesticide.

❸ A robin eats the worm.

❹ After the robin flies away, it is eaten by a bird of prey.

Figure 6.8D ● The robins' place in a food chain

DDT does not react with air or water. Since it is such a stable compound, the amount of DDT in the world is steadily increasing. We know that it kills insects and that, in sufficiently large amounts, it kills fish and birds. A human being may eat fish and birds contaminated with DDT over a long period of time. We do not know what the results of the buildup of DDT will be. There is no direct evidence that DDT causes illness in humans. It has been found, however, that cancer victims in the USA have twice as much DDT in their fatty tissues as the rest of the population. In the USA and Europe, DDT is now banned. DDT is still used for essential jobs, for example, fighting malaria. Safer insecticides are used for agricultural purposes. One choice is organic compounds of phosphorus, which are broken down in time to safe compounds.

Other methods of controlling insects

Scientists are looking for other ways of controlling insects. Some methods are described as **biological control**. One such method is to breed **predator insects**. These are insects which kill the insects that are eating the crops. Of course, the predator insects must not eat the crops! Another idea is to use radioactivity to sterilise male insects so that they cannot breed. Insects give out aromatic chemicals called **pheromones** when they want to attract a mate. Research workers have worked out a method of extracting pheromones from insects and using the extract to bait a trap. The insects which are lured into the trap are then killed.

Great Spruce Bark beetles lay their eggs under the bark of spruce trees. When the larvae hatch, they feed on the tree's nutrients, thus killing the tree. Belgian beetles are predators of Great Spruce Bark beetles. Belgian beetles also lay their eggs under the bark of spruce trees. When the larvae of the Belgian beetles hatch, they eat the larvae of Great Spruce Bark beetles, but they do not harm the tree.

SUMMARY

It would benefit the environment if we could find ways of cutting down on the use of insecticides. Research is being carried out on the use of predator insects, sterilisation and trapping.

CHECKPOINT

1 In the spring of 1988, North Africa was threatened by the worst plague of locusts for 30 years. The locust control programme financed by the United Nations uses short-lived insecticides such as fenitrothion. These do not last long enough to lie in wait for advancing swarms of locusts. An area sprayed with fenitrothion is safe for locusts after only three days. With long-lasting insecticides, such as dieldrin, strips of desert can be sprayed to prevent the advance of locusts. Experts say that dieldrin would do no harm to a desert environment. However, dieldrin has been shown to cause genetic damage. Which would you use – fenitrothion or dieldrin – if you were in charge of the locust control programme? Explain your views.

2 The Mediterranean fruit fly is an insect pest. Large numbers of male flies were irradiated by a colbalt-60 source (see Topic 4.1) and then released. They mated but they produced no offspring.
(a) Why were fruit farmers pleased with the use of radioactivity?
(b) What is the advantage of this technique over the use of chemical pesticides?

6.9 Fungicides and molluscicides

LOOK AT LINKS
for **bilharzia**
See Topic 6.2.

Fungi cause the expensive diseases of mildew and potato-blight. To prevent these diseases, crops are sprayed with fungicides. Many copper salts are fungicides, e.g. copper(II) sulphate.

Molluscs, including slugs and snails, eat crops and spread diseases. Molluscicides are being used to fight them. Bilharzia is a disease which affects people in tropical countries. It is spread by water snails.

IT'S A FACT

In 1985 nearly 10 000 tonnes of pesticide were applied to farmland in the UK. More than 800 000 tonnes were used worldwide.

Advantages of using pesticides
• They kill pests quickly.
• Pesticides increase food production.
• They are easy to store and use.
• There is such a large variety of chemicals available that most pests can be eliminated.

SUMMARY

Fungicides and molluscicides are used to kill harmful fungi and molluscs.

Disadvantages

- Pests can develop resistance to a particular pesticide.
- The spray can be carried by the wind and so affect wildlife.
- Pesticides can seep into the soil and drain into rivers and lakes, where they will harm wildlife (see Figure 6.9A).
- Pesticides can enter food chains and build up from one feeding level to the next. They can reach a toxic level at the top of a food chain.

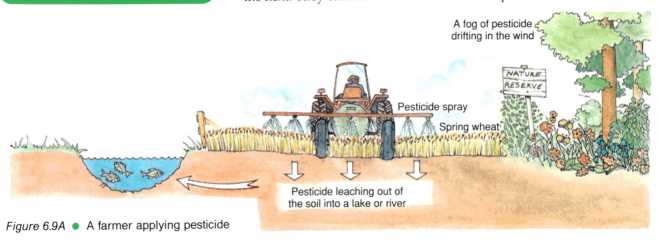

A fog of pesticide drifting in the wind

NATURE RESERVE

Pesticide spray

Spring wheat

Pesticide leaching out of the soil into a lake or river

Figure 6.9A ● A farmer applying pesticide

CHECKPOINT

❶ (a) How do chemical insecticides help farmers?
   (b) What did farmers use as insecticides before DDT?
   (c) In what way is DDT better than the insecticides which were used previously?
   (d) Why has the use of DDT on farms been banned?
   (e) For what purpose is DDT still used? Why is it in such demand?

❷ Why did Rachel Carson's book get people worried?

❸ Do you think we should give up using chemical insecticides?

❹ DDT was found in the flesh of eagles. They are birds of prey: they do not eat insects. How did the insecticide get into the eagles' flesh?

6.10 Energy

An ecosystem consists of a community of organisms living in a particular part of the physical (non-living) environment. Ecosystems are more or less self-supporting. The animals, plants, soil, water, etc. which make up a farm are an ecosystem.

Farms are ecosystems which are managed so as to produce as much food as possible. The amount of food produced can be assessed in terms of the energy content of the food produced. For example, the energy content of beef is 1050 kJ/100 g; the energy content of oatmeal is 170 kJ/100 g. The amount of food produced depends on the amount of energy which enters the farm ecosystem and also on the efficiency with which the energy is converted into the energy content of plant and animal tissue. The energy which enters the system is the **input**, and the energy content of the food produced is the **output**.

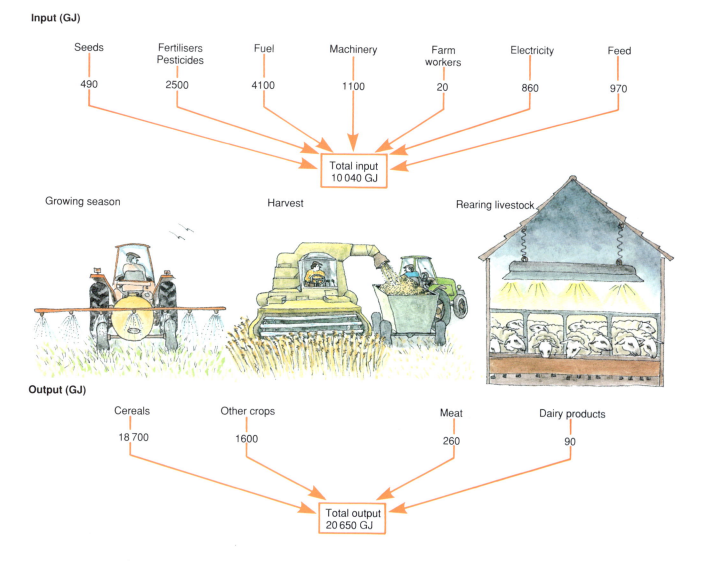

Input (GJ)

Seeds	Fertilisers Pesticides	Fuel	Machinery	Farm workers	Electricity	Feed
490	2500	4100	1100	20	860	970

Total input
10 040 GJ

Growing season Harvest Rearing livestock

Output (GJ)

Cereals	Other crops	Meat	Dairy products
18 700	1600	260	90

Total output
20 650 GJ

Figure 6.10A ● Energy input and output on a 640 hectare farm in the 1980s. Note that fuel is used to power machinery, in the generation of electricity and in the manufacture of fertilisers, pesticides, etc. In total, fuel accounts for nearly 99% of energy input. Manual work accounts for less than 0.2% of energy input.

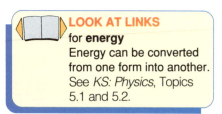

LOOK AT LINKS
for **energy**
Energy can be converted from one form into another.
See *KS: Physics*, Topics 5.1 and 5.2.

LOOK AT LINKS
Photosynthesis converts the energy of sunlight into the energy of chemical bonds in sugars
See Topic 11.1.

Figure 6.10A shows the input and output of a typical intensive farm of 640 hectares in southern England. The farm is chiefly arable but raises some livestock. Figure 6.10B shows the output and input on a farm of the same size and type in the 1820s. The figures have been worked out from historical records. The unit is the gigajoule, GJ (1 GJ = 10^9 J).

The values in Figures 6.10A and B do not include the energy of sunlight as input. It is assumed to be the same today as it was in the 1820s. The values in the two figures are summarised in Table 6.3.

Table 6.3 ● A comparison of intensive and traditional farms. (Energy values are in GJ.)

Energy	Intensive farm	Traditional farm	Ratio:	Intensive farm Traditional farm
Input	10 000	245	41	
Output	20 700	3440	6	
Ratio: Output/Input	2	14		

Input (GJ)

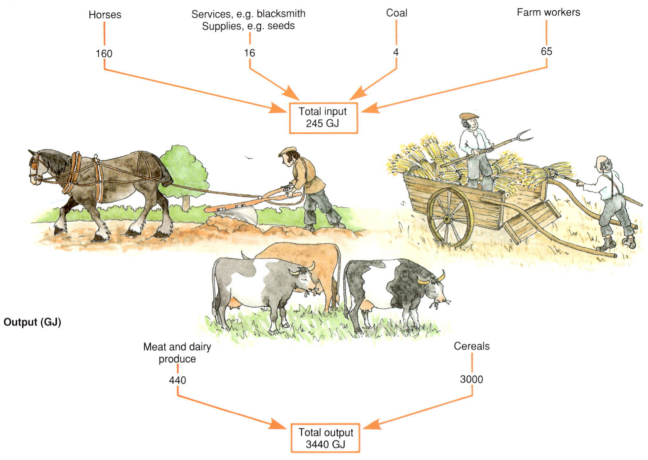

Horses	Services, e.g. blacksmith Supplies, e.g. seeds	Coal	Farm workers
160	16	4	65

Total input
245 GJ

Output (GJ)

Meat and dairy produce	Cereals
440	3000

Total output
3440 GJ

Figure 6.10B ● Energy input and output on a 640 hectare farm in the 1820s. Note that a horse uses 8 MJ/hour at work, while a farm worker uses 0.8 MJ/hour (MJ = megajoule, 1 MJ = 10^6 J). More than 98% of the work is done by horses and people.

The increase in productivity (output per hectare) from the 1820s to the 1980s was achieved with a decreasing number of farm workers per hectare. Today's farm worker produces 60 times more food than a farm worker in the 1820s. As you see from Table 6.3, the sixfold increase in output per hectare is achieved by means of a 40-fold increase in the input of energy. This energy comes almost entirely from oil. Each year a farm worker uses the energy from over 11 tonnes of oil to enable him or her to produce so much food per worker. This brackets the energy demands of agriculture with those of the steel industry and the ship building industry. When the world's supplies of oil dwindle, the **energy crisis** will affect intensive farming.

CHECKPOINT

❶ (a) How much greater is the output of the intensive farm compared with that of the traditional farm?

(b) How is the increase in productivity achieved? Name three factors.

(c) How many times more energy has to be supplied to the intensive farm to achieve the increased productivity?

(d) To achieve the same level of productivity (the same output of energy), how many times more energy would have to be supplied to the intensive farm than to the traditional farm?

6.11 Hunger

Much of the food which intensive farming methods produce in the UK is in excess of what the country needs. The reason why farmers continue to grow more and more is that the Government guarantees to pay farmers high prices for their produce. Crops, meat and dairy produce which are not eaten have to be stored. They form the very expensive grain, beef and butter 'mountains' which are stored by the European Community (EC). Food production in Western Europe as a whole has steadily increased since 1950. The population has remained almost constant. With so much food available, we eat more than we need, yet in poorer countries millions of people do not have enough food. Figure 6.11A compares Western Europe with Africa, where the population is increasing at a fast rate.

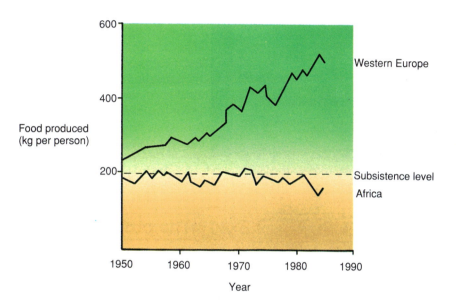

Figure 6.11A ● Food production in Western Europe and in Africa. (Source: US Department of Agriculture) Note the subsistence level, the minimum food needed to stay healthy. Millions of people in Africa are below this level. Note the gap in food production between Western Europe and Africa. Why has it increased since 1950?

What is the solution to the problem of countries which cannot feed themselves adequately? Should we ship over supplies of food from Europe? Shipping over massive food supplies is of course necessary in times of famine. It is, however, not a satisfactory way of life for a country or a region to be completely dependent on food supplies from outside.

Should we help such countries to develop intensive farming? The dependence on oil makes intensive farming an unwise choice for poorer countries. High-technology machinery is expensive to buy and expensive to run. If spare parts are not available or if local people are not trained in repair work, the machinery may lie idle. On the other hand, infertile lands in many parts of the world have been improved by the application of traditional, non-intensive systems. Sinking wells and digging irrigation channels have been very successful in increasing food crops. Planting trees and shrubs to prevent erosion has helped to save deforested areas. The **cross-breeding** of plants has resulted in new varieties of crops. Planting new varieties of crops which can survive a lack of rainfall has helped to increase yields in infertile areas.

LOOK AT LINKS
for **cross-breeding**
See Topic 19.

115

 **AIR POLLUTION**

7.1 Smog

As you read through this topic, think about the 15 000 or 20 000 litres of air that you breathe in each day. Obviously you hope that it is clean air. Unfortunately the air that most of us breathe is polluted. In this topic, we shall look into what can be done to reduce the pollution of air.

Four thousand people died in the great London smog of December 1952. Smog is a combination of smoke and fog. Fog consists of small water droplets. It forms when warm air containing water vapour is suddenly cooled. The cool air cannot hold as much water vapour as it held when it was warm, and water condenses. When smoke combines with fog, fog prevents smoke escaping into the upper atmosphere. Smoke stays around, and we inhale it. Smoke contains particles which irritate our lungs and make us cough. Smoke also contains the gas sulphur dioxide. This gas reacts with water and oxygen to form sulphuric acid, H_2SO_4. This strong acid irritates our lungs, and they produce a lot of mucus which we cough up.

The Government did very little about the cause of smog until 1956. Then there was another killer smog. A private bill brought by a Member of Parliament (the late Mr Robert Maxwell, the newspaper owner) gained such widespread support that the Government was forced to act. The Government introduced its own bill, which became the Clean Air Act of 1956. The Act allowed local authorities to declare smokeless zones. In these zones, only low-smoke and low-sulphur fuels can be burned. The Act banned dark smoke from domestic chimneys and industrial chimneys. People started using natural gas and electricity.

7.2 The problem

All the dust and pollutants in the air pass over the sensitive tissues of our lungs. Any substance which is bad for health is called a **pollutant**. The lung diseases of cancer, bronchitis and emphysema are common illnesses in regions where air is highly polluted. From our lungs, pollutants enter our bloodstream to reach every part of our bodies. The main air pollutants are shown in Table 7.1.

Hyperbook
(data base)

Use *Hyperbook* to read about:
- acid rain,
- the greenhouse effect,
- pollution generally.

With the computer, try moving from one article to another and perhaps printing out any particularly interesting documents.

Table 7.1 ● The main pollutants in air (Emissions are given in millions of tonnes per year in the UK.)

Pollutant	Emission	Source
Carbon monoxide, CO	100	Vehicle engines and industrial processes
Sulphur dioxide, SO_2	33	Combustion of fuels in power stations and factories
Hydrocarbons	32	Combustion of fuels in factories and vehicles
Dust	28	Combustion of fuels; mining; factories
Oxides of nitrogen, NO and NO_2	21	Vehicle engines and fuel combustion
Lead compounds	0.5	Vehicle engines

In this topic, we shall look at where these pollutants come from, what harm they do and what can be done about them.

7.3  Dispersing air pollutants

LOOK AT LINKS
for **convection currents**
Why does warm air rise?
See *KS: Physics*, Topic 6.5.

The surface of the Earth absorbs energy from the Sun and warms up. The Earth warms the lower atmosphere. The air in the upper atmosphere is cooler than the air near the Earth. **Convection currents** carry warm air upwards. Cold air descends to take its place (see Figure 7.3A). In this way, the warm dirty air from factories and motor vehicles is carried upwards and spread through the vast upper atmosphere.

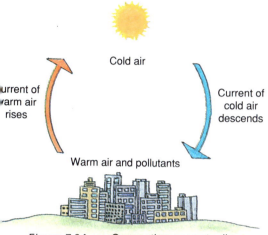

Figure 7.3A ● Convection currents disperse pollutants

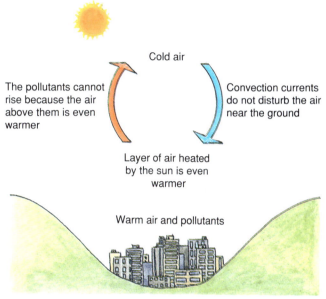

Figure 7.3B ● A temperature inversion traps pollutants

SUMMARY

Pollutants are carried upwards by rising currents of warm air. A temperature inversion stops the dispersal of pollutants. Temperature inversions occur in places with a hot climate and still air.

A low-lying area surrounded by higher ground tends to have still air. If an area like this has a hot climate, it is possible for the Sun to warm a layer of air in the upper atmosphere (Figure 7.3B). If the Sun is very hot, this layer of air may become warmer than that near the ground. There is a **temperature inversion**. The air near the ground is no longer carried upwards and dispersed. Pollutants accumulate in the layer of still air at ground level, and the city dwellers are forced to breathe them.

7.4  Sulphur dioxide

Pollution

Use a word-processing system or better still a desktop publishing package to design and print a poster or news-sheet on the topic of pollution.

● *Where does sulphur dioxide come from?*

Worldwide, 150 million tonnes of sulphur dioxide a year are emitted. Almost all the sulphur dioxide in the air comes from industrial sources. The emission is growing as countries become more industrialised. Half of the output of sulphur dioxide comes from the burning of coal. Most of the coal is burned in power stations. All coal contains between 0.5 and 5 per cent sulphur.

Sulphur	+	Oxygen	→	Sulphur dioxide
(coal)		(air)		
$S(s)$	+	$O_2(g)$	→	$SO_2(g)$

Industrial smelters, which obtain metals from sulphide ores, also produce tonnes of sulphur dioxide daily.

SUMMARY

Sulphur dioxide causes bronchitis and lung diseases. The Clean Air Acts have reduced the emission of sulphur dioxide from low chimneys. Factories, power stations and metal smelters send sulphur dioxide into the air. In the upper atmosphere, sulphur dioxide reacts with water to form acid rain.

● What harm does sulphur dioxide do?

Sulphur dioxide is a colourless gas with a very irritating smell. Inhaling sulphur dioxide causes coughing, chest pains and shortness of breath. It is poisonous; at a level of 0.5%, it will kill. Sulphur dioxide is thought to be one of the causes of bronchitis and lung diseases.

● What can be done about it?

After the Clean Air Acts of 1956 and 1968, the emission of sulphur dioxide and smoke from the chimneys of houses decreased. At the same time, the emission of sulphur dioxide and smoke from tall chimneys increased. Tall chimneys carry sulphur dioxide away from the power stations and factories which produce it (see Figure 7.4A). Unfortunately it comes down to earth again as acid rain (see Figure 7.5B).

Figure 7.4A ● Smoking chimneys

CHECKPOINT

❶ The emission of sulphur dioxide from low chimneys decreased between 1955 and 1975 from 1.7 million tonnes a year to 0.6 million tonnes a year. In the same period, the emission of sulphur dioxide from tall chimneys increased from 1.4 million tonnes to 3.0 million tonnes.
  (a) Explain why there was a decrease in sulphur dioxide emission from low chimneys.
  (b) Who benefited from the decrease in emission from low chimneys?
  (c) Explain why tall chimneys are not a complete answer to the problem of sulphur dioxide emission.

7.5 Acid rain

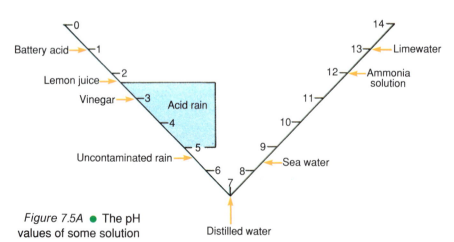

Figure 7.5A ● The pH values of some solution

Rain water is naturally weakly acidic. It has a pH of 5.4. Carbon dioxide from the air dissolves in it to form the weak acid, carbonic acid, H_2CO_3. What we mean by acid rain is rain which contains the strong acids, sulphuric acid and nitric acid. Acid rain has a pH between 2.4 and 5.0 (see Figure 7.5A).

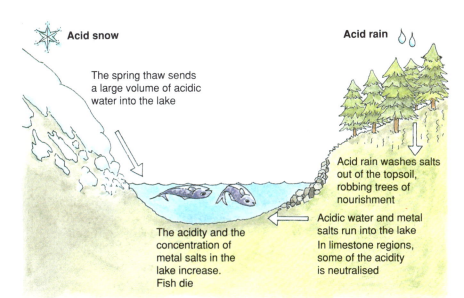

Acid snow

The spring thaw sends a large volume of acidic water into the lake

The acidity and the concentration of metal salts in the lake increase. Fish die

Acid rain

Acid rain washes salts out of the topsoil, robbing trees of nourishment

Acidic water and metal salts run into the lake
In limestone regions, some of the acidity is neutralised

Figure 7.5B ● Where acid rain comes from

How do sulphuric acid and nitric acid get into rain water? Tall chimneys emit sulphur dioxide and other pollutant gases, such as oxides of nitrogen. Air currents carry the gases away. Before long, the gases react with water vapour and oxygen in the air. Sulphuric acid, H_2SO_4, and nitric acid, HNO_3, are formed. The water vapour with its acid content becomes part of a cloud. Eventually it falls to earth as acid rain or acid snow which may turn up hundreds of miles away from the source of pollution (see Figure 7.5B).

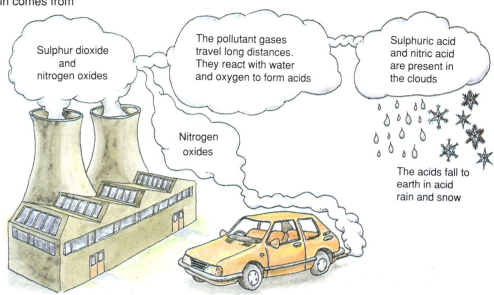

Sulphur dioxide and nitrogen oxides

The pollutant gases travel long distances. They react with water and oxygen to form acids

Sulphuric acid and nitric acid are present in the clouds

Nitrogen oxides

The acids fall to earth in acid rain and snow

Figure 7.5C ● Where acid rain goes

Acid rain which falls on land is absorbed by the soil (see Figure 7.5C). At first, the nitrates in the acid rain fertilise the soil and encourage the growth of plants. But acid rain reacts with minerals, converting the metals in them into soluble salts. The rain water containing these soluble salts of calcium, potassium, aluminium and other metals trickles down through the soil into the subsoil where plant roots cannot reach them. In this way, salts are leached out of the topsoil, and crops are robbed of nutrients. One of the salts formed by acid rain is aluminium sulphate. This salt damages the roots of trees. The damaged roots are easily attacked by viruses and bacteria, and the trees die of a combination of malnutrition and disease. The Black Forest is a famous beauty spot in Germany which makes money from tourism. About half the trees there are now damaged or dead. Pollution is an important issue in Germany. The Green Party is a major political party. It is campaigning for the reduction of pollution. There are dead forests also in Czechoslovakia and Poland. In 1987, the UK Forestry Commission reported that damage to spruce and pine trees is as widespread in the UK as in Germany. One third of British trees are damaged, but not everyone agrees that the cause of the damage is acid rain (see Figure 7.5D)

IT'S A FACT

The most acidic rain ever recorded fell on Pitlochrie in Scotland; it had a pH of 2.4.

● SCIENCE AT WORK ●

It is possible to monitor the emission from the chimneys of factories and power stations from a distance. A van can carry equipment using infra-red radiation to detect sulphur dioxide. The readings are automatically recorded by a computer.

IT'S A FACT

In 1979, 31 European countries, including the UK, signed the Convention on Long Range Transboundary Pollution. All the nations agreed to stop exporting pollution across their borders. In 1986, the UK joined the **30 per cent club**. All the nations in the 'club' agreed to reduce their emissions of sulphur dioxide by 30 per cent.

IT'S A FACT

Together, the USA, Canada and Europe send 100 million tonnes of sulphur dioxide into the air each year. Nine tenths of this comes from burning coal and oil.

● SCIENCE AT WORK ●

It cost £40 million to fit FGD to Drax power station in Yorkshire. Fidlers' Ferry in Wales is also fitted with FGD.

The acidic rain water trickles through the soil until it meets rock. Then it travels along the layer of rock to emerge in lakes and rivers. Lakes are more affected by acid rain than rivers are. They become more and more acidic, and the concentrations of metal salts increase. Fish cannot live in acidic water. Aluminium compounds, e.g. aluminium hydroxide, come out of solution and are deposited on the gills. The fish secrete mucus to try to get rid of the deposit. The gills become clogged with mucus, and the fish die. An acid lake is transparent because plants, plankton, insects and other living things have perished.

Thousands of lakes in Norway, Sweden and Canada are now 'dead' lakes. One reason why these countries suffer badly is that acidic snow piles up during the winter months. In the spring thaw, the accumulated snow melts suddenly, and a large volume of acidic water flows into the lakes. Acid rain is partially neutralised as it trickles slowly through soil and over rock. Limestone, in particular, keeps damage to a minimum by neutralising some of the acidity. There is not time for this partial neutralisation to occur when acid snow melts and tonnes of water flow rapidly down the hills and into the lakes.

The UK is affected too. In 1982, lakes and rivers in south-west Scotland had become so acidic that the water companies started treating the lakes with calcium hydroxide (lime). The aim is to neutralise the acidic water and revive stocks of fish. In Wales, the water company has for some years poured tonnes of powdered limestone into acidic lakes. A number of lakes are 'dead' and the fish in many others are threatened.

Figure 7.5D ● The effects of acid rain

● What can be done about acid rain?

There are three main methods of attacking the problem of acid rain. They all cost money, but then the damage done by acid rain costs money too.

❶ Low-sulphur fuels can be used. Crushing coal and washing it with a suitable solvent reduces the sulphur content by 10 to 40 per cent. The dirty solvent must be disposed of without creating pollution on land or in rivers. Oil refineries could refine the oil which they sell to power stations. The cost of the purified oil would be higher, and the price of electricity would increase.

❷ Flue gas desulphurisation, FGD, is the removal of sulphur from power station chimneys after the coal has been burnt and before the waste gases leave the chimneys. As the combustion products pass up the chimney, they are bombarded by jets of wet powdered limestone. The acid gases are neutralised to form a sludge. The method will remove 95 per cent of the acid combustion products. FGD can be fitted to existing power stations. One of the products is calcium sulphate, which can be sold to the plaster board industry and to cement manufacturers.

LOOK AT LINKS
The reactions of acids with limestone (calcium carbonate) and metals are described in
KS: Chemistry, Topics 19.3 and 12.1.

❸ Pulverised fluidised bed combustion, PFBC, uses a new type of furnace. The furnace burns pulverised coal (small particles) in a bed of powdered limestone. An upward flow of air keeps the whole bed in motion. The sulphur is removed during burning. The PFBC uses much more limestone than the FGD method: one power station needs 1 million tonnes of limestone a year (4 times as much as the FGD method). The PFBC method also produces a lot more waste material, which has to be dumped.

CHECKPOINT

❶ What is the advantage of building a power station close to a densely populated area? What is the disadvantage?

❷ Why do power stations and factories have tall chimneys? Are tall chimneys a solution to the problem of pollution? Explain your answer.

❸ Why does acid rain attack (a) iron railings (b) marble statues and (c) stone buildings?

❹ (a) Why does Sweden suffer badly from acid rain?
(b) Why do lakes suffer more than rivers from the effects of acid rain?

❺ A country decides to increase the price of electricity so that the power stations can afford to use refined low-sulphur fuel oil. In what ways will the country actually *save* money by reducing the emission of sulphur dioxide?

7.6 Carbon monoxide

● *Where does it come from?*

Worldwide, the emission of carbon monoxide is 350 million tonnes a year. Most of it comes from the exhaust gases of motor vehicles. Vehicle engines are designed to give maximum power. This is achieved by arranging for the mixture in the cylinders to have a high fuel to air ratio. This design leads to incomplete combustion. The result is the discharge of carbon monoxide, carbon and unburnt **hydrocarbons**.

LOOK AT LINKS
for **hydrocarbons**
See *KS Chemistry*, Topic 15.7.

LOOK AT LINKS
for **haemoglobin**
See Topic 14.2.

● *What harm does carbon monoxide do?*

Oxygen combines with **haemoglobin**, a substance in red blood cells. Carbon monoxide is 200 times better at combining with haemoglobin than oxygen is. Carbon monoxide is therefore able to tie up haemoglobin and prevent it combining with oxygen. A shortage of oxygen causes headache and dizziness, and makes a person feel sluggish. If the level of carbon monoxide reaches 0.1% of the air, it will kill. Carbon monoxide is especially dangerous in that, being colourless and odourless, it gives no warning of its presence. Since carbon monoxide is produced by motor vehicles, it is likely to affect people when they are driving in heavy traffic. This is when people need to feel alert and to have quick reflexes.

● *What can be done?*

Soil contains organisms which can convert carbon monoxide into carbon dioxide or methane. This natural mechanism for dealing with carbon monoxide cannot cope in cities, where the concentration of carbon monoxide is high and there is little soil to remove it. People are trying out a number of solutions to the problem.

SUMMARY

Carbon monoxide is emitted by vehicle engines. It is poisonous. Catalytic converters fitted in the exhaust pipes of cars reduce the emission of carbon monoxide.

- Vehicle engines can be tuned to take in more air and produce only carbon dioxide and water. Unfortunately, this increases the formation of oxides of nitrogen.
- Catalytic converters are fitted to the exhausts of many cars. The catalyst helps to oxidise carbon monoxide in the exhaust gases to carbon dioxide.
- New fuels may be used in the future. Some fuels, e.g. alcohol, burn more cleanly than hydrocarbons (see Topic 20.2).

CHECKPOINT

❶ (a) What are the products of complete combustion of petrol?
(b) What harm do these products do?
(c) What conditions lead to the formation of carbon monoxide?
(d) What harm does it do?

❷ How does carbon monoxide act on the body?

❸ Which types of people are likely to breathe in carbon monoxide? Is there anything they can do to avoid it?

❹ A family was spending the weekend in their caravan. At night, the weather turned cold, so they shut the windows and turned up the paraffin heater. In the morning, they were all dead. What had gone wrong? Why did they have no warning that something was wrong?

7.7 Smoke, dust and grit

LOOK AT LINKS
for **smog**
See Topic 7.1.

LOOK AT LINKS
for **electrostatic attraction**
How do electrostatic precipitators work?
See *KS: Physics*, Topic 20.1.

SUMMARY

Particles of smoke and dust and grit are sent into the air by factories, power stations and motor vehicles. Dirt damages buildings and plants. It pollutes the air we breathe; mixed with fog, it forms smog.

Millions of tonnes of smoke, dust and grit are present in the atmosphere. Dust storms, forest fires and volcanic eruptions send matter into the air. Human activities such as mining, land-clearing and burning coal and oil add to the solid matter in the air.

● *What harm do particles do?*

Particles darken city air by scattering light. Smoke increases the danger of smog. Solid particles fall as grime on people, clothing, buildings and plants.

Sunlight which meets dust particles is reflected back into space and prevented from reaching the Earth. Some scientists believe that the increasing amount of dust in the atmosphere is serious. A fourfold increase in the amount of dust would make the Earth's temperature fall by about 3 °C. This would affect food production.

● *How can particles be removed?*

Industries use a number of methods. These include:
- using sprays of water to wash out particles from their waste gases
- passing waste gases through filters,
- electrostatic precipitators, which remove dust particles from waste gases by **electrostatic attraction**.

7.8 Metals

Many heavy metals and their compounds are serious air pollutants. 'Heavy' metals are metals with a density greater than 5 g/cm^3.

Mercury

Earth-moving activities, such as mining and road-making, disturb soil and rock and allow the mercury which they contain to escape into the air. Mercury vapour is also released into the air during the smelting of many metal ores and the combustion of coal and oil. Both mercury and its compounds cause kidney damage, nerve damage and death.

Lead

● *Where does it come from?*

The lead compounds in the air all come from human activity. Vehicle engines, the combustion of coal and the roasting of metal ores send lead and its compounds into the air. Unlike the other pollutants in exhaust gases, lead compounds have been purposely added to the fuel. Tetraethyl lead, TEL, is added to improve the performance of the engine.

● *What harm does it do?*

Lead compounds settle out of the air on to plant crops, and contaminate our food. The level of lead in our environment is high: some areas still have lead plumbing; old houses may have peeling lead-based paint. City dwellers take in lead from many sources. Many people have blood levels of lead which are nearly high enough to produce the symptoms of lead poisoning. Symptoms of mild lead poisoning are headache, irritability, tiredness and depression. Higher levels of lead cause damage to the brain, liver and kidneys. Scientists have suggested that behaviour disorders such as hooliganism and vandalism may be due in part to lead poisoning.

● *What can be done?*

This type of pollution can be remedied. We can stop adding lead compounds to petrol. Research chemists have found other compounds which can be used to improve engine performance. Vehicles made in the UK after autumn 1990 are adjusted to run on lead-free petrol. Most petrol stations now stock lead-free petrol. The USA, Germany and Japan use lead-free petrol because the catalytic converters fitted to their cars are 'poisoned' by lead compounds.

Figure 7.8A ● City dwellers breathe exhaust gases

SUMMARY

Heavy metals are serious air pollutants. Levels of mercury and lead and their compounds in the air are increasing.

CHECKPOINT

❶ Name the pollutants which come from motor vehicles.

❷ Name the pollutants which can be reduced by fitting catalytic converters into vehicle exhausts. What effect will this modification have on the price of cars?

❸ Catalytic converters will only work with lead-free petrol. When TEL is no longer added to petrol, motorists will have to use high octane (4 star) fuel. What effect will this have on the cost of motoring?

❹ What effect does the use of TEL have on the air, apart from its effect on catalytic converters?

❺ In which ways will the control of pollution from vehicles cost money? In which ways will a reduction in the level of pollutants in the air save money? (Consider the effects of pollution on people and materials.) Will the expense be worthwhile?

7.9 Chlorofluorohydrocarbons

The ozone layer

There is a layer of ozone, O_3, surrounding the Earth. It is 5 km thick at a distance of 25–30 km from the Earth's surface. The ozone layer cuts out some of the ultraviolet light coming from the Sun. Ultraviolet light is bad for us and for crops. Long exposure to ultraviolet light can cause skin cancer. This complaint is common in Australia among people who spend a lot of time out of doors. If anything happens to decrease the ozone layer, the incidence of skin cancer from exposure to ultraviolet light will increase. An excess of ultraviolet light kills **phytoplankton**, the minute plant life of the oceans which are the primary food on which the life of an ocean depends.

LOOK AT LINKS
for **phytoplankton**
See Topic 2.

● The problem
Ozone is a very reactive element. If the upper atmosphere becomes polluted, ozone will oxidise the pollutants. Two pollutants are accumulating in the upper atmosphere. One is the **propellant** from aerosol cans (see Figure 7.9A).

When the pressure is released, the propellant liquid vaporises and forces the polish out of the can

BUG BRITE

Mixture of propellant and useful liquid, e.g. polish or insecticide, under pressure

Figure 7.9A ● An aerosol can

Many of the propellants are chlorofluorohydrocarbons (CFCs). They are very unreactive compounds. They spread through the atmosphere without reacting with other substances and drift into the upper atmosphere. There they meet ozone, which oxidises CFCs and in doing so is converted into oxygen.

Ozone + CFC ➜ Oxygen + Oxidation products

Another pollutant found at this height is nitrogen monoxide, NO. It comes from the exhausts of high-altitude aircraft, such as Concorde. Ozone oxidises nitrogen monoxide to nitrogen dioxide:

Ozone + Nitrogen monoxide ➜ Oxygen + Nitrogen dioxide
$O_3(g)$ + $NO(g)$ ➜ $O_2(g)$ + $NO_2(g)$

SCIENCE AT WORK
Microcomputers and sensors are taken up in aircraft to keep watch on the ozone layer. The sensor detects the thickness of the ozone layer and the microcomputer records the measurements.

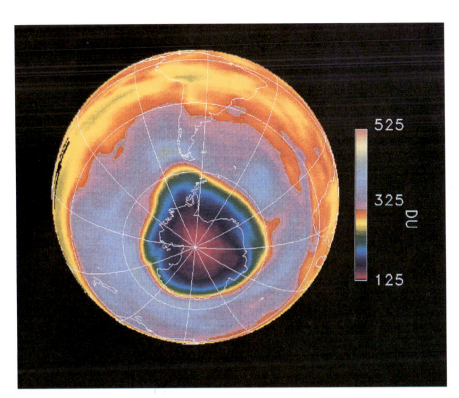

Figure 7.9B ● The ozone hole

What should be done?

Is it happening? Is the ozone layer becoming thinner? In June 1980 the British Antarctic Expedition discovered that there was a gap in the ozone layer over Antarctica during certain months. In 1987, research workers in the US confirmed that there was a thinning of the ozone layer which was 'large, sudden and unexpected... far worse than we thought'. In 1988 a team of scientists working in the Arctic Ocean discovered that the ozone layer over Northern Europe was thinner than it had been.

Knowing that the danger had appeared over more populated regions of the globe, spurred many countries to take action. At a meeting in Montreal in 1987 many countries agreed to reduce their use of CFCs by 50% by the year 2000. Since that date, many countries have agreed to speed up their programme of phasing out CFCs. Aerosols containing CFCs have been banned in the USA since 1988. In 1988 many makers of toiletries in the UK agreed to stop using CFCs by the end of 1989. They are now using spray cans with different propellants, which they label 'ozone-friendly', or pump-action cans. There is more of a problem with the CFCs used as refrigerants, in air conditioners, in the manufacture of polyurethane foam and as solvents. Chemists are now finding stable compounds to replace CFCs. In the USA, Du Pont Chemicals have agreed to stop using CFCs after the year 2000. In the UK, ICI chemists are working hard to find substitutes to enable ICI to do the same.

SUMMARY

The ozone layer protects animals and plants from ultraviolet radiation. As it reacts with pollutants in the upper atmosphere, the ozone layer is becoming thinner. CFCs and nitrogen monoxide from high altitude planes are the culprits. The use of CFCs is being reduced.

CHECKPOINT

❶ Look round your kitchen, bathroom and garage. How many products in aerosol cans do you buy? How convenient is it to have each of these products in an aerosol can? What inconvenience would you suffer if aerosol cans were banned? How many of the aerosol cans are labelled 'ozone-friendly'? What does this mean?

❷ Speaking on 23 February 1988, the Prince of Wales announced that he had banned aerosols from his household. He said that some members of his household had difficulty in finding a suitable alternative hairspray.
  (a) What concern led the Prince of Wales to take this step?
  (b) What properties must the propellant in the hairspray possess to work effectively and to be safe in use?
  (c) What substitute can you suggest for an aerosol hairspray?

❸ How does their lack of chemical reactivity make CFCs (a) useful and (b) dangerous?

7.10 The greenhouse effect

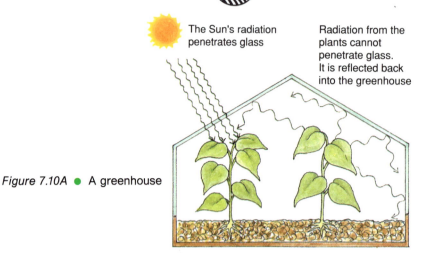

The Sun's radiation penetrates glass

Radiation from the plants cannot penetrate glass. It is reflected back into the greenhouse

Figure 7.10A ● A greenhouse

The Sun is so hot that it emits high-energy radiation. The Sun's rays can pass easily through the glass of a greenhouse. The plants in the greenhouse are at a much lower temperature. They send out infra-red radiation which cannot pass through the glass. The greenhouse therefore warms up (Figure 7.10A).

LOOK AT LINKS
Infra-red radiation and light are electromagnetic waves.
See *KS: Physics*, Topic 6.7

IT'S A FACT

Worldwide, 16 thousand million tonnes of carbon dioxide are formed each year by the combustion of fossil fuels!

IT'S A FACT

Other gases which contribute to the greenhouse effect are methane, oxides of nitrogen from vehicle exhausts and CFCs from aerosols and refrigerators.

Radiant energy from the Sun falls on the Earth and warms it. The Earth radiates heat energy back into space as infra-red radiation. Unlike sunlight, infra-red radiation cannot travel freely through the air surrounding the Earth. Both water vapour and carbon dioxide absorb some of the infra-red radiation. Since carbon dioxide and water vapour act like the glass in a greenhouse, their warming effect is called the greenhouse effect. Without carbon dioxide and water vapour, the surface of the Earth would be at $-40\,°C$. Most of the greenhouse effect is due to water vapour. The Earth does, however, radiate some wavelengths which water vapour cannot absorb. Carbon dioxide is able to absorb some of the radiation which water vapour lets through (see Figure 7.10B).

The surface of the Earth has warmed up by $0.75\,°C$ during the last century. The rate of warming up is increasing. Unless something is done to stop the temperature rising, there is a danger that the temperature of the Arctic and Antarctic regions might rise above $0\,°C$. Then, over the course of a century or two, polar ice would melt and flow into the oceans. If the level of the sea rose, low-lying areas of land would disappear under the sea.

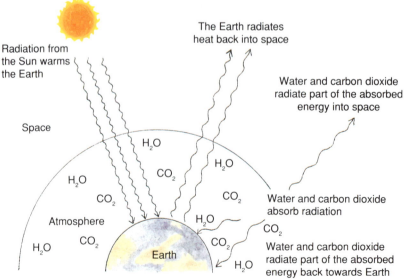

The Earth radiates heat back into space

Radiation from the Sun warms the Earth

Space

Water and carbon dioxide radiate part of the absorbed energy into space

H_2O

CO_2

H_2O

H_2O

CO_2

Atmosphere

CO_2

H_2O

CO_2

H_2O

Earth

CO_2

H_2O

CO_2

Water and carbon dioxide absorb radiation

Water and carbon dioxide radiate part of the absorbed energy back towards Earth

Figure 7.10B ● The greenhouse effect

IT'S A FACT

Every month, an area of tropical forest equal to the size of Belgium is felled and burned in South America.

One reason for the increase in the Earth's temperature is that we are putting too much carbon dioxide into the air. The combustion of coal and oil in our power stations and factories sends carbon dioxide into the air. The second reason is that we are felling too many trees. In South America, huge areas of tropical forest have been cut down to make timber and to provide land for farming. In many Asian countries, forests have been cut down for firewood. The result is that worldwide there are fewer trees to take carbon dioxide from the air by photosynthesis. The percentage of carbon dioxide is increasing, and some scientists calculate that it will double by the year 2000 (see Figure 7.10C). This would raise the Earth's temperature by 2 °C.

SUMMARY

Carbon dioxide and water vapour reduce the amount of heat radiated from the Earth's surface into space and keep the Earth warm. Their action is called the greenhouse effect. The percentage of carbon dioxide in the atmosphere is increasing, and the temperature of the Earth is rising. If it continues to rise, the polar ice caps could melt.

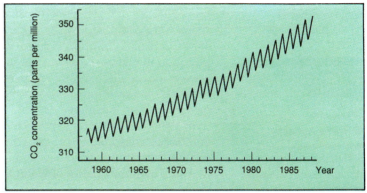

Figure 7.10C ● Atmospheric concentration of carbon dioxide. (These measurements were made at the Mauna Loa Observatory in Hawaii)

CHECKPOINT

❶ The amount of carbon dioxide in the atmosphere is slowly increasing.
(a) Suggest two reasons why this is happening.
(b) Explain why people call the effect which carbon dioxide has on the atmosphere the 'greenhouse effect'.
(c) Why are some people worried about the greenhouse effect?
(d) Suggest two things which could be done to stop the increase in the percentage of carbon dioxide in the atmosphere.

❷ *Selima* Did you hear what Miss Sande said about the greenhouse effect making the temperature of the Earth go up?
Joshe I don't know what she's worried about. We wouldn't be here at all if it weren't for the greenhouse effect.
(a) What does Joshe mean by what he says? What would the Earth be like without the greenhouse effect?
(b) Is Joshe right in thinking there is no cause for worry?

❸ The burning of fossil fuels produces 16 000 million tonnes of carbon dioxide per year. Carbon dioxide is thought to increase the average temperature of the air. It is predicted that the effect of this increase in temperature will be to melt some of the ice at the North and South Poles. Describe the effects which this could have on life for people in other parts of the world.

TOPIC 8 ● WATER POLLUTION

8.1 Pollution by industry

> What's the problem? We need clean drinking water – nothing could be more important. We need industry, and industry needs to dispose of waste products; in the process our water is polluted.

SUMMARY

The National Rivers Authority controls pollution of inland rivers. It does not regulate the discharge of pollutants into tidal rivers, estuaries and the sea. The estuaries in the UK are heavily polluted by industry and by sewage.

IT'S A FACT

B & N Chemicals in 1981 filled the town of Haverhill in Suffolk with chemical smells. The River Stow was so polluted that the water works was unable to draw water from the river. The company was prosecuted and convicted. The fine was a mere £325.
British Tissues was prosecuted in 1983 after discharging far more waste than it was allowed into the River Don. The fine was £750. It was obviously cheaper to pay fines than to treat the waste.

SUMMARY

Many industrial firms do not keep their discharges of wastes within the limits set by law.

Controls

You will notice that many industrial firms are on river banks. These firms can get rid of waste products by discharging them into rivers. Until 1989, the quantities of waste which industries were allowed to discharge into rivers were controlled by the water authority of each region. Under the 1974 Control of Pollution Act, the water authorities had power to control pollution in inland rivers but not in tidal rivers, estuaries and the sea (except for the discharge of radioactive waste). In spite of the Act, more than 2800 km of Britain's largest rivers are too dirty and lacking in oxygen to keep fish alive.

In 1989 the UK Water Privatisation Bill became law. The water authorities were sold to private companies, and are now run for profit as other industries are. The Government set up a National Rivers Authority to watch over the quality of water and prosecute polluters.

● *Estuaries*

Many of the worst polluters discharge into coastal waters and estuaries. The oil refineries, chemical works, steel plants and paper mills on coasts and estuaries can pour all the waste they want into estuaries and the sea. In the 1930s, fishermen could make a living in the Mersey. Now, it is too foul to keep fish alive. One reason is the discharge of raw sewage into the Mersey. The other is that too many firms pour waste into the estuary. There is unemployment in Merseyside, and the Government does not want to make life difficult for industry in the area. The industries on the banks of the Mersey have been given permission to fall below the standards of the Control of Pollution Act (see Figure 8.1A).

Other estuaries, such as the Humber, the Tees, the Tyne and the Clyde, are also polluted by industry.

Figure 8.1A ● The Mersey

FIRST THOUGHTS

Why has it taken so long for industry to react to the tragedy of Minamata?

LOOK AT LINKS
for **food chains**
See Topic 2.

Mercury and its compounds

A well-known case of industrial pollution is the tragedy of Minamata, a fishing village on the shore of Minamata Bay in Japan. A plastics factory started discharging waste into the bay in 1951. By 1953, a thousand people in Minamata were seriously ill. Some were crippled, some were paralysed, some went blind, some became mentally deranged, and some died. The cause of the disease was found to be the mercury compounds which the plastics factory discharged into the Bay. Although the level of mercury compounds in the Bay water was low, mercury was concentrated through the **food chain** (see Figure 8.1B). The level of mercury in the fish in the Bay was high, and fishers and their families became ill through eating the fish.

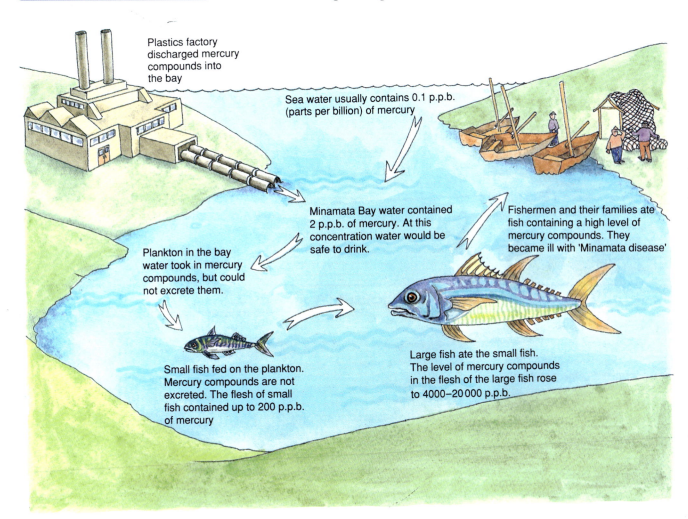

Plastics factory discharged mercury compounds into the bay

Sea water usually contains 0.1 p.p.b. (parts per billion) of mercury

Minamata Bay water contained 2 p.p.b. of mercury. At this concentration water would be safe to drink.

Fishermen and their families ate fish containing a high level of mercury compounds. They became ill with 'Minamata disease'

Plankton in the bay water took in mercury compounds, but could not excrete them.

Small fish fed on the plankton. Mercury compounds are not excreted. The flesh of small fish contained up to 200 p.p.b. of mercury

Large fish ate the small fish. The level of mercury compounds in the flesh of the large fish rose to 4000–20 000 p.p.b.

Figure 8.1B ● The food chain which led to the Minamata disease

Other countries have experienced the results of mercury pollution. In 1967, many lakes and rivers in Sweden were found to be so contaminated by mercury that fishing had to stop. In 1970, high mercury levels were found in hundreds of lakes in Canada and the USA. As late as 1988, the ICI plant on Merseyside discharged more mercury than the permitted level. Now that the danger is known, the polluting plants have taken care to reduce spillage of mercury. The danger is still there, however. Mercury from years of pollution lies in the sediment at the bottom of lakes. Slowly it is converted by bacteria into soluble mercury compounds. These may get into food chains.

SUMMARY

Mercury and its compounds are poisonous. If mercury gets into a lake or river, it is converted slowly into soluble compounds. These are likely to accumulate in fish and may be eaten by humans.

CHECKPOINT

❶ When a car engine has an oil change, the waste oil is sometimes poured down the drain. What is wrong with doing this?

❷ Does it matter whether rivers are clean and stocked with fish or foul and devoid of life? Explain your answer.

❸ The Minamata tragedy happened when Japan was building up its industry after the war. In spite of Japan's experience, Sweden, Canada and the USA found an excess of mercury in their lakes twenty years later. Why had they not learned from Japan's mistake?

8.2 Thermal pollution

FIRST THOUGHTS

What's wrong with warming up the water?

SUMMARY

Thermal pollution means warming rivers and lakes. It reduces the concentration of oxygen dissolved in the water.

Industries use water as a coolant. A large nuclear power station uses 4000 tonnes of water a minute for cooling. River water is circulated round the power station, where its temperature increases by 10 °C, and is returned to the river. If the temperature of the river rises by many degrees, the river is **thermally polluted**. As the temperature rises, the solubility of oxygen decreases. At the same time, the **biological oxygen demand** increases (see Topic 20.6). Fish become more active at the higher temperature, and need more oxygen. The bacteria which feed on decaying organic matter become more active and use more oxygen.

8.3 Pollution by sewage

FIRST THOUGHTS

The population of the UK is increasing. One result is the need for more sewage works. What happens when a country does not keep up with this need?

IT **Inside Science** (program)

Use this software to deal with an accidental pollution spillage.

IT **Life and Death of a River** (data base)

Explore a case of river pollution and see what can be done to clean it up. Use this with either the *Key* or *Keysheet* package.

Figure 8.3A ● British beaches and the EC standard

In Topic 20.6 you will read how sewage is treated before it is discharged into rivers or the sea. Unfortunately, as some water companies do not have enough plants to treat all their area's sewage, they discharge some raw sewage into rivers and estuaries. The Mersey receives raw sewage from Liverpool and other towns. In Sussex, sewage treatment works are inadequate and sewage is discharged into the sea. This creates nasty results at bathing beaches around the country.

SUMMARY

During the 1980s, the United Nations set a target of safe water and sanitation for all by 1990. The aim was to provide wells and pumps, kits for disinfecting water and hygienic toilets. The sum needed was £25 billion, slightly more than the world spends on its armies in one month. The target was not reached by 1990, but the work is continuing.

LOOK AT LINKS
for **sewage treatment**
See Topic 20.6.

The quality of the water at dozens of Britain's bathing beaches fails to meet standards set by the European Community (EC). Many British beaches have more coliform bacteria and faecal bacteria in the water than the EC standard.

The Third World

Of the four billion people in the world, two billion have no toilets, and one billion have unsafe drinking water. In Third World countries (the developing countries) three out of five people have difficulty in obtaining clean water. Some Third World communities have to use a river as a source of drinking water as well as for disposal of their sewage. Bacteria are present in faeces, and they infect the water. Many diseases are spread by contaminated water. They include cholera, typhoid, river blindness, diarrhoea and schistosomiasis. Four-fifths of the diseases in the Third World are linked to dirty water and lack of sanitation. Five million people each year are killed by water-borne diseases (see Figure 8.3B).

Figure 8.3B ● Their water supplies

FIRST THOUGHTS

Farmers need to use fertilisers. What happens when a crop does not use all the fertiliser applied to it? There can be pollution, as this section explains.

8.4 Pollution by agriculture

Fertilisers

A lake has a natural cycle. In summer, algae grow on the surface, fed by nutrients which are washed into the lake. In autumn the algae die and sink to the bottom. Bacteria break down the algae into nutrients. Plants need the elements carbon, hydrogen, oxygen, nitrogen and phosphorus. Water always provides enough carbon, hydrogen and oxygen; plant growth is limited by the supply of nitrogen and phosphorus. Sometimes farm land surrounding a lake receives more fertiliser than the crops can absorb. Then the unabsorbed nitrates and phosphates in the fertiliser wash out of the soil into the lake water. When fertilisers wash into a lake, they upset the natural cycle. The algae multiply rapidly to produce an **algal bloom**. The lake water comes to resemble a cloudy greenish soup (see Figure 8.4A). When the algae die, bacteria feed on the dead material and multiply. The increased bacterial activity consumes much of the dissolved oxygen. There is little oxygen left in the water, and fish die from lack of oxygen. The lake becomes difficult for boating because masses of algae snag the propellers. The name given to this accidental fertilisation of lakes and rivers is **eutrophication**.

Many parts of the Norfolk Broads are now covered with algal bloom. The tourist industry centred on the Broads would like to see them restored to their former condition.

Use an oxygen sensor and probe to measure the percentage of oxygen in air. The same probe can be used to measure dissolved oxygen in water. You'll need to connect the sensor to a computer or a datalogger.

Lough Neagh in Northern Ireland is the UK's biggest inland lake. It supplies Belfast's water and it also supports eel-fishing. Algae now block the filters through which water flows to the water treatment plant. Eels and other fish are in danger as the level of dissolved oxygen falls. The problem is being tackled by removing phosphates from the treated sewage which enters the lough. Treatment with a solution containing aluminium ions and iron(III) ions precipitates phosphates. Each year, this stops 60 tonnes of phosphorus in the form of phosphates from entering Lough Neagh. This pollution is unnecessary. Detergents without phosphates would leave laundry only a little less sparkling white, but would not pollute our rivers and lakes.

Fertiliser which is not absorbed by crops can be carried into the ground water (the water in porous underground rock). Ground water provides one third of Britain's drinking water. The EC has set a maximum level of nitrates in drinking water at 50 mg/l (12 p.p.m. of nitrogen in the form of nitrate). Four out of the ten water companies in England and Wales have drinking water which exceeds this nitrate level. In 1989, the EC decided to prosecute the UK for falling below EC water standards.

Figure 8.4A ● Algal bloom

SUMMARY

When a crop receives more fertiliser than it can use, nitrates and phosphates wash into lakes and rivers. There, they stimulate the growth of weeds and algae. When the plants die, bacterial decay of the dead material uses oxygen. The resulting shortage of dissolved oxygen kills fish.

The level of nitrates in ground water, from which we obtain much of our drinking water, is rising.

Pollution can be reduced by reducing the application of fertilisers and by omitting phosphates from detergents.

There are two health worries over nitrates. Nitrates are converted into nitrites (salts containing the NO_2^- ion). Some chemists think that nitrites are converted in the body into nitrosoamines. These compounds cause cancer. The other worry is that nitrites oxidise the iron in haemoglobin. The oxidised form of haemoglobin can no longer combine with oxygen. The extreme form of nitrite poisoning is 'blue baby' syndrome, in which the baby turns blue from lack of oxygen. Babies are more at risk than adults because babies' stomachs are less acidic and assist the conversion of nitrates into nitrites.

The level of nitrites in drinking water permitted by the EC is 0.1 mg/l. Some parts of London have nitrite levels which are higher than this. The UK Government has agreed to bring the UK into line with the rest of Europe. To install nitrate-stripping equipment would cost £200 million. *Should the Government reduce the use of fertilisers? How? Should the Government introduce a tax on fertilisers or ration fertilisers?*

Pesticides

FIRST THOUGHTS

What are the 'drins'? Why is the EC worried about the level of drins in UK water?

Other pollutants which must worry us are the pesticides dieldrin, endrin and aldrin (sometimes called the 'drins'). They cause liver cancer and affect the central nervous system. The EC sets a maximum level of 5×10^{-9} g/l for 'drins'. Half the water in the UK exceeds this level. The danger with the 'drins' is that fish take them in and do not excrete them. The level of 'drins' in fish may build up to 6000 times the level in water.

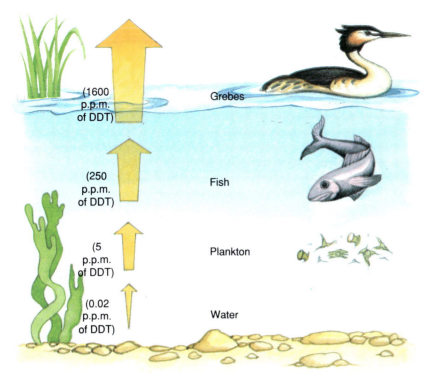

(1600 p.p.m. of DDT) Grebes

(250 p.p.m. of DDT) Fish

(5 p.p.m. of DDT) Plankton

(0.02 p.p.m. of DDT) Water

Figure 8.4B ● A food chain in Clear Lake, California

Figure 8.4B shows what happened when DDT, another powerful insecticide, was used to spray Clear Lake in California to get rid of mosquitoes. It is another example of pollutants being concentrated by a food chain (see p. 110).

SUMMARY

Pesticides are serious pollutants, especially when they can be concentrated through a food chain.

CHECKPOINT

1 Groups of settlers in North America always built their villages on river banks, and discharged their sewage into the river. How did the river dispose of the sewage? Why can this method of sewage disposal not be used for larger settlements?

2 British Tissues make toilet paper, paper towels, paper handkerchiefs, etc. They use a lot of bleach on the paper, and this bleach is one of the chemicals which the firm has to dispose of. Can you suggest how the firm could reduce the problem of bleach disposal?

3 (a) Why do some lakes develop a thick layer of algal bloom?
(b) Why is algal bloom less likely to occur in a river?
(c) What harm does algal bloom do to a lake that is used as (i) a reservoir (ii) a fishing lake (iii) a boating lake?

4 The concentration of nitrates in ground water is rising. Explain:
(a) why this is happening,
(b) why some people are worried about the increase.

5 Water companies can tackle the problem of high nitrate levels by:
• blending water from high-nitrate sources with water from low-nitrate sources
• closing some sources of water
• treating the water with chemicals
• ion exchange
• microbiological methods

(a) Say what you think are the advantages and disadvantages of each of these methods.
(b) Which do you think would be the most expensive treatments? How will water companies be able to pay for the treatment?
(c) Suggest a different method of reducing the level of nitrates in ground water.
(d) Say who would pay for the method which you mention in (c) and how they would find the money.

8.5 Pollution by lead

Lead water pipes have been used for centuries. What is wrong with them?

Lead was used to make water pipes in Roman times. Until the 1950s, lead water pipes were used widely in the UK. Slowly, lead dissolves in water. Lead and its compounds are poisonous. Since 1950, copper has been used for water pipes. In cold water systems, plastic pipes are used. In hard water areas, lead pipes are safe because a layer of insoluble lead carbonate, calcium carbonate and magnesium carbonate builds up and acts as a protective barrier which stops lead dissolving. In soft water areas, the pH of the water may be less than 5, and lead dissolves more rapidly. Water companies make the water more alkaline by adding calcium hydroxide. During the 1983 strike of UK water workers, the treatment of many water supplies with calcium hydroxide stopped. Within two weeks, the level of lead in the tap water of some houses with lead pipes had risen from 40 µg/l to over 800 µg/l. In the North-West, 600 000 houses have lead pipes. Some parts of Scotland have water which contains more lead than the EC standard.

SUMMARY

Some parts of the UK have lead water pipes. Lead compounds are toxic. In hard water areas, the deposition of insoluble scale in the pipes stops lead dissolving. In other areas, the solution of lead is reduced by keeping the water alkaline.

CHECKPOINT

1 (a) How does lead get into our drinking water?
  (b) What harm does it do?
  (c) In what ways can the amount of lead in drinking water be reduced?

8.6 Pollution by oil

Huge oil tankers sail the sea. Sometimes one has an accident – not often,.but often enough to cause a pollution disaster.
The first big oil spill off the UK coast was when the oil tanker, the *Torrey Canyon*, sank off Cornwall in 1967. The pollution fouled beaches in Cornwall and killed thousands of sea birds. Now, tankers of up to 550 000 tonnes are in use, and accidents can happen on a much larger scale.

Prince William Sound was a beautiful unspoiled bay in Alaska. It was home to a huge variety of sea animals. Death struck over an area of 1300 square kilometres in 1989. The supertanker *Exxon Valdez* left Valdez with a cargo of oil from Alaska. Only 40 km out of port, the tanker hit a submerged reef and 60 million litres of crude oil leaked from her tanks. Fish, sea mammals and migrating birds from all parts of the American continent perished in the giant oil slick.

Sea birds dive to obtain food. If the sea is polluted by a discharge of oil, sea birds may find themselves in an oil slick when they surface. Then oil sticks to their feathers and they cannot fly. They drift on the surface, becoming more and more waterlogged until they die of hunger and exhaustion. Thirty five thousand sea birds died in the *Exxon Valdez* disaster.

Oil spills at sea are the results of capsizings, collisions and accidental spills during loading and unloading at oil terminals. There is another source. After a large tanker has unloaded, it may have 200 tonnes of oil left in its tanks. While in port, the tanker is flushed out with water sprays, and the cleaning water is collected in a special tank, where the oil separates. Some captains save time by flushing out their tanks at sea and pumping the wash water overboard. This is illegal. Maritime nations have tried to set up standards to stop pollution of the seas, but several

Figure 8.6A ● The *Torrey Canyon*

Figure 8.6B ● The *Exxon Valdez* accident

Figure 8.6C ● Ten thousand sea otters perished in the *Exxon Valdez* spill. Some ate poisoned fish. Others drowned when their fur became clogged with oil

Figure 8.6D ● Burning the spilled oil from a smaller scale accident

Figure 8.6E ● Another victim

135

═══ SCIENCE AT WORK ═══

Bacteria can be used to clean out a tanker's storage compartment. The empty tank is filled with sea water, nutrient, air and bacteria. When the tanker reaches its destination, the tank contains clean water, a small amount of recoverable oil and an increased number of bacteria. The bacteria can be used as animal feed.

nations have not signed the agreements. Enforcing agreements is very difficult as it is impossible to detect everything that happens at sea.

Various methods have been tried for the removal of oil from the surface of the sea.

- **Dispersal**   Chemicals are added to emulsify the oil. The danger is that they may be toxic to marine life.

- **Sinking**   Oil may be treated with sand and other fine materials to make it sink. A danger is that the sunken oil may cover and destroy the feeding areas of marine creatures.

- **Burning**   Burning oil is dangerous as a fire can spread rapidly over the sea. Research has been done on safe methods of burning oil, but they leave 15 per cent of the oil behind as lumps of tar.

- **Absorbing**   Absorbents do not work well in the open sea. They provide the best way of cleaning a beach or preventing an oil spill from reaching the shore.

- **Skimming off**   The method of surrounding an oil spill with a line of booms to prevent it spreading and then pumping oil off the surface has been used with some success.

- **Solidifying**   Scientists at British Petroleum have discovered chemicals which will solidify oil spills. The chemicals must be sprayed on to the oil slick from the air. They convert the oil into a rubber-like solid which can be skimmed off the surface in nets.

- **Bacteria**   There are bacteria which will decompose petroleum. A mixture of bacteria (of the correct strain) and nutrients is sprinkled on to the spill from the air.

▼ SUMMARY

Spillage of oil from large tankers is a source of pollution at sea. It kills marine animals and washes ashore to pollute beaches.

CHECKPOINT

 (a) What are the causes of oil spills at sea?
   (b) What damage do they do?
   (c) Who pays to clean up the mess?
   (d) Suggest what can be done to stop pollution of the sea by oil.

? THEME QUESTIONS

● **Topic 5**

1 Equal amounts of four soil samples, taken from different locations, were shaken up with water in graduated cylinders and left to settle. The results are shown below.

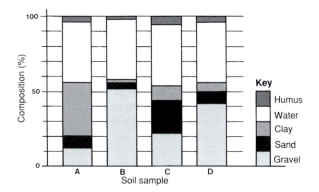

(a) Which soil had (i) the highest humus content (ii) the lowest gravel content?

(b) What was the percentage of clay in (i) sample A (ii) sample B? (iii) Explain how the addition of sand and humus would improve the texture and properties of sample A.

(c) What was the percentage of sand in (i) sample C (ii) sample D? (iii) Which of these two samples was taken from a good vegetable garden? Give your reasons.

(d) (i) Soil B contained the smallest number of earthworms. Try to explain this.
(ii) Soil B was also unsuitable for the growth of root crops such as carrots and turnips. Suggest an explanation for this.

2 Complete the table below to highlight the differences between a clay soil and a sandy soil.

Property	Clay soil	Sandy soil
Particle size		
Crumb structure		
Aeration and drainage		
Mineral content		
How can it be improved?		

3 (a) Name five other constituents of soil apart from rock particles.

(b) What is a loam?

(c) (i) What is humus and how is it formed?
(ii) Describe two ways in which humus can be renewed in the soil.

(d) Irrigation can vastly increase food production in hot regions of the world, but what are the drawbacks it can bring?

(e) Outline the advantages and disadvantages of each of the following agricultural practices
(i) monoculture (ii) the use of large fields.

4 A classroom contains 36 pupils. The doors and windows are closed for half an hour. Answer these questions about the air at the end of the half hour.

(a) Will the air temperature be higher or lower? Explain your answer.

(b) Will the air be more or less humid (moist)? Explain your answer.

(c) Will the percentage of carbon dioxide in the air be higher or lower? Explain your answer. Say how the change in carbon dioxide content will affect the class.

● **Topic 6**

5 (a) What type of plants can use atmospheric nitrogen as a nutrient?

(b) Why can't most plants use atmospheric nitrogen in this way?

(c) Name the nitrogen-containing compounds that are built by plants.

(d) Name one natural process that produces nitrogen compounds that enter the soil and are used by plants.

(e) Explain why there is not enough nitrogen from natural sources to support the growth of crop after crop on the same land.

6 (a) Explain what is meant by the **nitrogen cycle**.

(b) Are we likely to run out of nitrogen?

(c) Say how the human race alters the natural nitrogen cycle (i) by taking nitrogen out of the cycle and (ii) by adding nitrogen.

(d) Explain how fertilisers have created a problem concerning nitrogen compounds.

(e) Suggest two ways in which this problem could be attacked.

7 An experiment was set up as shown in the diagram. A sample of unsterilised soil was suspended in flask A and sterilised soil in flask B. Both flasks contained potassium hydroxide solution. At the start of the experiment, the levels of the coloured liquid were equal in both arms of the U-tube.

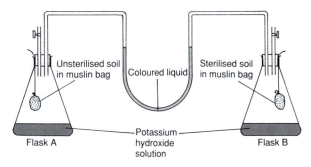

(a) What change is likely to have taken place in the levels of the coloured liquid after 48 hours?

(b) Explain fully your answer.

137

8 Explain each of the following statements:
  (a) Prolonged use of artificial fertilisers is bad for the soil.
  (b) Extensive use of fertilisers on arable land can lead to the pollution of waterways.
  (c) Growing clover improves the fertility of the soil.

9 The diagram below shows the nitrogen cycle. The labelled arrows represent different processes.

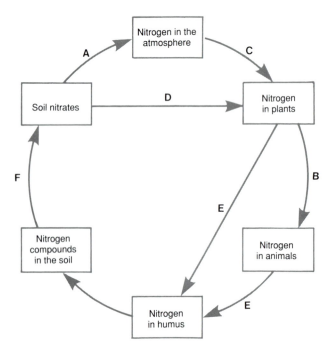

  (a) Name the processes A–F, e.g. A = denitrification.
  (b) (i) Why is process C useful to plants?
      (ii) Process C takes place in clover but not in grass. Why is this so?
  (c) What organisms are responsible for process E?
  (d) In what form is nitrogen passed in process B?

10 Many pesticides are potentially harmful because they are not biodegradable and can build up two lethal limits in animals. The food chain shows the build up of a particular pesticide.

$$\text{Aphid} \rightarrow \text{Lacewing larva} \rightarrow \text{Blue tit} \rightarrow \text{Buzzard}$$
$$5 \qquad\qquad 500 \qquad\qquad 5000 \qquad 2500$$

  (Pesticide levels shown in relative amounts)

  (a) How did the pesticide enter the food chain in the first place?
  (b) Which animal in the food chain is likely to receive a lethal amount first?
  (c) The most widespread insecticide use in Britain used to be DDT. Why do you think that its use has been banned?
  (d) One of the sub-lethal effects of DDT was that it affected the shell-producing gland in many birds of prey such as the peregrine falcon. It resulted in eggs being laid with thinner than normal shells. How did this lead to the death of many young?
  (e) An alternative method of keeping down pest numbers is biological control. Explain how it works.

● *Topic 7*

11 When petrol burns in a car engine, carbon dioxide and carbon monoxide are two of the products.
  (a) Write the formula of (i) carbon dioxide (ii) carbon monoxide.
  (b) Explain the statement 'Carbon monoxide is a product of incomplete combustion.'
  (c) Red blood cells contain haemoglobin. What vital job does haemoglobin do in the body?
  (d) If people breathe in too much carbon monoxide, it may kill them. How does carbon monoxide cause death?
  (e) Explain how people can be poisoned accidentally by carbon monoxide.
  (f) What precautions can people take to make sure that carbon monoxide is not formed in their homes?
  (g) The blood of people who smoke contains more carbon monoxide than the blood of non-smokers. Can you explain why?

12 (a) What is the difference between oxygen and ozone?
  (b) What converts oxygen into ozone?
  (c) What converts ozone into oxygen?
  (d) What is the ozone layer? Where is it?
  (e) Why is the ozone layer becoming thinner?
  (f) Why does the decrease in the ozone layer make people worry?

● *Topic 8*

13 (a) Name three pollutants that are produced by power stations.
  (b) For one of the pollutants, describe the kind of cleaning system that can be used to stop the pollutant being discharged into the air.
  (c) How will the cost of electricity be affected by (i) installing the cleaning system and (ii) stocking the chemicals consumed in running the cleaning system?

14 Many industrial plants take water from a river and then return it at a higher temperature. What harm can this do? What name is given to this practice?

15 Bacteria in river water are able to convert many pollutants into harmless products. What, then, is the harm in dumping waste into rivers?

16 What happens when plastics are dumped in lakes and rivers?

17 Water normally contains 10 p.p.m. of oxygen. It takes 4 g of oxygen to oxidise completely 1 g of oil. A garage does an oil change on a car, and pours the 4 kg of dirty oil down the drain. What mass of water will be stripped of its dissolved oxygen by oxidising the oil from this car?

18 Why is the pollution of estuaries so common? Who gains from being able to pollute estuaries? Who loses from this pollution?

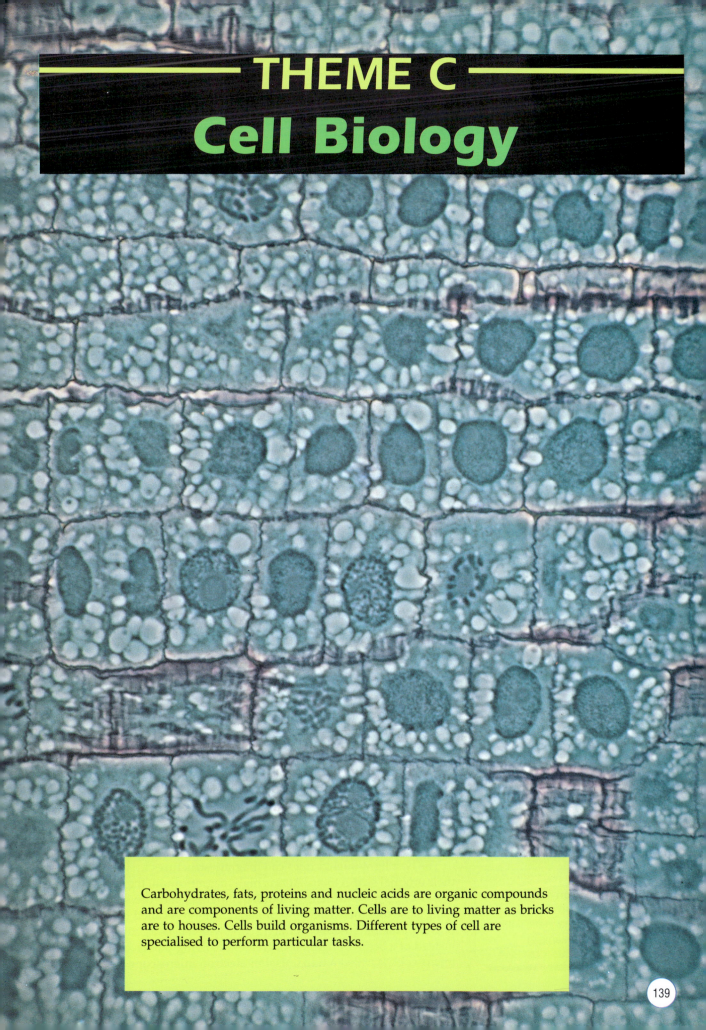

THEME C
Cell Biology

Carbohydrates, fats, proteins and nucleic acids are organic compounds and are components of living matter. Cells are to living matter as bricks are to houses. Cells build organisms. Different types of cell are specialised to perform particular tasks.

TOPIC 9

THE CELL

FIRST THOUGHTS

9.1 Seeing cells

Cells were first seen in 1665 when the British scientist Robert Hooke looked at cork under his microscope. His discovery is one example of the way in which an advance in technology – in this case the microscope – leads to scientific discoveries. Modern microscopes reveal the structure of cells in great detail.

Figure 9.1A ● Robert Hooke's drawing of the outline of cork cells (from *Micrographia*, 1665)

Figure 9.1A shows what Robert Hooke saw when he looked at thin strips of cork with his microscope. He called the little box-like shapes **cells** because they reminded him of the small rooms (called cells) in which monks live. The scientists of the day did not appreciate the importance of Hooke's discovery. A century and a half later, in 1838, two German biologists, Mathias Schleiden and Theodor Schwann, put forward a **cell theory of life**. Their theory has been altered slightly in the light of later discoveries. Modern cell theory states:

- All living things are made of cells.
- New cells are formed when old cells divide into two.
- All cells are similar in structure and function (the way they work), but not identical.
- The structure of an organism depends on the way in which the cells are organised.
- The functions of an organism (the way it works) depend on the functions of its cells.

The more we know about the structure and function of cells, the better we shall be able to understand how a whole organism functions.

Microscopes

Most cells are too small to be seen with the naked eye. Without microscopes, we should know very little about them. Figure 19.1B shows a modern light microscope. A lamp lights the specimen, which the observer views through two **magnifying lenses**. The total magnification is given by

Total magnification = Magnifying power x Magnifying power
of specimen   of eyepiece   of objective lens

The best light microscopes can magnify structures up to 1500 times their original size (x1500). At this high magnification, the image becomes less clear because the lenses cannot distinguish between small structures lying close together. This limit of magnification is called the resolving power of the microscope. A microscope which uses a beam of electrons instead of a beam of light has a greater resolving power. Such a microscope is called a **transmission electron microscope** (see Figure

RESOURCE ACTIVITY PACK

LOOK AT LINKS
for **magnifying lenses**
See *KS: Physics*, Topic 13.4.

LOOK AT LINKS
for **electron microscopes**
Learn more about the effect of magnetic fields on electron beams in *KS: Physics*, Topic 22.4.

Observer

Eyepiece lens - usually magnifies x10 (though x7 and x5 eyepieces are available)

Focusing knob

Objective lens poised over specimen - may be low power (x5), medium power (x10 to x 20) or high power (x40 to x100)

Revolving turret - houses up to four objective lenses

Cover slip - a very thin glass sheet over specimen

Specimen mounted on slide

Glass slide

Stage - the platform which holds the slide

Lamp - only present in some microscopes

Hole in stage

Mirror - reflects light through the hole in the stage, through the specimen and into the observer's eye

Light path

Figure 9.1B ● A light microscope

9.1C). It can magnify up to 500 000 times (x500 000) without loss of clarity. The transmission electron microscope enables you to see the structure of cells in great detail; that is, the **fine structure**. As we cannot see electrons, the electron beam is used to give a fluorescent image on a screen, which can be photographed.

Figure 9.1C ● A transmission electron microscope (TEM)

Some of the photographs in this book give a three-dimensional view of living things, for example, Figure 9.1H overleaf. These photographs are taken with a **scanning electron microscope**. The beam of electrons inside a scanning electron microscope does not pass through the specimen: the beam is reflected by the surface of the specimen and therefore shows the surface in great detail.

SUMMARY

The invention of the light microscope led to the discovery of cells. The electron microscope has revealed the fascinating world inside the cell and opened up the study of cell biology.

LOOK AT LINKS
Look at Figure 1.5A. It summarises the differences between plant cells, animal cells and bacterial cells.

Cell structure

Plant cells, animal cells and bacterial cells are different from one another. However, they also have features in common. Figures 9.1D–I illustrate the differences and the similarities. The drawings have been made from views of cells under the transmission electron microscope.

Figure 9.1D ● A typical animal cell. Features of different cells viewed under the electron microscope have been combined to form this generalised picture.

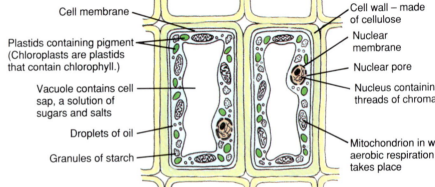

Endoplasmic reticulum a network of channels running through the cytoplasm from the nucleus to the surface of the cell; sections can be seen throughout the cell

Oil droplet

Cell membrane

Mitochondrion one of the 'power houses' of the cell in which energy is released by aerobic respiration

Nucleus containing threads of chromatin (made of DNA and protein)

Ribosomes – bead-like structures on the endoplasmic reticulum and in the cytoplasm (Proteins are made in the ribosomes.)

Nucleolus

Nuclear membrane

Pore in nuclear membrane

Glycogen granule

Cytoplasm

Vacuole – small, not present in all animal cells

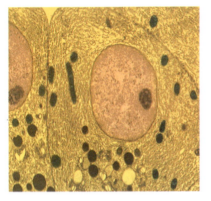

Figure 9.1E ● Animal pancreas cell (x700)

Cell membrane

Plastids containing pigment (Chloroplasts are plastids that contain chlorophyll.)

Vacuole contains cell sap, a solution of sugars and salts

Droplets of oil

Granules of starch

Cell wall – made of cellulose

Nuclear membrane

Nuclear pore

Nucleus containing threads of chromatin

Mitochondrion in which aerobic respiration takes place

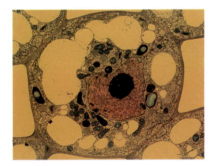

Figure 9.1G ● Plant palisade cell (x500)

Figure 9.1F ● A generalised plant cell

Chromosome – a single, coiled strand of DNA. There is no distinct nucleus though this is called the nuclear zone. (Some bacteria have more than one chromosome.)

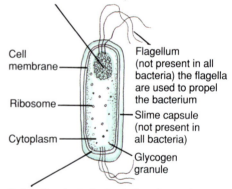

Cell membrane

Ribosome

Cytoplasm

Flagellum (not present in all bacteria) the flagella are used to propel the bacterium

Slime capsule (not present in all bacteria)

Glycogen granule

Cell wall protects the bacterium, has a shape characteristic of the bacterium. Made of lipid, protein and a carbohydrate other than cellulose.

Figure 9.1I ● A bacterium. The cells of most bacteria are only just visible under a light microscope, but the transmission electron microscope shows their structure in detail.

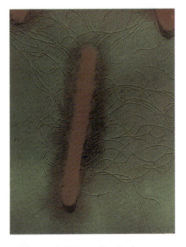

Figure 9.1H ● Bacterium (x3000)

SUMMARY

Plant cells, animal cells and bacterial cells have many features in common. There are also differences between these three types of cell. Many cells have special features which enable them to carry out specific tasks in an organism.

❶ Compare the structures of the three types of cell in Figures 9.1E, G and I. Then copy and complete the table below. Put a tick in each space if the structure is present. Add any comments needed to distinguish between the structures. As an example chromatin has been filled in for you.

Structure in cell	Bacterium	Plant cell	Animal cell
Cell membrane			
Cell wall			
Cytoplasm			
Nucleus			
Nuclear membrane			
Chromatin	✓ one chromosome	✓ many chromosomes	✓ many chromosomes
Endoplasmic reticulum			
Mitochondria			
Ribosomes			
Chloroplast			
Vacuole			

9.2 Movement into and out of cells

Cells need a non-stop supply of water and the substances dissolved in it to stay alive. This is why there is constant movement of solutions inside cells and also into and out of cells. After reading this section you will be able to explain what is happening to the plant in Figure 9.2A.

Diffusion

Diffusion is the movement of a substance through a solution or through a gas. In a solution, diffusion happens when there is a higher concentration of a substance in one part of the solution than in another. Then the substance diffuses. It spreads out until the concentration is the same throughout the solution. Diffusion is very important in the solutions inside cells.

Figure 9.2A ● (a) A plant wilting through lack of water
(b) The same plant recovering after watering

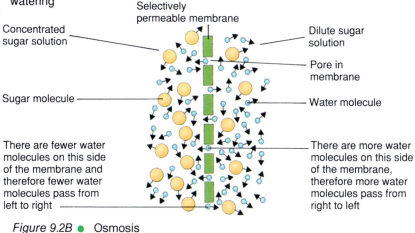

Concentrated sugar solution

Selectively permeable membrane

Dilute sugar solution

Pore in membrane

Sugar molecule

Water molecule

There are fewer water molecules on this side of the membrane and therefore fewer water molecules pass from left to right

There are more water molecules on this side of the membrane, therefore more water molecules pass from right to left

Figure 9.2B ● Osmosis

Osmosis

Movement of substances take place through the cell membrane which separates the cell contents from the surroundings. Cell membranes allow some substances to pass through and stop other substances from passing: they are **selectively permeable** or **semi-permeable**. In general, substances pass through a membrane if their particles are smaller than the pores in the membrane (see Figure 9.2B).

When two solutions are separated by a selectively permeable membrane, water passes through the membrane in both directions. The net flow of water is from the more dilute solution to the more concentrated solution. The flow continues until the two concentrations are equal. The flow of water from a more dilute solution to a more concentrated solution is called **osmosis**.

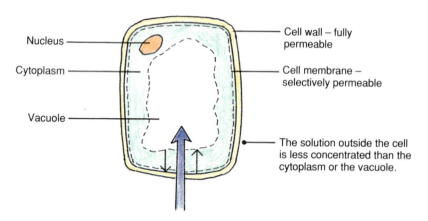

Nucleus

Cytoplasm

Vacuole

Cell wall – fully permeable

Cell membrane – selectively permeable

The solution outside the cell is less concentrated than the cytoplasm or the vacuole.

Water flows through the cell wall and cell membrane into the cytoplasm and into the vacuole. The increased pressure of water in the vacuole presses the cytoplasm against the cell wall. When the cell contains as much water as it can hold, the cell is fully **turgid**

Figure 9.2C ● A turgid cell

● *Turgor and plasmolysis*

Osmosis takes place between the cytoplasm and the solution outside the cell. The changes happening inside the cells due to osmosis bring about visible changes in the plant. When a plant cell is placed in a less concentrated solution, water passes through the cell wall and membrane, through the cytoplasm and into the vacuole. The cell becomes **turgid** (see Figure 9.2C).

Sometimes plant cells may be placed in a more concentrated solution, although this does not often happen in nature. Then water passes out of the vacuole, out of the cytoplasm, out through the cell membrane and cell wall into the solution outside the cell. The pressure of the vacuole on the cytoplasm decreases until the cytoplasm pulls away from the cell wall. The cell becomes limp, rather like a partly blown-up balloon. Cells in this condition are described as **plasmolysed** (see Figure 9.2D).

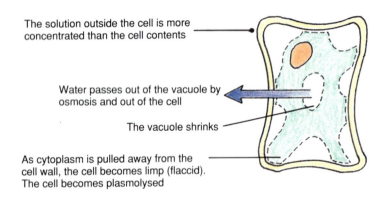

The solution outside the cell is more concentrated than the cell contents

Water passes out of the vacuole by osmosis and out of the cell

The vacuole shrinks

As cytoplasm is pulled away from the cell wall, the cell becomes limp (flaccid). The cell becomes plasmolysed

Figure 9.2D ● A plasmolysed cell

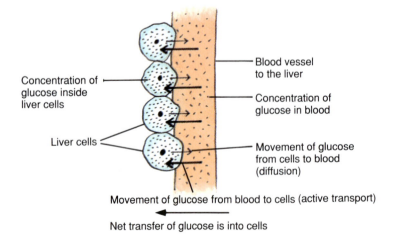

Concentration of glucose inside liver cells

Liver cells

Blood vessel to the liver

Concentration of glucose in blood

Movement of glucose from cells to blood (diffusion)

Movement of glucose from blood to cells (active transport)

Net transfer of glucose is into cells

Figure 9.2E ● Active transport

Sometimes substances move across the cell membrane from a region where they are present in low concentration to a region where they are present in high concentration. That is, they move in the reverse direction to normal diffusion. Such movement is called **active transport**. By this means, cells may build up stores of substances which would otherwise be spread out by diffusion (see Figure 9.2E). Active transport requires more energy than normal diffusion.

❶ Refer to Figure 9.2A. Explain fully, with the aid of a diagram, what has happened in the plant cells to produce the change from photograph (a) to photograph (b).

❷ The following three conversations take place in a kitchen. Give scientific explanations for all three observations.
(a) *Michael*   Why are you putting that celery in a jug of water?
   Kate   It will keep crisper in water.
(b) *Jonathan*   I've sprinkled sugar on my strawberries and it's gone all pink.
   Beth   That's because it makes the juice come out.
(c) *Jasper*   I cut some chips earlier and put them in salt water to keep.
   Miranda   They seem to have shrunk!

❸ (a) Explain what is meant by a semi-permeable or selectively permeable membrane.
(b) The figure opposite shows a bag made from visking tubing (a semi-permeable membrane) filled with a 15% salt solution (15 g of salt per 100 g of water). Draw sketches to show the changes that will happen if the beaker contains (i) distilled water (ii) 7% salt solution (iii) 15% salt solution (iv) 30% salt solution.

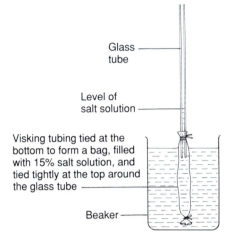

Glass tube

Level of salt solution

Visking tubing tied at the bottom to form a bag, filled with 15% salt solution, and tied tightly at the top around the glass tube

Beaker

❹ Peas are sometimes preserved by drying them.
(a) How much water do dried peas take up when they are soaked in water? Plan an experiment to find out. You should plan a quantitative experiment – that is to measure the exact quantity (e.g. mass) of water absorbed by a known mass of dried peas.
(b) What makes the cells of a pea absorb water when the pea is placed in water? Describe what happens inside the cells. Use scientific words to describe the difference between the cells in the dried pea and the cells in the soaked pea.
(c) Water does not flow into a dried pea indefinitely. What makes the flow of water stop?

❺ Explain the difference between osmosis and active transport. Give one example of the importance of active transport.

IT'S A FACT

More than 30% of the energy released by cellular respiration is used in the active transport of substances into cells.

9.3 The nucleus

An imaginary journey

Figure 9.3A ● Journey into the nucleus

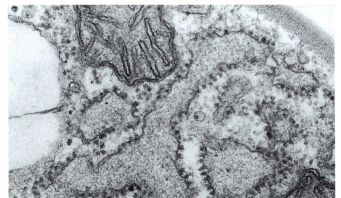

(a) The rope-like endoplasmic reticulum extends throughout the cell. Note the clusters of ribosomes (0.02 µm wide) and the mitochondrion at the top of the photo (10 µm long, 1 µm across).

(b) The nucleus lies ahead surrounded by the nuclear membrane. The pits are nuclear pores (0.005 µm across).

Imagine you are shrinking! You become smaller and smaller until you can pass through one of the pores in the cell membrane. Once inside the cytoplasm, you set off in search of the nucleus. You know that the endoplasmic reticulum connects the cell membrane to the nucleus so you follow it through the cytoplasm (see Figure 9.3A(a)). Your journey takes you past sausage-shaped mitochondria and globular ribosomes, until you come up against a wall. It is the nuclear membrane, pitted with holes (see Figure 9.3A (b)). Hoping that these are nuclear pores, you plunge through one of them. You are inside the nucleus! It looks like an explosion in a spaghetti factory. Everywhere you can see coiled strands of the nucleic acids, DNA and RNA. These are the molecules that control the activities of the whole cell.

Making proteins

LOOK AT LINKS
for **proteins**
See Topic 10.4.

A vital activity which takes place in cells is the manufacture of **proteins**. To make one molecule of a certain protein, hundreds or thousands of amino acid molecules must combine in the right order. The substance responsible for getting the order right is **DNA**. Present in the nucleus, DNA carries instructions – called a **code** – for combining amino acids in the right order.

DNA contains the bases adenosine, thymine, guanine and cytosine (A, T, G and C). The base uracil (U) is found in RNA instead of thymine. There are about twenty amino acids. The instructions needed to assemble one amino acid in its correct place in the protein molecule are contained in a row of three bases: a **codon** (see Figure 9.3B).

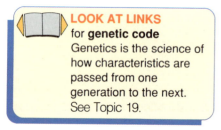

LOOK AT LINKS
for **genetic code**
Genetics is the science of how characteristics are passed from one generation to the next.
See Topic 19.

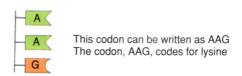

This codon can be written as AAG
The codon, AAG, codes for lysine

Figure 9.3B ● A row of three bases forms a codon

Each amino acid has a different codon, GGC codes for (gives instructions for) the amino acid glycine, and AAG codes for lysine. A length of DNA which codes for the whole of one protein is called a **gene**. This is why the DNA code is called the **genetic code**.

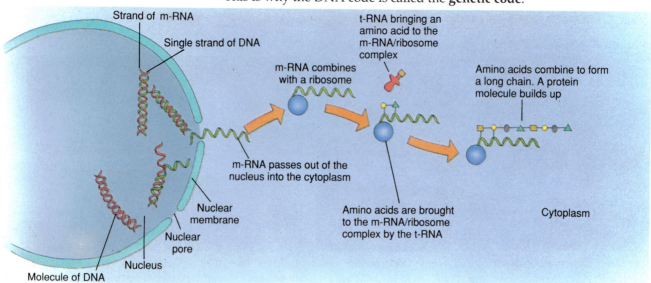

Figure 9.3C ● How DNA controls protein synthesis

DNA is present in the nucleus. Proteins are synthesised on the ribosomes in the cytoplasm. *How do the instructions for making proteins travel from DNA to the ribosomes?* DNA employs a messenger to take instructions to where they are needed. This messenger is a substance called messenger-RNA (m-RNA).

Messenger-RNA forms in the nucleus after a DNA double helix unwinds. Each single strand of DNA attracts bases, which combine to form a strand of m-RNA. Because of the way it is assembled, m-RNA is a complement of one strand of DNA. It can travel through the nuclear pores into the cytoplasm where it bonds to a ribosome. Another type of RNA is involved at this stage. Transfer-RNA (t-RNA) brings amino acids to the m-RNA–ribosome complex. The amino acids become attached to the m-RNA–ribosome complex and then combine with one another to form a long chain of amino acids: a protein molecule. The order in which amino acids combine is the result of the code on m-RNA, which is a copy of the code on one of the strands of DNA (see Figure 9.3C).

The chemical reactions which take place in a cell are catalysed by enzymes, and enzymes are proteins. In controlling which proteins are made in a cell, DNA controls enzyme production and therefore the whole of the structure and all the functions of the cell.

SUMMARY

The nucleus of a cell contains DNA. The sequence of bases in DNA is a 'code' of instructions for synthesising proteins. Protein synthesis takes place in the cytoplasm. DNA employs messenger-RNA to carry the code out of the nucleus to the ribosomes in the cytoplasm. Amino acids are sorted into the correct order by the messenger-RNA on the ribosomes. The amino acids combine to form a protein. By controlling protein synthesis, DNA controls the whole life of the cell.

CHECKPOINT

❶ Explain the difference between a codon and a gene.

❷ Copy the following sequence of bases.

A A T C C T G A C T A G

How many codons are contained in this section of DNA? Assume the first codon begins at the left hand end and that codons do not overlap. Mark off the codons on the sequence. Assuming that each codon codes for one amino acid, how many amino acids are represented in this section of DNA?

❸ (a) Where in the cell is protein made?
(b) Why is making new protein essential for the cell?

9.4  The dividing cell

FIRST THOUGHTS

The cells which make up living things die due to accidents and age. To survive, an organism must be able to make new cells as fast as its old cells die. To grow, an organism must be able to make new cells at an even faster rate.

LOOK AT LINKS
for **meiosis**
In meiosis each daughter cell receives only half a set of chromosomes. Sex cells (eggs and sperm) are produced by meiosis. Why? See Topic 15.1.

New cells are formed when old cells divide into two. The cells which divide are called **parent cells**, and the new cells are called **daughter cells**. Two overlapping processes take place:
• The nucleus divides into two. There are two ways of dividing: **mitosis** and **meiosis**. We shall deal only with mitosis in this section.
• The cytoplasm divides.

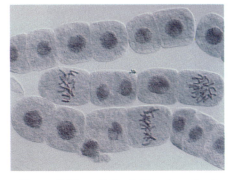

Figure 9.4A ● Cells dividing by mitosis

DNA replication

Before mitosis takes place, new DNA is made inside the nucleus. The process is called **replication** because each chromosome makes an exact copy – a replica – of itself.

Figure 9.4B ● DNA replication

(a) The double helix unwinds to form two single strands

(b) A new strand of DNA forms alongside each of the original strands of DNA. Each new strand combines with an old strand to form a new molecule of DNA. In this way, two new molecules of DNA are formed. Each is a replica of the original.

IT'S A FACT

A chromosome contains about 10 000 times its own length of DNA. Two metres of DNA can coil up sufficiently tightly to cram into one cell nucleus.

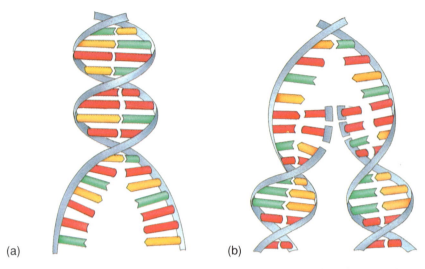

(a)   (b)

Hundreds of thousands of nucleotide bases may add each second to the growing strand of DNA. Occasionally the wrong base adds by mistake. Then the new DNA formed is slightly different from the original. This change is called a **mutation**.

SUMMARY

The nucleus contains chromosomes. A chromosome consists of DNA wound round a core of protein and folded tightly into a compact structure. Genes are lengths of DNA molecules which code for proteins.

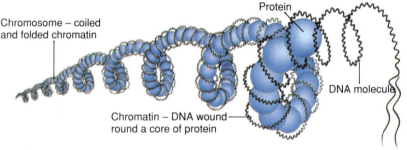

Figure 9.4C ● The structure of a chromosome

Chromosomes consist of folded strands of DNA coiled round a protein core (see Figure 9.4C). The DNA part of the structure controls the **inheritance** of characteristics.

The stages in mitosis

Before mitosis takes place, new DNA is made in the nucleus. Each chromosome then divides into a pair of identical **chromatids** (see Figure 9.4D). The first sign of cell division by mitosis is when the chromosomes can be seen more easily under the microscope. This happens because chromatin coils up to form shorter and thicker chromosomes.

Figure 9.4D ● Chromosomes

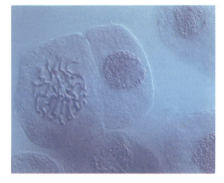

(a) Chromosomes become visible in the nucleus as chromatin coils up

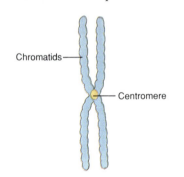

(b) A chromosome: a pair of chromatids joined at a point called the centromere

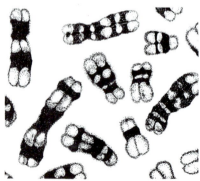

(c) A group of chromosomes. Note the differing positions of the centromeres

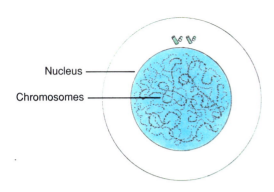

Nucleus

Chromosomes

Parent cell

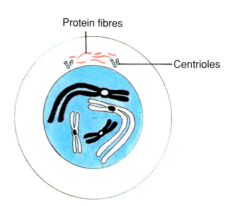

Protein fibres

Centrioles

Stage 1 Two bodies called centrioles move to opposite ends of the cell. Protein fibres form round each of them

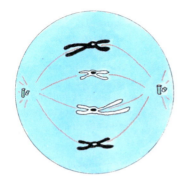

Stage 2 The protein fibres arrange themselves between the centrioles into a structure called the spindle. The nuclear membrane has disappeared. Pairs of chromatids arrange themselves on the equator of the spindle. *NOTE*: the spindle of a plant cell forms without the help of centrioles.

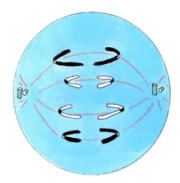

Stage 3 The centromeres divide into two. The new centromeres separate and move along the spindle, one to each end of the cell. They carry their chromatids with them.

Stage 4 The chromatids are now the new chromosomes. They gather into two bunches. A nuclear membrane forms around each bunch.

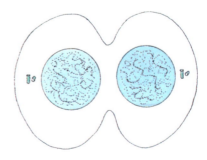

Stage 5 Two daughter cells have been formed. Each has the same number of chromosomes as the parent cell.

Figure 9.4E ● The stages of mitosis. (Only four chromosomes are shown in detail.)

● *Division of the cytoplasm*
Let us look in more detail at Stage 4 of Figure 9.4E. Once the new nuclei have formed, the cytoplasm begins to divide. Mitochondria and chloroplasts divide. Other cell structures are built up from materials in the cytoplasm. The new structures are distributed more or less equally between the two daughter cells.

IT'S A FACT

When cells are grown on dishes in a solution of all the substances they need to live, the cells divide. They spread out until they form a complete layer over the bottom of the dish. At this point, the cells are touching neighbouring cells and, in the case of normal cells, cell division stops. This is called contact inhibition (stopping by touching). Cancer cells are different. They continue to divide even after they are all touching. In the body, masses of cancer cells are called tumours. By multiplying faster than normal cells, cancer cells destroy healthy tissue. Many cancers can be cured if they are detected early, but left untreated cancer can endanger a person's life.

Division of the cytoplasm happens differently in animal cells and plant cells (see Figures 9.4F and G).

Figure 9.4F ● The start of division of cytoplasm in an animal cell (in this case a frog's egg). A furrow develops. It pinches the cell membrane in. As the furrow deepens, the parent cell divides into two daughter cells

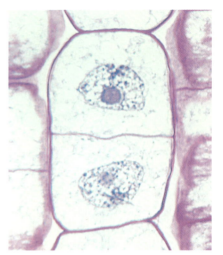

Figure 9.4G ● Division of the cytoplasm in a plant cell. A thin, slab-like structure called the cell plate develops across the middle of the spindle. As it extends outwards, the cell plate meets the sides of the cell dividing the cytoplasm into two daughter cells.

SUMMARY

In mitosis, cells divide into two. A full set of chromosomes is passed to each daughter cell. The chromosomes in each daughter cell are identical to those of the parent cell.

● **The importance of mitosis**

Mitosis divides the chromosomes of the parent cell into two identical groups. Daughter cells are therefore genetically the same as the parent cell and one another.

Why do you think that keeping the genes constant from generation to generation is so important? (Hint: Remember that mitosis is the method by which living things repair damage and also grow. For example, old skin cells are always replaced by new skin cells. As roots grow through the soil, root tip cells divide by mitosis to form new root tissue.)

CHECKPOINT

❶ (a) Why do the cells of a tissue need to undergo mitosis?
(b) What effect does mitosis have on the chromosomes of the parent cell?
(c) What is the relationship between the chromosomes of the parent cell and the chromosomes of the daughter cells?
(d) What is the importance of this relationship for the health of the tissue?

❷ (a) In everyday language, what is a 'replica'?
(b) What is formed by the replication of DNA?
(c) Refer to Figure 9.4B, which shows the replication of DNA. Explain briefly what is happening.

❸ Refer to Figure 9.4C. Explain the relationship between
(a) DNA and chromatin,
(b) chromatin and a chromosome.

❹ Refer to Figure 9.4D. Explain the relationship between
(a) a chromosome and a chromatid,
(b) a chromatid and a centromere.

9.5 Cells, tissues and organs

FIRST THOUGHTS

All cells have certain features in common (see Topic 9.1). This section deals with the differences in shape and structure between different types of cell.

Do you remember that some cells exist as independent organisms? You can see that the organisms in Figures 1.5C and 1.5D are made of single cells. Other cells cannot survive on their own; they exist as members of multicellular organisms. The bodies of multicellular organisms are made of different types of cell. Each type of cell is specialised to perform a particular biological task.

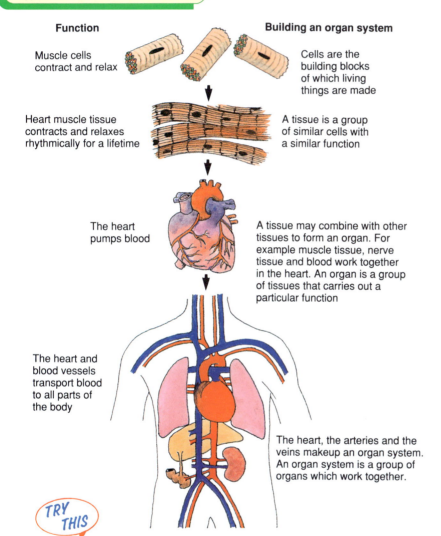

Function

Muscle cells contract and relax

Heart muscle tissue contracts and relaxes rhythmically for a lifetime

The heart pumps blood

The heart and blood vessels transport blood to all parts of the body

Building an organ system

Cells are the building blocks of which living things are made

A tissue is a group of similar cells with a similar function

A tissue may combine with other tissues to form an organ. For example muscle tissue, nerve tissue and blood work together in the heart. An organ is a group of tissues that carries out a particular function

The heart, the arteries and the veins makeup an organ system. An organ system is a group of organs which work together.

Tissues and organs

A group of similar cells makes a **tissue**. Different tissues together make up an **organ**, and different organs are combined in an **organ system**. Figure 9.5A shows how one organ system, the human heart and its blood vessels, is organised.

Types of cell

The human body is constructed from more than two hundred different types of cell. Fewer types of cell form the bodies of flowering plants (e.g. foxglove), fungi (e.g. field mushroom) and simple animals. The fresh water animal *Hydra* has only seven different types of cell.

Figure 9.5A ● The human heart and blood vessels: an organ system

TRY THIS

Cell shape and specialisation

The appearance of cells often helps us to understand what they do. Study the cells in Figure 9.5B. You will see them again in Figures 9.5C, D and E. Briefly describe how the shape of each one helps you to understand what it does.

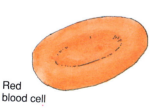

Red blood cell

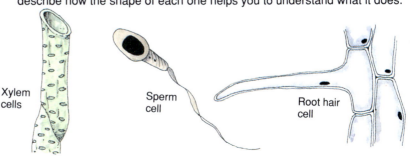

Xylem cells

Sperm cell

Root hair cell

Figure 9.5B ● Cells (not to scale)

Figures 9.5C, D and E show cell types from different kingdoms. Notice the features that all the cells have in common, such as cytoplasm, cell membranes and nuclei. Notice also how specialisation for particular biological tasks makes each type of cell different.

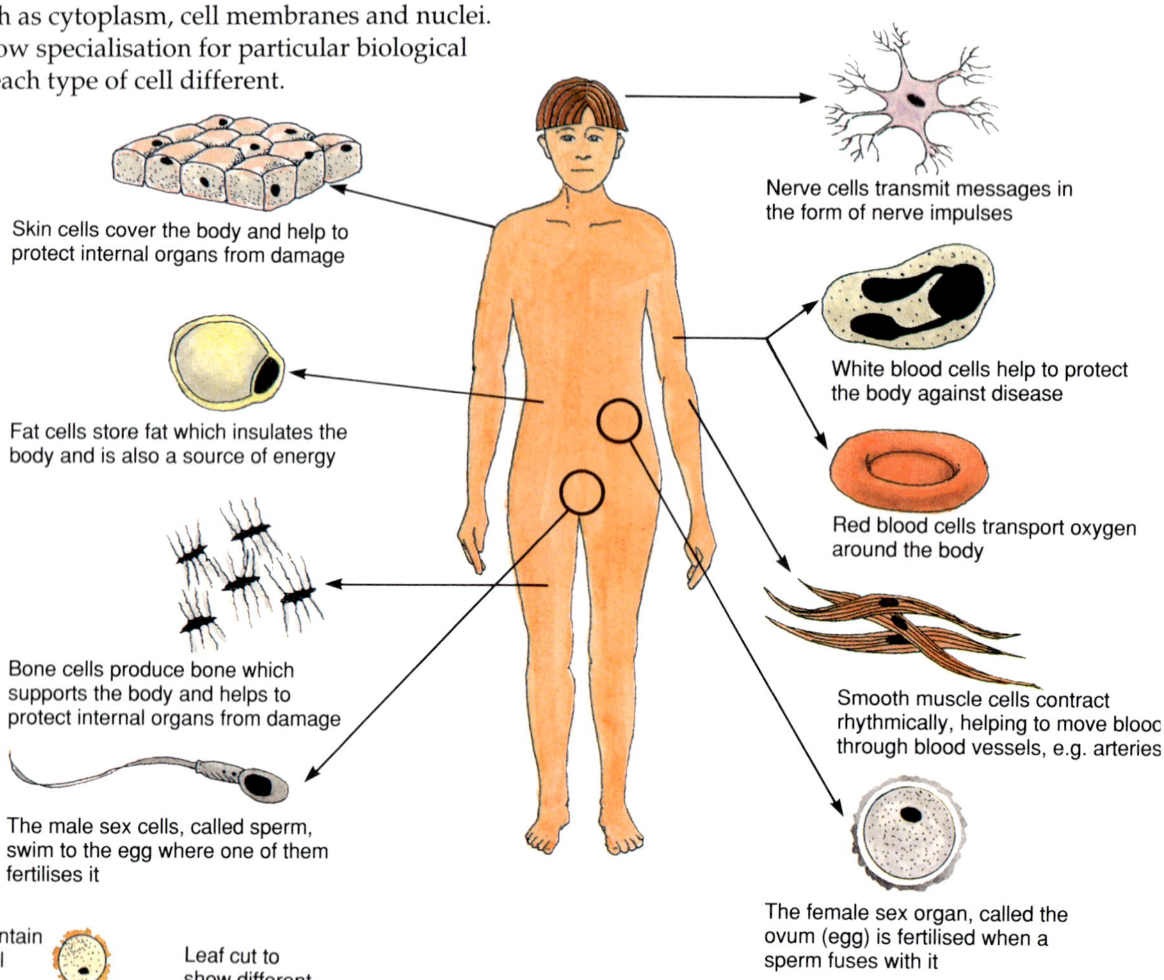

Skin cells cover the body and help to protect internal organs from damage

Fat cells store fat which insulates the body and is also a source of energy

Bone cells produce bone which supports the body and helps to protect internal organs from damage

The male sex cells, called sperm, swim to the egg where one of them fertilises it

Nerve cells transmit messages in the form of nerve impulses

White blood cells help to protect the body against disease

Red blood cells transport oxygen around the body

Smooth muscle cells contract rhythmically, helping to move blood through blood vessels, e.g. arteries

The female sex organ, called the ovum (egg) is fertilised when a sperm fuses with it

Figure 9.5C ● Animal kingdom – human

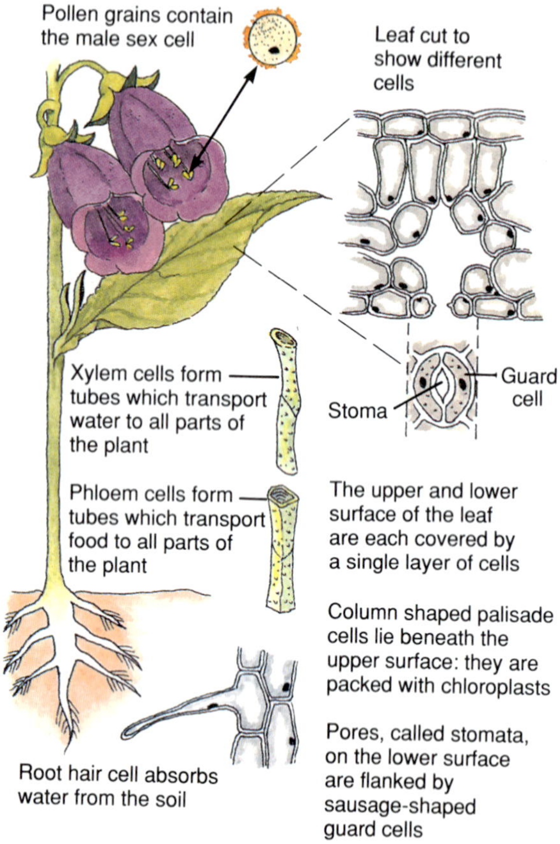

Pollen grains contain the male sex cell

Leaf cut to show different cells

Xylem cells form tubes which transport water to all parts of the plant

Phloem cells form tubes which transport food to all parts of the plant

Root hair cell absorbs water from the soil

Guard cell

Stoma

The upper and lower surface of the leaf are each covered by a single layer of cells

Column shaped palisade cells lie beneath the upper surface: they are packed with chloroplasts

Pores, called stomata, on the lower surface are flanked by sausage-shaped guard cells

Figure 9.5D ● Plant kingdom – foxglove

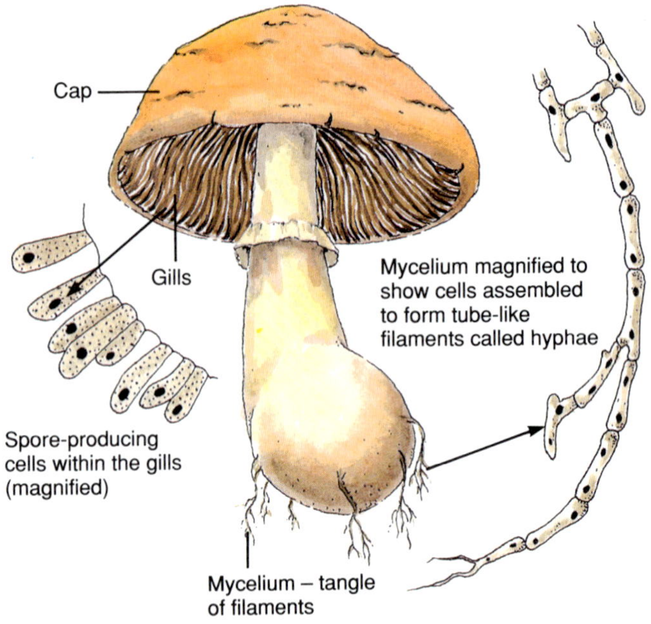

Cap

Gills

Spore-producing cells within the gills (magnified)

Mycelium – tangle of filaments

Mycelium magnified to show cells assembled to form tube-like filaments called hyphae

Figure 9.5E ● Kingdom Fungi – field mushroom

152

Cell size

Surface area to volume ratio

Figure 9.5F shows three cubes of different sizes. A cube has six faces.

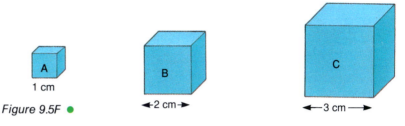

Figure 9.5F ●

Copy and complete the table

	Cube A	Cube B	Cube C
Surface area of one face	1 cm x 1 cm = 1 cm^2		
Surface area of cube	6 x 1 cm^2 = 6 cm^2		
Volume of cube	1 cm x 1 cm x 1 cm = 1 cm^3		
Ratio: surface area/ volume	6 cm^2/1 cm^3 = 6/cm		

Say whether the surface area/volume ratio of a cube increases or decreases as the cube increases in size.

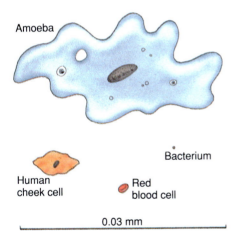

Amoeba

Bacterium

Human cheek cell

Red blood cell

0.03 mm

Figure 9.5G ● A range of cells drawn to scale

Cells are not cube-shaped, but your calculations on cubes apply to any shape of cell. What you have found out is that when a cell grows the surface area of the cell membrane increases more slowly than the volume of the cell (because surface area increases with the square of the side, and volume increases with the cube of the side). As the cell grows, it needs to take in more food and gases. But the area of the cell membrane is not increasing at the same rate as the volume. After the cell reaches a certain size, the surface area becomes insufficient to meet the needs of the larger volume. At this point, the cell either dies or divides into daughter cells.

Cell size depends on the rate at which it uses materials. Cells with a fast turnover of materials are usually smaller than cells with a low turnover.

Figure 9.5G gives an idea of the range of size of cells. Comparing the different sizes is rather like comparing the difference in size between a mouse and a whale. The range is enormous! However, the size of most cells falls somewhere in between: invisible to the naked eye but easily seen under a light microscope.

CHECKPOINT

❶ Copy and complete the following paragraph. You may use the words in the list, once, more than once or not at all.

(a)　　an organ　organs　cells　types　tissues

Living things are made of _____. Groups of similar _____ with similar functions form _____ that can work together as _____. A group of _____ working together form _____ system.

❷ (a) What effect does the division of a cell into small daughter cells have on the surface area/volume ratio?
(b) Why are cells with a high material turnover usually smaller than those with a low material turnover?

TOPIC 10

THE CHEMICALS OF LIFE

10.1 Elements and compounds

Life – what a mystery it is! Are the compounds in living things different from those in the rest of the Universe? Do they have some mysterious property of life?

All matter is made of chemical elements and their compounds. The compounds in living organisms are made up of the same elements as the rest of the universe. They obey the same laws of physics and chemistry. The elements that living organisms need come from the environment; from the Earth and its atmosphere. There are twenty of them. Six of them together make up 95% (by mass) of living matter. In order of abundance, they are carbon, hydrogen, nitrogen, oxygen, phosphorus and sulphur. A list of the symbols, CHNOPS, will help you to remember them.

What, then, makes something composed of ordinary chemicals live? That is a difficult question! Part of the answer lies in the chemistry of carbon compounds. Living things are made of water, salts and carbon compounds.

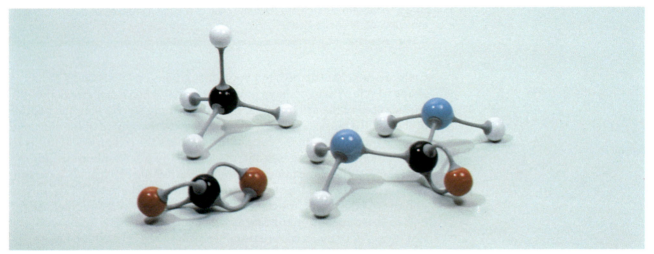

(a) Methane, CH_4, released by cattle

(b) Carbon dioxide, CO_2, in exhaled breath

(c) Urea, $CO(NH_2)_2$, excreted in urine

Figure 10.1A ● Models of the molecules of some simple carbon compounds

An atom of carbon can form four covalent bonds. Figure 10.1A shows some simple carbon compounds which occur in living things. Atoms of carbon are able to combine to form long chains. Many of the carbon compounds in living things have much larger molecules than those shown in Figure 10.1A. Often such compounds are formed by small molecules joining up to form large molecules. The carbon compounds in living things are called **organic compounds**. Some examples are listed in Table 10.1.

LOOK AT LINKS
for **organic compounds**
There is more about organic chemistry, the chemistry of carbon compounds, in *KS: Chemistry*, Topics 30 and 31.

Table 10.1 ● Organic compounds

Substances with small molecules which can combine to form large molecules		Substances with large molecules which are formed as a result
Sugars	→	Starch, cellulose, chitin Topic 10.2
Fatty acids and glycerol	→	Lipids (Topic 10.3)
Amino acids	→	Proteins (Topic 10.4)
Nucleotides	→	Nucleic acids (Topic 10.5)

FIRST THOUGHTS

10.2  Carbohydrates

What do sugar and starch, the cell walls of plants and the exoskeletons of insects have in common? This section will tell you!

LOOK AT LINKS
for **respiration**
See Topic 12.

Sugars

Carbohydrates are compounds containing carbon, hydrogen and oxygen only. Carbohydrates do a vital job in all living organisms: they provide energy. When carbohydrates react with oxygen, carbon dioxide and water are formed and energy is released. This reaction takes place slowly inside cells. Energy is released in a controlled way and used by the cells for all their activities. The process by which living cells release energy is called **cellular respiration**. Most of our energy is obtained from the respiration of carbohydrates (usually glucose), although living organisms also respire fats, oils and proteins.

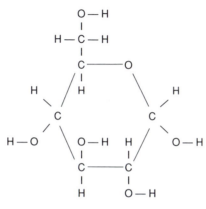

(b) The formula for glucose in full

(c) The formula in a shorthand form

Carbohydrates include sugars, starches and cellulose. Two of the simplest carbohydrates are the sweet-tasting sugars **glucose** and **fructose**. They both have the formula $C_6H_{12}O_6$, but the structures (the arrangement of atoms) of their molecules are different. The structure of a molecule of glucose is shown in Figure 10.2A. As you can see, six of the atoms are joined to form a ring. Sugars with one ring of atoms are called **monosaccharides** (mono = one). There are two kinds of monosaccharides: **hexoses**, which contain 6 carbon atoms, e.g. glucose and fructose, and **pentoses**, which contain 5 carbon atoms, e.g. ribose, $C_5H_{10}O_5$.

(a) Model of a glucose molecule

Figure 10.2A ● Glucose

RESOURCE
-ACTIVITY-
PACK

Two molecules of glucose can combine to form one molecule of the sugar maltose, $C_{12}H_{22}O_{11}$ (see Figure 10.2B).

Glucose → Maltose + Water
$2C_6H_{12}O_6(aq)$ → $C_{12}H_{22}O_{11}(aq)$ + $H_2O(l)$

Maltose is a **disaccharide**: its molecules contain two sugar rings. Sucrose (table sugar) is another disaccharide. It is formed from glucose and fructose.

Glucose + Fructose → Sucrose + Water

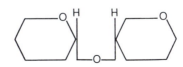

Figure 10.2B ● The formula for maltose in shorthand form

Starch, glycogen and cellulose

Starch (a stored food substance in plant cells), glycogen (a stored food substance in animal cells) and cellulose (of which plant cell walls are composed) are all carbohydrates, although they do not taste sweet. Starch and glycogen are slightly soluble in water; cellulose is insoluble. These compounds are **polysaccharides**. Polysaccharides are carbohydrates whose molecules contain a large number (hundreds) of sugar rings (see Figure 10.2C).

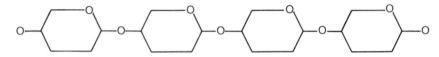

Figure 10.2C ● Part of a starch molecule

In starch, glycogen and cellulose, the molecules are long chains of glucose rings. Polysaccharides differ in the length and structure of their chains (see Figure 10.2D).

As starch and glycogen are only slightly soluble they can remain in the cells of an organism without being dissolved out and therefore make good food stores (see Figures 10.3E and F). Cells convert starch and glycogen into glucose, which is soluble. Glucose is oxidised in the cells to release energy.

Some polysaccharides are building materials. Cellulose fibres make the walls of plant cells (see Figure 10.2G). The tough framework of fibres makes the plant cell wall rigid. Another polysaccharide that provides a strong structure is chitin. It forms the exoskeleton round insects' bodies which protects their insides (see Figure 10.2H).

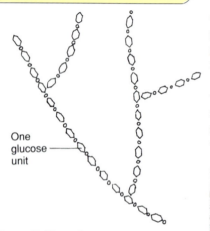

Figure 10.2D ● A model of part of a starch molecule

One glucose unit

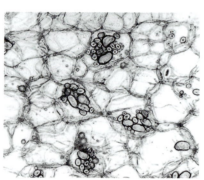

Figure 10.2E ● A thin section of potato stained with iodine. Note the grains of starch which have been stained black

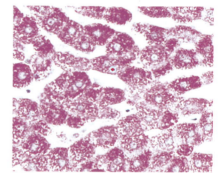

Figure 10.2F ● A thin section of liver. Note the glycogen stained pink

Figure 10.2G ● Plant cell wall. (x10 000) Note the frame work of cellulose fibres

Figure 10.2H ● A chitin-impregnated exoskeleton

SUMMARY

Carbohydrates:
- Monosaccharides
  - sugars, e.g. glucose and fructose with a six-carbon ring and ribose with a five-carbon ring
- Disaccharides
  - made up of two monosaccharides
  - sugars such as sucrose and maltose
- Polysaccharides
  - made up of many monosaccharides
  - starch and glycogen, store energy in plants and animals
  - cellulose, strengthens plant cell walls
  - chitin, strengthens insect exoskeletons

CHECKPOINT

1. Runners in the London marathon ate pasta the evening before. Why did they think this would give them energy?

2. Explain why rice, bread and potatoes are 'high-energy foods'.

3. Give two examples of the importance of cellulose and chitin as building materials in living things.

4. Glucose, maltose and starch are carbohydrates. Answer the following questions about them.
   (a) Which one tastes sweet?
   (b) Which two are very soluble in water?
   (c) List the elements that make up all three.
   (d) The formula of glucose is $C_6H_{12}O_6$. What is the ratio number of atoms of hydrogen: number of atoms of oxygen?
   (e) The formula of maltose is $C_{12}H_{22}O_{11}$. What is the ratio number of atoms of hydrogen: number of atoms of oxygen?
   (f) Using your answers to (d) and (e), try to explain where the name 'carbohydrate' comes from.
   (g) The 'shorthand' formula for glucose is

   Draw the shorthand formulas for maltose and starch.

10.3 Lipids

FIRST THOUGHTS

Have you seen brands of margarine and cooking oil described as 'high in polyunsaturates' and wondered what it means? This section will tell you.

What are lipids?

Lipids is a name for **fats** and **oils**. Lipids are made of carbon, hydrogen and oxygen. A difference between fats and oils is that most fats are solid at room temperature and most oils are liquid at room temperature. Fats and oils have a 'greasy' feel and are insoluble in water.

Fats and oils are important as:

- **Sources of energy**. Fats and oils are stores of energy. In cellular respiration, lipids provide about twice as much energy per gram as carbohydrates.
- **Insulation**. Mammals have a layer of fat under the skin. This layer helps to keep the animals warm. Whales and seals which live in Antarctic waters have an especially thick layer of fat, called blubber, to protect them from extreme cold.

Figure 10.3A ● Fur seals

- **Protection**. Delicate organs, such as the kidneys, are protected by a layer of hard fat.
- **Food**. Vitamins A, D and E are soluble in fats and oils. Foods containing lipids provide animals with these essential vitamins.

Fats and oils are also used to build cells. Look at the magnified picture of a **cell membrane** in Figure 10.3B. You can see two 'tramlines', which scientists think are composed of molecules of a **phospholipid** – a fat containing phosphorus.

LOOK AT LINKS
for **cell membranes**
See Topic 9.1.

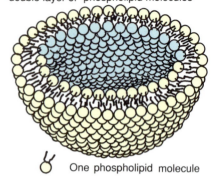

A magnified view of the 'tramlines' shows a double layer of phospholipid molecules

One phospholipid molecule

(a) A photograph of a cell membrane taken through an electron microscope. Part of the cell membrane shows as two 'tramlines'. (x250 000)

(b) The diagram shows how scientists interpret the photograph

Figure 10.3B ● Cell membrane

SUMMARY

Fats and oils are together called lipids. They are sources of energy. A layer of fat insulates the bodies of mammals from the cold. Fats form part of the structures of cells.

LOOK AT LINKS
The importance of fats and oils in the diet is discussed in Topic 11.

Saturated and unsaturated fats and oils

Do you watch your diet? Many people are concerned about having too much fat in their diet, particularly **saturated fats**. Let us consider what the term saturated means here.

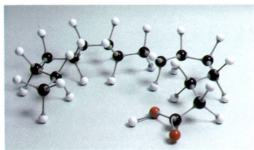

(c) Model of a molecule of fatty acid (hexadecanoic acid)

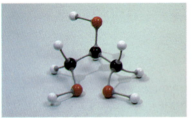

(a) Model of a molecule of glycerol

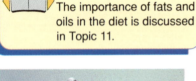

(b) The formula of glycerol

(d) The formula of hexadecanoic acid. Can you see how the acid gets its name? (hexa = six, deca = ten)

Figure 10.3C ● Glycerol and hexadecanoic acid

Fats and oils are compounds of fatty acids and glycerol (see Figure 10.3C). One molecule of glycerol can combine with three molecules of a fatty acid to form a molecule of a triglyceride and three molecules of water. **Lipids** are mixtures of triglycerides.

Glycerol + Fatty acid ➜ Triglyceride + Water

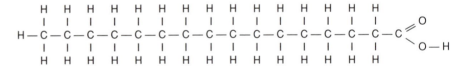

Like other organic compounds, fatty acids may be saturated or unsaturated. If there is one double bond between carbon atoms, the

LOOK AT LINKS
The place of fats and oils in a healthy diet is discussed in Topic 11.

compound is described as **unsaturated**. If there is more than one double bond between carbon atoms, the acid is described as **polyunsaturated**. Below is part of the carbon chain in a saturated compound. This compound is described as **saturated** because it cannot combine with any more hydrogen atoms.

$$
\begin{array}{ccccccccc}
H & H & H & H & H & H & H & H & H \\
| & | & | & | & | & | & | & | & | \\
-C & -C & -C & -C & -C & -C & -C & -C & -C- \\
| & | & | & | & | & | & | & | & | \\
H & H & H & H & H & H & H & H & H
\end{array}
$$

Below is part of the carbon chain in an unsaturated compound. This compound is described as unsaturated because it can add more hydrogen atoms across the double bond.

$$
\begin{array}{ccccccccc}
H & H & H & H & H & H & H & H & H \\
| & | & | & | & | & | & | & | & | \\
-C & -C & -C & -C & -C & -C & =C & -C & -C- \\
| & | & | & | & | & & & | & | \\
H & H & H & H & H & & & H & H
\end{array}
$$

Below is part of the carbon chain in a polyunsaturated compound

$$
\begin{array}{ccccccccc}
H & H & H & H & H & H & H & H & H \\
| & | & | & | & | & | & | & | & | \\
-C & -C & -C & -C & =C & -C & -C & =C & -C- \\
| & | & | & & & | & & & | \\
H & H & H & & & H & & & H
\end{array}
$$

The fats which contain glycerol combined with saturated fatty acids are called saturated fats. Fats which contain glycerol combined with unsaturated fatty acids are called unsaturated fats or polyunsaturated fats, depending on the number of double bonds in a molecule. Animal fats contain a large proportion of saturated compounds and are solid. Plant oils contain a large proportion of unsaturated compounds and have lower melting points. Many scientists believe that eating a lot of saturated fat increases the risk of heart disease.

SUMMARY

Lipids (fats and oils) are mixtures of compounds of glycerol with organic acids. These compounds can be saturated (with only single bonds between carbon atoms) or unsaturated (with a double bond between carbon atoms) or polyunsaturated (with more than one carbon–carbon double bond per molecule).

CHECKPOINT

❶ (a) Say which are the fats and which are the oils.
   soft margarine, butter, lard, hard margarine, cooking oil
(b) What is the main physical difference between fats and oils?

❷ Refer to Figure 10.3A. What feature helps the seals to keep warm?

❸ Briefly explain why fats and oils are better stores of energy than carbohydrates are.

❹ Briefly explain the difference between saturated and unsaturated fats.

❺ The manufacturers of *Flower* margarine claim that their product encourages healthy eating because it is 'low in saturated fats'.
(a) What do they mean by 'low in saturated fats'?
(b) Name a product which contains much saturated fat.
(c) What particular benefit to health is claimed for products such as *Flower*?

❻ The following equation represents the formation of a fat.

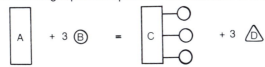

(a) Name the substances A and D.
(b) Name one substance which could be B.
(c) Name one fat and one oil which could be C.

10.4 Proteins

FIRST THOUGHTS

Muscle, skin, hair and haemoglobin: what do they have in common? You can find out in this section.

Proteins are compounds which contain carbon, hydrogen, oxygen, nitrogen and sometimes sulphur. Proteins are vitally important because:
- proteins are the materials from which new tissues are made. If organisms are to grow and if they are to repair damaged tissues, they need proteins
- enzymes are proteins. None of the reactions which take place in animals and plants would take place rapidly enough without enzymes
- hormones are proteins which control the activities of organisms
- proteins can be oxidised in respiration to provide energy

Every protein molecule is made from a large number of molecules of **amino acids**. The simplest amino acid is **glycine** (see Figure 10.4A).

LOOK AT LINKS
for **protein**
Which foods provide protein in our diet?
See Topic 11.

$$
\begin{array}{cc}
H & H \\
| & | \\
N - C - C & {\Large\diagup}^{O} \\
| & | \quad\quad O-H \\
H & H
\end{array}
$$

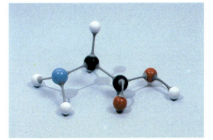

Figure 10.4A ● Model of a glycine molecule

There are about 20 amino acids. They have the general formula

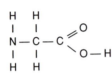

R is a different group of atoms in each amino acid; it can be —H, —CH₃, —CH₂OH and many other groups. In the following equations, each shape represents a different amino acid.

Two amino acids can combine to form a **peptide** and water

The peptide which is formed can combine with more amino acids. Many different amino acids can combine to form a long chain

Peptides are molecules with up to 15 amino acids.
Polypeptides are molecules with 15–100 amino acids.
Proteins have still larger molecules. Part of a protein molecule is shown below.

IT'S A FACT

Fifteen million children die each year of starvation and disease. You will have seen pictures of children with swollen abdomens. They are suffering from a disease called kwashiorkor. They are getting some food but are starved of protein.

LOOK AT LINKS
for **making protein**
See Topic 9.

Dr Max Perutz began work on the structure of haemoglobin in 1936. It took over 20 years to find out that the molecule contains 574 amino acid groups, arranged in four chains. With a total of 10 000 atoms, the molecule has a mass 65 000 times that of a hydrogen atom.

LOOK AT LINKS
for **kwashiorkor**
See Topic 11.

Essential and non-essential amino acids

Animals are able to make some amino acids in their bodies, but they cannot make all the amino acids they need. They must obtain the ones they cannot make from their diet. The amino acids that animals need to obtain from their diet are called **essential amino acids**. The amino acids which animals can make themselves are called **non-essential amino acids**.

A protein which contains all the essential amino acids is called a **first-class protein**. Such proteins are found in lean meat, fish, cheese and soya beans. A protein which lacks essential amino acids is called a **second-class protein**. Such proteins are found in flour, rice and oatmeal. In parts of the world where rice is the staple diet, kwashiorkor is common.

The chain of amino acid groups in a protein is folded and twisted into a certain three-dimensional shape. The part that a protein plays in a living organism depends on the shape of its molecules (see Figures 10.4B, C, D). The shape is decided by the order (sequence) of amino acid groups in the protein molecule.

Figure 10.4B ● Cross section of a human hair. Hair, nails, claws and feathers consist chiefly of the protein keratin. (x800)

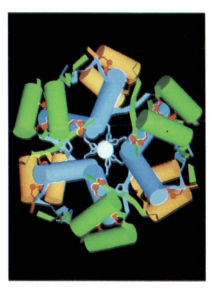

Figure 10.4C ● Insulin. In 1953, Dr Frederick Sanger of the UK reported the complete sequence of amino acids in the protein insulin. This was the first time that a protein structure had been worked out. Insulin controls blood sugar levels. Its molecule is one of the smallest protein molecules.

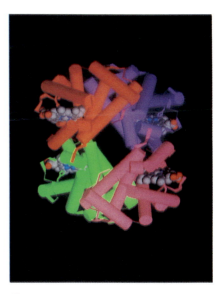

Figure 10.4D ● Haemoglobin. In 1959, Dr Max Perutz worked out the shape of the haemoglobin molecule. Haemoglobin is the red pigment in blood which transports oxygen round the body. The haemoglobin molecule is larger than the insulin molecule.

SUMMARY

Many parts of living organisms are made of proteins. The molecules of proteins are long chains of amino acids. The molecules of peptides and polypeptides are shorter chains of amino acids.

LOOK AT LINKS
The reactions which take place in a living cell are called the **metabolism** of the cell.
See Topic 10.2.

Enzymes, an important group of proteins

Many chemical reactions take place inside a cell. The speeds at which the cell reactions take place are controlled by **catalysts** called **enzymes**. Enzymes are proteins. There are thousands of enzymes in living things. We say that enzymes are **specific**; this means that each enzyme catalyses a certain chemical reaction or a type of chemical reaction (see Figure 10.4E). The substance which the enzyme helps to react is called its **substrate**. The enzyme amylase increases the speed at which starch reacts with water to form sugars.

Starch + Water ➔ Sugars

The enzyme has a group of atoms called the **active site**. Part of the substrate molecule fits into the active site. This fit has been described as 'like a lock and key'.

Substrate (part of a starch molecule)

Active site

Amylase molecule

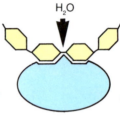

Chemical bonds form between the substrate and the active site. These bonds make it easier for a molecule of water to attack the molecule of starch.

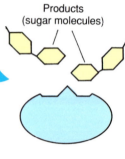

The starch molecule is attacked by a water molecule. Weakened by its bonds to the enzyme the starch molecule is hydrolysed to form two molecules of sugar.

H_2O

Products (sugar molecules)

Figure 10.4E ● How the enzyme amylase catalyses the breakdown of starch. Notice that the enzyme is unchanged at the end of the reaction and able to catalyse the breakdown of more starch

SUMMARY

Enzymes are an important group of proteins. They catalyse the reactions which take place in living cells.

CHECKPOINT

❶ What is the difference between (a) a peptide and a polypeptide (b) a polypeptide and a protein?

❷ (a) To what group of chemical compounds do enzymes belong?
(b) What vital job do enzymes do?
(c) Give the names of three substrates on which enzymes work and the names of the products that are formed.

10.5 Nucleic acids

FIRST THOUGHTS

When two scientists found out the structure of DNA, there was great excitement in the scientific world, and they won Nobel prizes. Why is DNA so important? What is so special about the structure? You can find out in this section.

RNA and DNA

The nucleic acids are ribonucleic acid (RNA) and deoxyribonucleic acid (DNA). DNA occurs in cell nuclei. The genes that living things inherit from their parents are lengths of DNA. The information that a cell needs to assemble molecules of amino acids in the correct order to make protein molecules is present in its DNA. RNA occurs in nuclei and cytoplasm. It transfers the information contained in DNA to the places in the cell where proteins are made.

RNA and DNA have large complex molecules made up from smaller molecules of compounds called **nucleotides** (see Figure 10.5A).

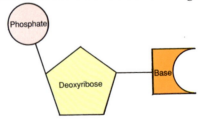

Phosphate

Deoxyribose

Base

Figure 10.5A ● A nucleotide in DNA. The five-carbon sugar deoxyribose occurs in DNA. The base is one of four: adenine, cytosine, guanine or thymine. (In RNA the sugar ribose occurs, and the base uracil is present instead of thymine.)

Nucleotides combine to form nucleic acids. The phosphate group of one nucleotide molecule combines with the sugar group of another. A strand of alternate sugar groups and phosphate groups is formed. The bases attached to the sugar groups stick out from the strand (see Figure 10.5B).

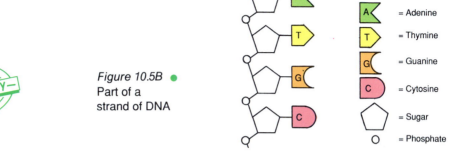

RESOURCE ACTIVITY PACK

Figure 10.5B ● Part of a strand of DNA

A= Adenine
T= Thymine
G= Guanine
C= Cytosine
⬠ = Sugar
○ = Phosphate

A DNA molecule consists of two such strands bonded together. The bases sticking out from one sugar–phosphate strand bond to bases sticking out from a second sugar–phosphate strand. In Figure 10.5C you will notice that C bonds to G and A bonds to T. This is always the case in DNA. (Remember, in RNA, the base uracil is present instead of thymine.)

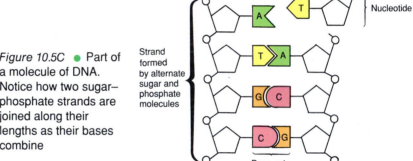

Nucleotide

Strand formed by alternate sugar and phosphate molecules

Base pairs

Figure 10.5C ● Part of a molecule of DNA. Notice how two sugar–phosphate strands are joined along their lengths as their bases combine

Who's behind the science

Erwin Chargaff, a biochemist, discovered that the amounts of adenine (A) and thymine (T) in DNA are equal and the amounts of guanine (G) and cytosine (C) are equal. This led to the idea of the one-to-one pairings of adenine–thymine and cytosine–guanine. Chargaff's rule, as it came to be known, helped Francis Crick and James Watson to construct their model of DNA.

As two strands of sugar–phosphate groups combine to form a molecule of DNA, the result resembles a ladder: the sugar–phosphate strands form the sides and base pairs form the rungs. The 'ladder' is in fact twisted into a spiral shape (see Figure 10.5D) so perhaps a spiral staircase is a better comparison. The shape is called a **double helix** – two intertwined spiral strands.

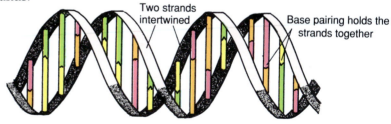

Two strands intertwined

Base pairing holds the strands together

Figure 10.5D ● The double helix: two connected spiral strands

CHECKPOINT

❶ (a) Name three types of group present in a nucleotide.
(b) Which sugar gives DNA its name?
(c) Briefly explain why DNA got the name 'the double helix'.
(d) The letters ATCG stand for bases. Name the four bases.
(e) Briefly explain the meaning of 'base pairs'.

The DNA story

Scientists recognised the vital importance of DNA in cells, and were therefore keen to discover its structure. Many scientists worked on the problem, using different techniques. Success came in 1953 when James Watson and Francis Crick combined all the evidence from the different research workers. They succeeded in building a model of DNA which fitted all the facts. This model was the double helix (see Figure 18.5D). So excited were Watson and Crick that they dashed out of the Cavendish Laboratory of Cambridge University, England, and ran along the street, looking for people to tell about their discovery.

The technique of X-ray crystallography was of prime importance. It involves passing a beam of X-rays through a crystal onto a photographic plate. X-rays affect photographic plates, and a pattern can be seen when the plate is developed. The pattern depends on the arrangement of the atoms in the crystal.

Rosalind Franklin of King's College, London, together with Maurice Wilkins, obtained the X-ray crystallography pictures that gave Watson and Crick the vital clues they needed to build the DNA model. Her work was essential to the discovery. She died in 1958. Whether, if she had lived longer, she would have shared the Nobel Prize that Watson, Crick and Wilkins later won, we cannot say.

Figure 10.5F ● Rosalind Franklin

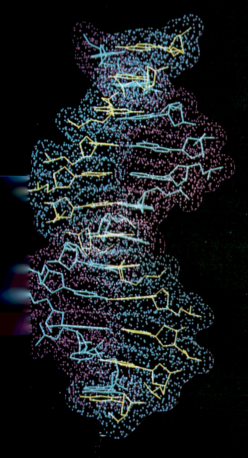

Figure 10.5E ● A computer graphics image of the DNA double helix looking

Figure 10.5G ● Maurice Wilkins with X-ray equipment

Figure 10.5H ● James Watson, an American biologist (left) and Francis Crick, a British physicist turned biologist (right) with their model of DNA at the Cavendish Laboratory, Cambridge. You can see the spiral pattern of the double helix. Notice the model sugar ring at Watson's left shoulder.

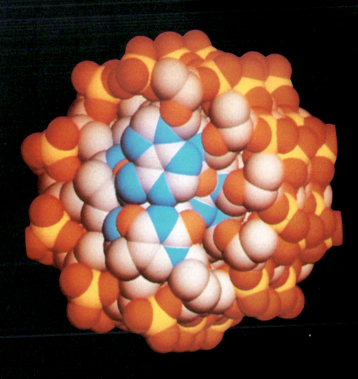

Figure 10.5I ● A space-filling computer graphics image of DNA showing an end-on view of its double helix structure

Figure 10.5J ● Six of the 1962 Nobel Prize winners on the eve of the award ceremony. From left to right: Maurice Wilkins, John Steinbeck (winner of the literature prize), John Kendrew, Max Perutz, Francis Crick and James Watson

Francis Crick, James Watson and Maurice Wilkins received the Nobel Prize from the King of Sweden for finding the structure of DNA in 1962. Also receiving prizes that year were Max Perutz for finding the structure of haemoglobin and John Kendrew for his work on the structure of myoglobin (the first protein to have its three dimensional structure worked out).

SUMMARY

The famous double helix, the structure of DNA, was worked out by James Watson and Francis Crick on the basis of the X-ray crystallography of DNA by Rosalind Franklin.

? THEME QUESTIONS

● Topic 9

1 Match each of the biological terms in Column A with the descriptions in Column B.

Column A	Column B
Ribosome	Contains genetic material
Cell wall	Contains cell sap
Vacuole	Site of aerobic respiration
Mitochondrion	A food reserve in animal cells
Nucleus	Site of photosynthesis
Starch grain	Site of protein synthesis
Cell membrane	Made of cellulose
Chloroplast	Controls the passage of substances into and out of the cell
Endoplasmic reticulum	Network of channels running through the cytoplasm
Glycogen	A food reserve in plant cells

2

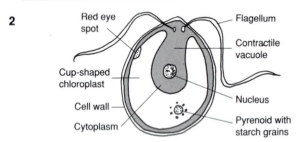

(a) Look carefully at the diagram and try to decide whether it is a plant cell or an animal cell. Give reasons for your choice.

(b) Which of the structures contains the genetic information?

(c) (i) This organism lives in freshwater. Explain why it constantly has to get rid of excess water from the cell.
(ii) Which structure enables it to do this?
(iii) What would happen to this cell if it were placed in sea water?

3 The diagram below shows a cell of yeast.

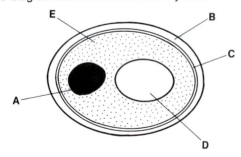

(a) Use the label letters **A**, **B**, **C**, **D**, and **E**, to identify the cell wall, the cell membrane, the nucleus, the vacuole and the cytoplasm.

(b) Give **two** ways in which this cell is different from a leaf mesophyll cell.

(c) Give **two** ways in which this cell is different from a cheek cell.

(LEAG)

4 The diagram shows the bacterium *Salmonella* which causes food poisoning.

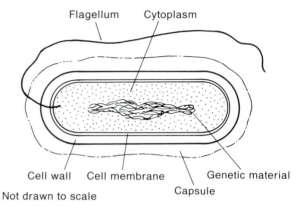

Not drawn to scale

(a) Give **one** similarity between the bacterial cell structure and the structure of a palisade cell in a leaf.

(b) How is the genetic material in bacterial cells organised?

(c) What method of asexual reproduction occurs in bacteria?

(d) The table shows the number of outbreaks of food poisoning caused by three types of bacteria in 1980. The table also shows the foods with which these outbreaks were linked.

Food	Bacteria		
	Salmonella	*Clostridium*	*Bacillus*
Beef	6	17	1
Chicken	36	8	2
Meat pies	18	13	1
Milk	13	0	0
Pork/ham	3	6	0
Rice	10	0	8

Use the information in the table to work out the following:
(i) the total number of outbreaks of food poisoning caused by *Clostridium*
(ii) the second most common food linked to outbreaks of food poisoning
(iii) the total number of outbreaks of food poisoning linked to beef
(iv) the percentage of outbreaks of food poisoning linked to beef caused by *Salmonella*. Show your working.

(e) Explain why it is unwise to keep foods such as meat pies in warm places for long periods.

(f) Food may be preserved by a number of different methods. These include pickling in vinegar (ethanoic acid), drying and putting food into concentrated sugar solution. Use your knowledge of biology to explain how each of these methods helps to preserve the food.

(LEAG)

5 Many hostile things have been done to environments in times of war.

For example, in 149 AD the Romans put salt in fields of wheat in order to destroy the plants and therefore make them useless for food.

If salt was put onto plants today and their cells compared with those of a healthy plant, the following would be seen.

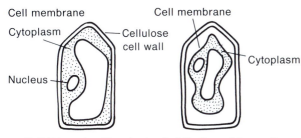

Cell **T** from healthy plant   Cell **S** after adding salt

(a) What is the difference in the position of the cytoplasm of cell **S** compared to cell **T**?

(b) Which liquid has moved out of the cell in order to produce this difference?

(c) Describe the process by which the liquid has moved out of the cell.

(MEG)

6 The diagram shows a cell with four chromosomes. Explain, using diagrams to aid your answer, how this cell can divide by mitosis.

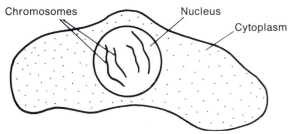

7 (a) Several different types of cell are shown below.

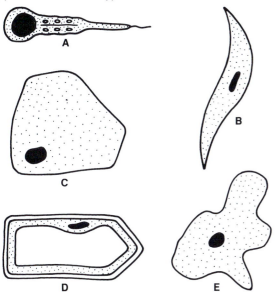

(i) What is the name of the large dark blob in each of these cells? State **one** reason why it is needed.

(ii) Which **one** of these cells is from a plant? How can you tell that this cell comes from a plant?

(b) (i) Where would you expect to find the genes in each cell?

(ii) Explain why genes are important.   (MEG)

8 The diagram shows a sycamore tree and some of the cells it contains.

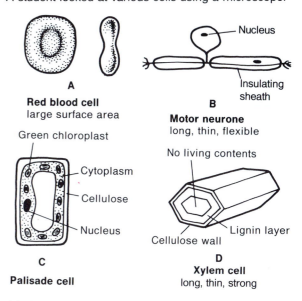

(a) Explain why xylem cells are hollow and strong.

(b) (i) In which part of the tree might palisade cells be found?

(ii) Explain the special purpose of palisade cells.

(c) (i) What is the job of root hair cells?

(ii) Explain how their shape helps them to do this job.   (MEG)

9 A student looked at various cells using a microscope.

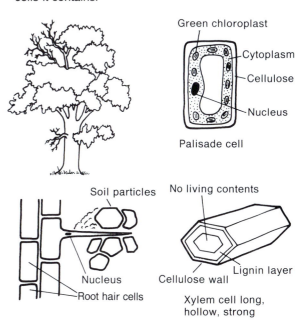

(a) Which cell can carry large amounts of water? Write down a reason for your choice.

(b) What is the special function of the nerve cell? How is it modified to carry out this function?

(MEG)

● *Topic 10*

10 Graph 1 shows how temperature affects an enzyme-controlled reaction.

Graph 1

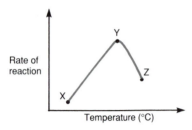

(a) Explain what is happening (i) between **X** and **Y** (ii) between **Y** and **Z**.
(b) At what temperature would you expect the reaction to be quickest?
(c) Suggest how high temperature might affect the active site of the enzyme.

Graph 2 shows how pH affects the rate of two different enzyme-controlled reactions.

Graph 2

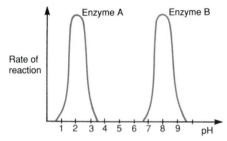

(d) Which enzyme is likely to be most active in (i) the stomach (ii) the duodenum?
(e) What does the graph tell you about the range of pH over which an enzyme is active?

11 Match the chemicals in Column A with their correct chemical composition in Column B and function in Column C.

Column A	Column B	Column C
Lipids	CHON	Growth and repair of cells
Nucleic acids	CHO	Solvent
Carbohydrates	HO	Store of energy
Water	CHON(SP)	Material of heredity
Proteins	CHO	Main source of energy

12 Match the chemicals in Column A with their functions in Column B.

Column A	Column B
Chitin	Pairs with cytosine
Cellulose	Makes up the cell membrane
Glycerol	Carries genetic information
Phospholipid	Same formula as glucose
DNA	Involved in protein synthesis
Fructose	Protein present in hair
Starch	Present in insect exoskeletons
Guanine	Forms a triglyceride with 3 fatty acids
RNA	Makes up plant cell walls

13 The diagram below represents part of a DNA molecule.

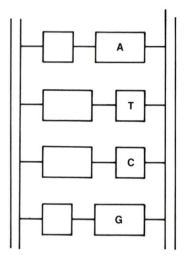

(a) Copy and complete the empty boxes with the letter of the correct base.
(b) Draw arrows on the diagram pointing to the chemical bonds that break when the DNA molecule replicates (reproduces itself).
(c) When in the life of a cell does the DNA replicate?

THEME D
Life Processes

What makes living things different from non-living things?

There is no easy answer to this question. The best we can hope to do is to find features shared by all living things that are absent from non-living things. These features are the characteristics of life. There are seven characteristics: nutrition, respiration, excretion, sensitivity, movement, growth and reproduction. *Life Processes* also examines the transport systems that plants and animals have developed to move food, oxygen, water and waste products around their bodies.

TOPIC 11  *NUTRITION*

11.1 **Green plants make food by photosynthesis**

> Life on Earth depends on light. Why? Because light energy is converted into the energy of the chemical bonds in food substances by photosynthesis, and all living things need food.

The Sun floods the Earth with light. Plant cells use light to help them to make food by **photosynthesis**. They trap the energy in sunlight and use it to convert carbon dioxide and water into sugars. A summary of the chemical reactions that take place is

catalysed by chlorophyll

Carbon dioxide + Water → Glucose + Oxygen

$$6CO_2(g) + 6H_2O(l) \rightarrow C_6H_{12}O_6(aq) + 6O_2(g)$$

How does photosynthesis get its name? 'Photo' means light (in Greek) and 'synthesis' means putting together. Photosynthesis takes place in the **chloroplasts** found in the cells of green plants. Chloroplasts contain the green pigment chlorophyll, which acts as a **catalyst** in the chemical reactions of photosynthesis.

Since the energy of the chemical bonds on the left hand side of the equation above is less than the energy of the chemical bonds on the right hand side, the reactants must take in energy before the reaction can happen. This energy is supplied by the sunlight that falls on the leaves of the plant.

LOOK AT LINKS
for **energy**
For how energy can be converted from one form into another
See *KS: Physics*, Topic 5.1.

LOOK AT LINKS
for **chloroplasts** and the structure of plant cells
See Topic 9.1.

LOOK AT LINKS
for **catalysts**
See *KS: Chemistry*, Topic 23.7.

Leaves, light, water and gases

Leaf cells are 'factories' which collect sunlight, water and carbon dioxide, and manufacture food. Figure 11.1A shows how leaves are adapted for photosynthesis and traces the path of water and carbon dioxide molecules from the air into a leaf, a leaf cell and a chloroplast.

Photosynthesis involves a series of chemical reactions inside chloroplasts. The reactions fall into two stages: *stage 1*, absorbing sunlight; *stage 2*, transforming the energy of sunlight (which is an unlimited supply) into the energy of chemical bonds (see Figure 11.1A). The sugars formed by photosynthesis are the source of energy for the plant. When animals eat plants, the food substances built up in the plants become available to them.

Different pigments in photosynthesis

Chlorophyll absorbs light of all colours except green. It therefore appears green. In fact there are several kinds of chlorophyll. The green **pigments**, chlorophyll A and chlorophyll B, are found in most plants. They strongly absorb red and blue light. Other pigments, called carotenoids, are also present. They are red, yellow and orange and absorb light of other colours. Together, the assortment of pigments enables a plant to absorb sunlight over a wide spectrum (Figure 11.1B).

LOOK AT LINKS
for **light** and **colour**
See *KS: Physics*, Topic 14.1.

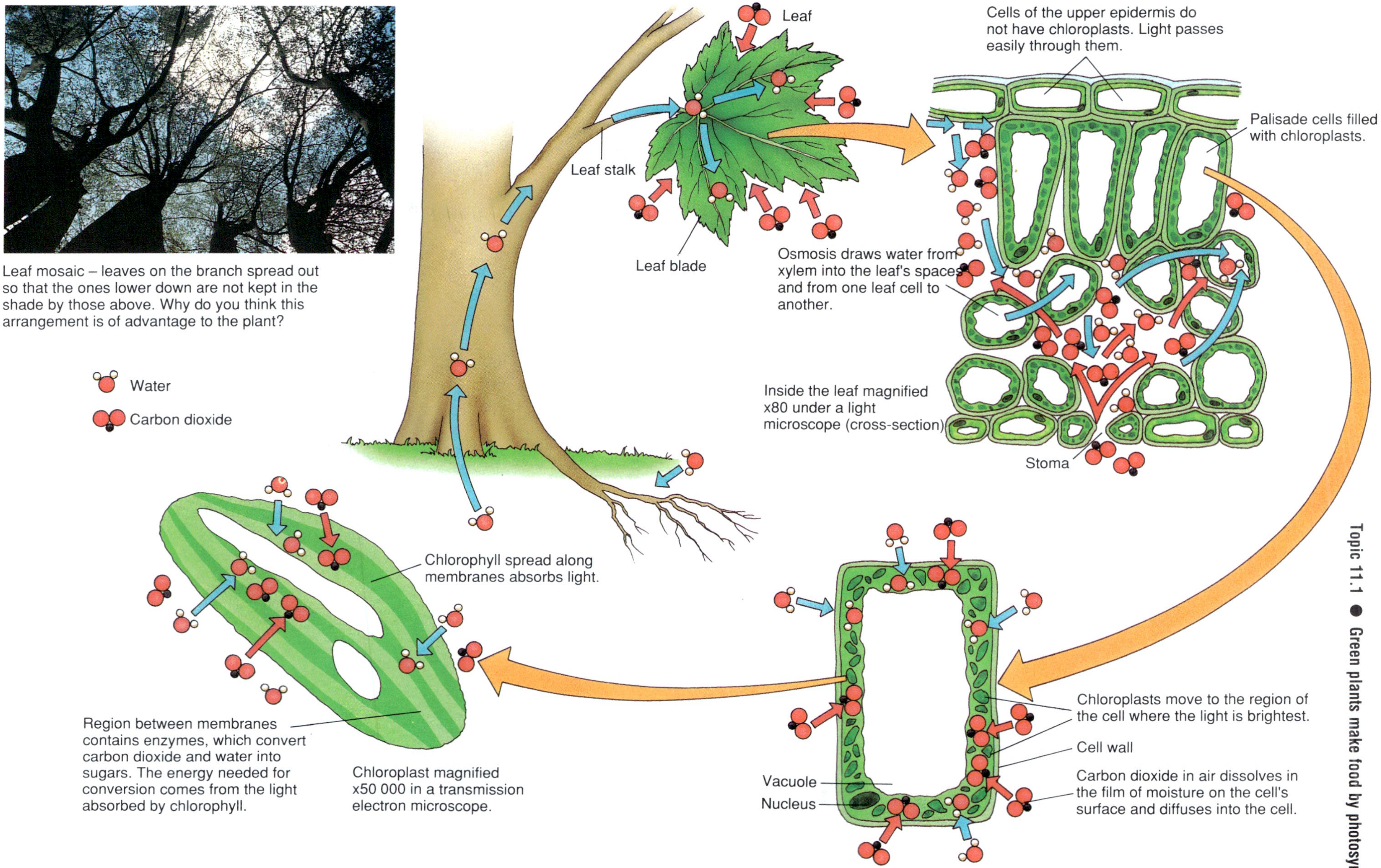

Palisade cells filled with chloroplasts.

Cells of the upper epidermis do not have chloroplasts. Light passes easily through them.

Chloroplasts move to the region of the cell where the light is brightest.

Cell wall

Carbon dioxide in air dissolves in the film of moisture on the cell's surface and diffuses into the cell.

Stoma

Leaf palisade cell magnified x200 under a light microscope.

Vacuole

Nucleus

Osmosis draws water from xylem into the leaf's spaces and from one leaf cell to another.

Inside the leaf magnified x80 under a light microscope (cross-section)

Leaf

Leaf blade

Leaf stalk

Chlorophyll spread along membranes absorbs light.

Chloroplast magnified x50 000 in a transmission electron microscope.

Region between membranes contains enzymes, which convert carbon dioxide and water into sugars. The energy needed for conversion comes from the light absorbed by chlorophyll.

Leaf mosaic – leaves on the branch spread out so that the ones lower down are not kept in the shade by those above. Why do you think this arrangement is of advantage to the plant?

Water

Carbon dioxide

Figure 11.1A ● The leaf and photosynthesis

171

LOOK AT LINKS
You can separate the
pigments in plant leaves by
chromatography.
See *KS: Chemistry*, Topic 11.9.

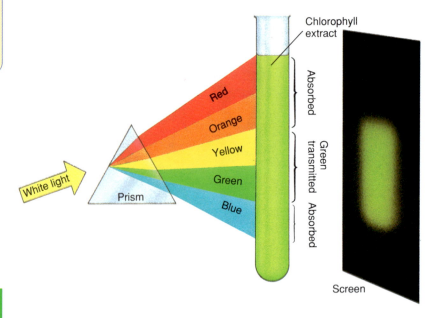

Figure 11.1B ● Spectrum of white light and chlorophyll

SUMMARY

Photosynthesis is the process by
which plants make food. Carbon
dioxide and water combine in the
presence of the catalyst,
chlorophyll, to form glucose and
oxygen. The reaction takes place
in sunlight. The energy of sunlight
is converted into the energy of the
chemical bonds in glucose.

In summer nearly all leaves are green, because the colours of the
carotenoids are masked by the chlorophylls. In autumn however, when
the production of chlorophylls tails off, the colours of the carotenoids
blaze through. Tomatoes, carrots and ripe fruits of different kinds get
their colours from carotenoids.

What conditions are best for photosynthesis?

Light intensity, temperature and supplies of carbon dioxide and water
together affect the rate at which plants make sugars by photosynthesis.
They are called **limiting factors** because if any one of them falls to a low
level, photosynthesis slows down or stops.

*Why do plants usually grow better
in a greenhouse?* Look at the
greenhouse in Figure 11.1C and
notice the bright lights, the water
sprinklers and the machine for
releasing carbon dioxide into the
atmosphere. The farmer can
control these conditions so that
photosynthesis occurs at
maximum efficiency, and the
plants grow quickly. With the
sunlight streaming through the
windows, no wonder the plants
are growing well! All the
conditions they need for
photosynthesis are just right.

Figure 11.1C ● The maximum efficiency greenhouse

Buttercups in the sun

Dog's mercury in the shade

● Light

Look at Figure 11.1D showing plants growing in different light conditions. The graph shows that in both cases the rate of photosynthesis increases as the light intensity increases. However, once a particular light intensity is reached (the compensation point) the rate of photosynthesis stays constant even if the light intensity increases further.

Notice that the compensation points for dog's mercury and the buttercup are not the same. *How has dog's mercury adapted to shady conditions?*

Light on a sunny summer's day is about 10 times brighter than the compensation point for most plants. Light, therefore, is not usually a limiting factor, except for plants growing in shady places.

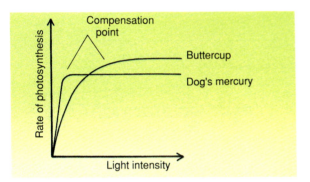

Figure 11.1D ● Light intensity as a limiting factor

● Carbon dioxide

Figure 11.1E shows two crops of lettuces grown in a glasshouse. One crop has had carbon dioxide added to the atmosphere; the other has not. *What is the percentage increase in yield of lettuces grown in a carbon dioxide enriched atmosphere?*

Increasing the amount of atmospheric carbon dioxide is only practical in a glasshouse. Outdoors, the level of carbon dioxide in the air is about 0.03%. The graph shows how the rate of photosynthesis increases with increasing carbon dioxide concentration.

The rate of photosynthesis stays constant after a certain concentration of carbon dioxide is reached. In the short term this limit point is about 0.5%, but over long periods the best level is about 0.1%. Above this level the amount of light often limits the plant's ability to use the extra carbon dioxide.

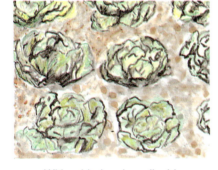

With added carbon dioxide
Mass of 10 lettuces = 1.1 kg

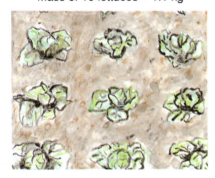

Without added carbon dioxide
Mass of 10 lettuces = 0.9 kg

Glasshouse crops like lettuces and tomatoes are commonly grown in an atmosphere of 0.1% carbon dioxide. The extra cost to the grower is more than offset by the increased crop yield.

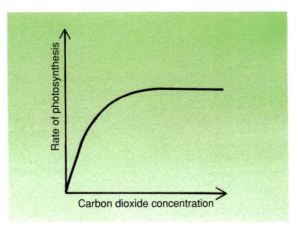

Figure 11.1E ● Carbon dioxide concentration as a limiting factor

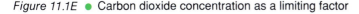

Figure 11.1F ● Turning the desert green

● **Warmth**

Most plants grow best in warm conditions. The higher the temperature, the faster the chemical reactions of photosynthesis – within limits! **Enzymes** are proteins that control chemical reactions (including photosynthesis) in living things. Temperatures below 0 °C and above 40 °C destroy proteins and reduce enzyme activity.

Plants which grow well at 20–25 °C grow in the UK; plants which thrive at 35–40 °C are found in tropical regions.

● **Water**

Water is one of the raw materials for photosynthesis. However, lack of water affects so many cell processes that it is impossible to single out its direct effect on photosynthesis. Figure 11.1F highlights the importance of water for plant growth. It makes farming possible – even in the desert!

CHECKPOINT

1 The diagram opposite shows a leaf and a small portion of it cut through lengthways.
   (a) How does the structure of the leaf enable it to collect light?
   (b) Name the parts labelled A, B, C and D.
   (c) Where in the leaf is most of the plant's food made by photosynthesis?
   (d) Where do gases pass into and out of the leaf?
   (e) How does C help the circulation of gases and water vapour?

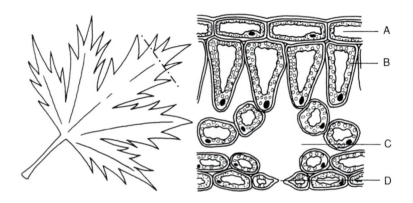

2 The diagram opposite represents the inside of a chloroplast.

   (a) What happens at A during photosynthesis?
   (b) How does the structure of A enable it to play a part in photosynthesis?
   (c) What happens at B during photosynthesis?
   (d) Briefly explain the relationship between what happens at A and what happens at B during photosynthesis.
   (e) Which gas is released into the environment during photosynthesis?

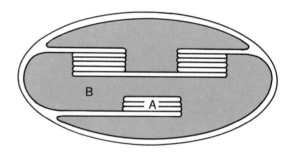

3 The table shows the number of bubbles of gas given off in one minute by water weed illuminated with light of different colours.

Colour of light	Number of bubbles given off in one minute
Blue	85
Green	10
Red	68

What relationship between colour of light and rate of photosynthesis do the results indicate? Suggest how results like this might help a gardener to grow bigger plants in a shorter time in a greenhouse.

④ The diagram below shows an experiment that was carried out to measure the rate of photosynthesis in water weed at 20 °C.

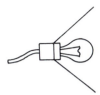 

The rate of photosynthesis was assessed at different light intensities by counting the number of bubbles of gas given off by the plant in a given time. The results are shown in the table.

Number of bubbles per minute	6	15	21	25	27	28	28
Light intensity (arbitrary units)	1	2	3	4	5	6	7

(a) Plot the results on a graph, putting 'Light intensity' on the horizontal axis and 'Number of bubbles' on the vertical axis.

Now answer the following questions, using the diagram and your graph to help you.

(b) At what light intensity did the plant produce 25 bubbles per minute?
(c) Suggest a better way of measuring the rate of photosynthesis than counting the bubbles.
(d) If the experiment was done at the following temperatures, what would be the effect on the rate of photosynthesis?
   (i)   2 °C
   (ii) 33 °C
   (iii) 65 °C
(e) What other factors affect the rate of photosynthesis?

11.2 Food – why do we need it?

FIRST THOUGHTS

We all have favourite foods. This section tells you why we need food and how our bodies use it.

Cars need good maintenance and fuel to keep them going. Without petrol they wouldn't get very far! Have you ever thought how we keep going? Food is the fuel that our bodies need if we are to keep active. But food is more than just a fuel. It also provides the raw materials that we need to build up our bodies and to keep them working properly. It is vital for:
• **Energy** – it is the fuel that keeps us going.
• **Growth** and repair of cells and tissue.
• **Metabolism** – all the complex chemical reactions that take place in the cells of our bodies.

Food contains water, and food from plants contains fibre. However, water and fibre are not usually thought of as nutrients, although fibre is an important part of our food and water is essential for life. Adults can survive for many weeks without nutrients but only for a few days without water.

As well as nutrients, water and fibre, food also has flavour, colour and texture. Small quantities of a range of substances give food these qualities. Cooks and food technologists aim to enhance them to make food more attractive.

IT'S A FACT

Water makes up about two-thirds of your body weight.

Energy nutrients

LOOK AT LINKS
for **carbohydrates, fats,**
and **cellular respiration**
See Topic 10 and Topic 12.

All the processes of life need **energy**. Every time you move, think, speak, read or blink, you use energy. Your body needs energy to grow, to resist disease, and to heal itself, and for all the other processes that make it work.

Energy for living comes from the **carbohydrates**, **fats** and **proteins** in food. It is stored in the chemical bonds of their molecules, and released in the oxidation reactions of cellular respiration. Carbohydrates, fats and proteins are known as the 'energy nutrients' and give foods their energy content.

The energy value of food is measured in the laboratory with an instrument called a bomb calorimeter (Figure 11.2A). Food is burnt and releases food energy as heat. The heat given off by the burning food is transferred to the surrounding water through the heat exchange coil. The change in water temperature is measured with the thermometer and used to work out the energy value of the food. *How is the heat transferred from the heated wire to the water?*

Looking at labels – Energy values

Food labels usually carry information on the energy content of the food. Look at the 'Nutrition information' or 'Nutrition – typical values' on a selection of food labels. The food's energy content is often called its **energy value**, and is usually given in kilojoules (kJ) per 100 g of food.

List your selection of foods and write down the energy value for each one.

- Which food on your list has the highest energy value and which the lowest?
- What nutrients would you find in the high energy food?
- Why are energy values given per 100 g of food?
- Work out the total amount of energy available from a container of one of the foods (the weight will be printed on the container).

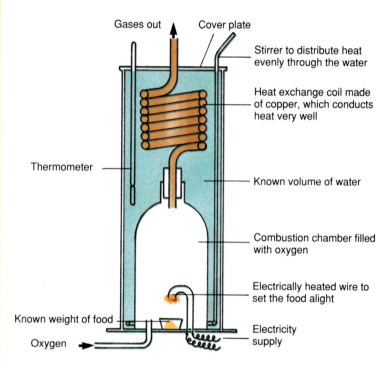

Figure 11.2A ● A bomb calorimeter

IT'S A FACT

If a small slice of full fat cheese is burned, it releases enough heat to make four cups of cool water too hot to touch.

LOOK AT LINKS
How to measure and work out how much energy is needed to heat 100 g of water by 1 °C is explained in *KS: Physics*, Topic 6.3.

Figure 11.2B ● Energy released from different foods

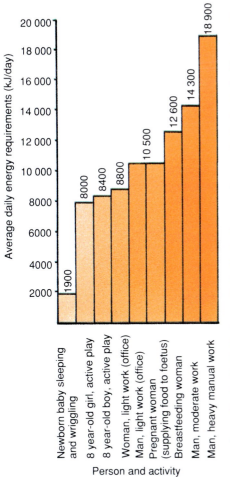

Figure 11.2C ● Average daily energy requirement of different people undertaking different activities

Our energy needs

The rate at which the body uses energy is called the **metabolic rate**. It is lowest (called the **basal metabolic rate**) when the body is at rest (sleeping). Breathing, the heartbeat, maintenance of body temperature, repair and replacement of cells and growth are some of the body activities that contribute to the basal metabolic rate. Any kind of activity increases the metabolic rate. Figure 11.2C shows the energy needed each day by people doing different things.

Measuring energy needs

The man in Figure 11.2D is running on a treadmill. The scientists are measuring the amount of energy he is using. The runner has not eaten and has rested for at least 12 hours before this test. *Why has he taken these precautions?* The amount of energy he uses in running is found by measuring how much oxygen he consumes. His nose is plugged so he breathes through the tubes in his mouth.

Figure 11.2E shows the balance between energy used and food energy needed. We are in 'energy balance' when the amount of energy we get from food equals the amount of energy our bodies use. Measure the height of each block on the right hand side of the see-saw. The total height represents the total amount of energy the man in Figure 11.2D used. *How much energy did he use for running? What is the answer as a fraction of the total energy he used?*

What will happen to the runner's weight if the energy see-saw drops to the left? The right?

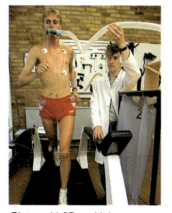

Figure 11.2D ● Using energy

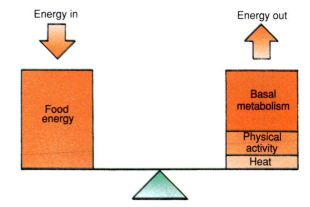

Figure 11.2E ● The energy balance

CHECKPOINT

❶ Look at Figure 11.2D. The runner uses 5 litres of oxygen to provide 84 joules of energy. In running 1 km in 6 minutes, he uses 25 litres of oxygen. How much energy does he use?

❷ Look at Figure 11.2E. Measure the height (in mm) of each part of the right hand block. Let the total height (in mm) represent your answer (in joules) to question 1. What height (in mm) represents the athlete's running? How much energy (J) is represented by the height of this block?

❸ What fraction is the energy used in running (answer to 2) of the total energy used by the athlete (answer to 1)?

❹ The basal metabolic rate varies with weight, age and sex. Women are generally lighter than men because muscle forms a lower proportion of their body weight, and they usually have a higher proportion of fat. Muscle as a proportion of body weight decreases with age. Think about the following statements:

* The heavier you are, the more energy you need.
* Young people need more energy than adults.
* Women need less energy than men.

Briefly explain each statement in the light of what you have read in this section and the background information given. Why do you think young people need more energy than adults? In what circumstances do you think the energy needs of a woman would increase sharply?

IT'S A FACT

Astronauts need more energy in the weightless conditions of space than when on Earth. They take in 12 600 kJ of food energy daily, yet still lose weight on space flights lasting more than two weeks. To help to prevent this, space-travellers are provided with high-energy foods. A typical meal prepared by NASA for astronauts aboard the space shuttle is cream of mushroom soup, mixed vegetables, smoked turkey and strawberries.

Nutrients for growth, repair and the control of metabolism

● Protein

Although protein is an 'energy nutrient' this is not its primary role in the body. Its most important use is for growth and repair. Muscle, blood and other body tissues are made of protein.

Amino acids are the 'building blocks' of protein.

Of the 20 naturally occurring amino acids, nine cannot be made by the body and must be supplied in food. They are called **essential amino acids**. The other eleven can be made by the body and are called **non-essential amino acids**.

Some proteins in food contain all of the essential amino acids the body needs for growth and repair. Other proteins lack one or two essential amino acids. Figure 11.2F shows the percentage and type of protein in different foods.

● Minerals

Some minerals are also important for growth and repair of the body. Others control metabolism. We need small amounts of them in our food for good health.

Minerals such as calcium sulphate and sodium chloride are present in the blood and tissue fluids.

Calcium is needed for making strong bones and teeth and for clotting blood. An adult needs about 1.1 g of calcium each day. Calcium deficiency can cause rickets (Figure 11.2G).

*Contain all the essential amino acids

Potatoes 2.1
*Milk 3.3
Peas 5.8
Rice 6.8
*Wholewheat flour 11.6
*Eggs 12.7
*Beef 18.1
*Cheese 25.4
*Soya beans 40.3

Percentage of protein

Type of food

Figure 11.2F ● The protein content of different foods

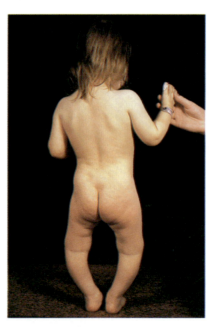

Figure 11.2G ● The badly bowed legs of a child with rickets

Iron is needed to make the blood protein **haemoglobin**. Insufficient iron in the diet causes anaemia. An adult needs about 16 mg of iron each day. Since only a small amount of iron is needed compared with calcium, iron is called a **trace element**. Minerals that are needed in larger amounts are called **major elements**. Diseases caused by mineral deficiency are called **deficiency diseases**. Table 11.1 summarises the information on the minerals we need for good health.

Table 11.1 ● Summary of minerals needed by humans – find out about beri-beri and goitre.

Mineral	Some sources	Importance in body	Deficiency disease
Major elements			
Calcium	Milk, cheese and other dairy products, bread	Making bones and teeth Blood clotting	Rickets (soft bones)
Sodium and Chlorine	Table salt, cheese, green vegetables	Keeping level and make-up of body fluids correct Transmission of nerve impulses	Cramp
Phosphorus	Most foods	Making bones and teeth Important in nucleic acids and energy release in cells	Rarely deficient
Sulphur	Dairy products, beans and peas	Part of vitamin B_1	Beri-beri
Potassium	Meat, potatoes, most fruit and green vegetables	Keeping level and make-up of body fluids correct Transmission of nerve impulses	Rarely deficient
Magnesium	Cheese, green vegetables, oats, nuts	Energy metabolism Calcium metabolism	Rarely deficient
Trace elements			
Iron	Liver, egg, meat, cocoa	Making haemoglobin	Anaemia
Fluorine	Water, tea, sea-food	Helps tooth enamel to resist decay	
Iodine	Fish, iodised table salt, water	Making the hormone thyroxin (which controls growth) in the thyroid gland	Goitre
Zinc	Meat, peas and beans	Protein metabolism Enzymes	Poor healing, skin complaints
Copper	Liver, peas and beans	Making haemoglobin Energy release	Rarely deficient
Cobalt	Meat, yeast, comfrey (a herb)	Part of vitamin B_{12}	Pernicious anaemia (failure to produce haemoglobin)

IT'S A FACT

Before iron tablets became available, people used to stick iron nails into oranges to make a tonic for the treatment of anaemia. The iron in the nails reacted with the ascorbic acid (vitamin C) in the orange, which was then squeezed and the iron-rich juice drunk as an iron tonic.

● **Vitamins**

We also need small amounts of vitamins for good health. A few vitamins are made in the body. The rest come from food or are made by the bacteria that live in our intestines. They are organic substances, which play an important role in the control of metabolism.

As different vitamins were discovered they were labelled alphabetically (A, B, C, etc.). However, in some cases a substance which was first thought to be a single vitamin later turned out to be several related substances, and numbers were added to the letter label (B_1, B_2, etc.). Table 11.2 lists the sources and functions of some of the vitamins needed by humans. Notice that some vitamins are soluble in fat and some in water. *Which vitamins are more likely to be stored in the body?*

Table 11.2 ● Summary of vitamins needed by humans. (The B vitamins help to release energy from food and to prevent anaemia. They promote healthy skin and muscle.)

Vitamin	Some sources	Importance in body	Deficiency disease
Fat soluble			
A	Liver, fish-liver oil, milk, dairy produce, green vegetables, carrots	Gives resistance to disease Protects eyes Helps you to see in the dark	Infections Poor vision in dim light
D	Fish-liver oil, milk, butter, eggs Made by the body in sunlight	Helps the body to absorb calcium from food	Rickets in children Brittle bones in adults
E	Milk, egg yolk, wheatgerm, green vegetables	Antioxidant, protects vitamins A, C, D, K and polyunsaturated fatty acids	Poorly understood in humans Causes sterility in rats
K	Green vegetables, pig's liver, egg yolk Produced by bacteria in gut	Helps to make blood clot	Spontaneous bleeding Long clotting time
Water soluble			
B_1	Whole cereals, wheatgerm, yeast, milk, meat	Helps body to oxidise food to release energy	Beri-beri Nervous disorders
B_2	Fish, eggs, milk, liver, meat, yeast, green vegetables	Helps body to oxidise food to release energy	Dry skin, mouth sores, poor growth
B_6	Eggs, meat, potatoes, cabbage	Helps to digest protein	Anaemia
B_{12}	Meat, milk, yeast, comfrey (herb)	Helps in the formation of red blood cells	Pernicious anaemia
C	Oranges, lemons and other citrus fruits, green vegetables, potatoes, tomatoes	Helps to bond cells together Helps in the use of calcium by bones and teeth	Scurvy (bleeding gums, and internal organs)

Deficiency diseases occur when the body lacks particular vitamins. They are easily cured by supplying the missing vitamins. Many sailors died of scurvy, a deficiency disease, on long sea voyages in the sixteenth and seventeenth centuries (see below). It causes bleeding in various parts of the body, particularly the gums (Figure 11.2H).

Who's behind the science

● **Scurvy**

In 1747 the naval doctor James Lind made the following report.

'On the 20th May I took twelve patients in the scurvy, on board the *Salisbury* at sea …. They all in general had putrid gums …. Two had each two oranges and one lemon given them every day …. The consequence was, that the most sudden and visible good effects were perceived from the use of the oranges and the lemons; one of those who had taken them being at the end of six days fit for duty.'

(From James Lind's account of how he treated scurvy aboard the HMS Salisbury.)

The other ten sailors received different treatments and did not recover.

By the early 1780s a daily ration of lemon juice was a compulsory part of a sailor's rations and scurvy was no longer a problem on board ship. Look at Table 11.2 to find out which vitamin is responsible for preventing scurvy. *Why do you think sailors on long sea voyages were particularly prone to scurvy?*

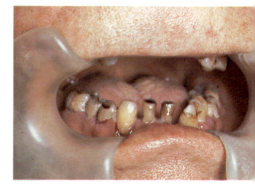
Figure 11.2H ● Scurvy

Vitamin D promotes absorption of calcium by the small intestine and incorporation of calcium into bones. A severe shortage of vitamin D reduces the calcium availability and may cause rickets in children (Figure 11.2G). Bones fail to grow properly and become soft, so when children with rickets start walking the bones bend with the weight of the body. Eating vitamin D-rich foods like fish-liver oil, butter, eggs and milk, prevents rickets. Bright sunlight also changes a chemical in the skin into vitamin D. In hot countries, where there is a lot of sunshine, people make most of the vitamin D they need in this way. If they move to cooler, cloudier climates, they need extra vitamin D in their food to prevent rickets. Children under five years old need about 0.1 mg of vitamin D daily; over fives need about 0.0025 mg daily. In adults the bones have stopped growing and lack of vitamin D causes a disease called osteomalacia. The bones lose calcium, become brittle and snap.

Water

Water makes up about two-thirds of your body weight. It is taken in either directly by drinking or indirectly as part of food.

Chemical reactions in the body also produce water. For example, the oxidation of carbohydrates and fats to release energy produces water.

Water is used in the body:
- as a solvent in which chemical reactions take place,
- as a solvent for waste matter which passes out of the body in solution,
- for transporting substances round the body (water is a major part of **blood** and **lymph**,
- as a means of keeping cool.

An adult needs about 2500 cm³ of water each day. The body loses and gains water through its different activities. Losses and gains roughly balance as Table 11.3 shows.

Table 11.3 ● Daily water balance sheet for an adult

Daily gains (cm³/day)		Daily losses (cm³/day)	
Drinks	1400	Urine	1500
Food	800	Faeces	100
Cellular respiration	300	Evaporation from lungs	350
		Sweat	550
Total	2500	Total	2500

Dietary fibre

Dietary fibre comes from plant foods. There are two types:
- Soluble fibre from fruit pulp, vegetables, oat bran and dried beans.
- Insoluble fibre from the cellulose of plant cell walls and the bran husk that covers wheat, rice and other cereal grains (Figure 11.2I). Wholemeal bread contains much more fibre than white bread because it is made from wholemeal flour – flour

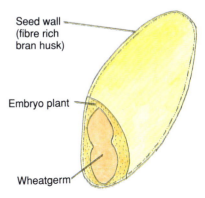

Figure 11.2I ● Section of a grain of wheat

SCIENCE AT WORK

Deficiency diseases such as rickets are rare in countries where vitamins are added to foods like margarine.

IT'S A FACT

Some desert animals such as the oryx take in all the water they need from their food and never drink.

IT'S A FACT

Fog rolling in from the coast of south-west Africa is virtually the only source of water in the Namib desert. When it is foggy the 'head standing beetle' creeps to the top of a sand dune and stretches its back legs, tilting its body forward, head down. The beetle drinks as fog condenses onto its body and runs down to its mouth.

RESOURCE ACTIVITY PACK

LOOK AT LINKS
for **plant cells**
See Topic 9.1.

TRY THIS

Try to dissolve pectin and cellulose in water. Compare your results. *Which is the soluble fibre and which is the insoluble one?*

LOOK AT LINKS
for more about the relationship between **cholesterol** and heart disease
See Topic 11.3.

made from the whole grain. In white flour the outer husks of the grain have been removed. The wheatgerm and the bran also contain most of the vitamins.

Soluble fibre dissolves in water to produce a gel. Insoluble fibre does not dissolve, but it 'holds' water and swells up if mixed with water.

The two types of dietary fibre have different effects. Insoluble fibre adds bulk to food. The muscles of the intestine can work against it and help food to pass through quickly. As a result, disease-causing substances produced by bacteria in the intestine and in the food do not remain in the intestine for very long.

Soluble fibre has the opposite effect. It slows down the passage of food through the intestine. It also seems to decrease the absorption of some minerals, and of glucose and **cholesterol**, into the body. Many people want to lower their cholesterol level. Perhaps this is why porridge is becoming more popular. With its high content of soluble fibre, porridge may reduce the absorption of cholesterol.

CHECKPOINT

❶ Why do we need water?

❷ Look at the amount of water needed daily by an adult (Table 11.3). What percentage of it is produced by 'cellular respiration'?

❸ (a) What is meant by 'fat-soluble vitamin' and 'water-soluble vitamin'?
(b) Which type is difficult to store in the body?

❹ What are the sources and functions of (a) vitamin C and (b) vitamin D?

❺ What can happen when we do not have enough (a) vitamin C and (b) vitamin D?

❻ Which foods are sources of calcium and iron?

❼ What can happen when a person does not have (a) enough calcium and (b) enough iron in his/her diet?

❽ Why is fibre important in the diet?

❾ Susan and Mary want to find out the energy values of sugar and butter. Their apparatus is shown in the diagram.

Susan set the sugar and butter alight and measured the rise in temperature of the water in each beaker after three minutes. All the sugar had burnt away but the butter was still alight.

Her results were: Rise in water temperature in beaker 1 = 3 °C
Rise in water temperature in beaker 2 = 5 °C

Mary also set alight the sugar and butter but she let all the food burn away before measuring the rise in temperature of the water in each beaker.

Her results were: Rise in water temperature in beaker 1 = 6 °C
Rise in water temperature in beaker 2 = 10 °C

(a) Susan and Mary both believe that butter contains more energy than sugar. Why do you think that Mary's experimental evidence is better than Susan's? You should find four reasons.
(b) What could Mary do to improve her accuracy?
(c) Since butter is mainly fat, Susan's and Mary's results might give them the idea that all fats contain more energy than sugar. Briefly explain how they could test this idea.

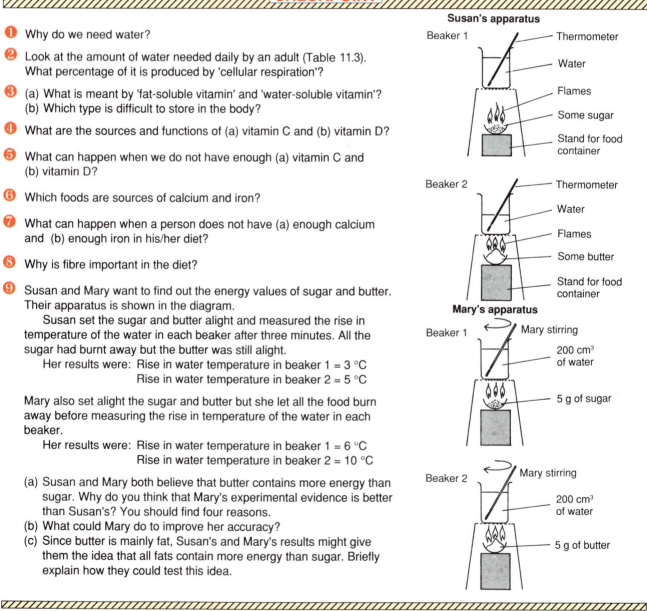

11.3  Diet and food

What is your diet? Is it healthy? This section will help you assess the food you eat and understand what makes a healthy diet.

Your diet is the food you eat and drink. It should contain nutrients, water and fibre in the correct amounts and proportions for good health. If it does, then your diet is said to be balanced or complete.

A healthy diet

By the time you are 70 years old you will have eaten about 30 tonnes of food. Advice about what you eat is big business. 'Experts' tell you that some foods are healthy and others are not. *Who is right? Who should you believe?*

If a diet consists of a single food then the food will be 'unhealthy' no matter what the food is, because no single food contains all the nutrients in the proportions we need for healthy living. For example, both beef and wholemeal bread lack vitamins A, C and D and are low in calcium. Beef also lacks dietary fibre, which wheat provides. Wheat lacks vitamin B_{12}, which beef provides. Together beef and bread provide more nutrients than either on its own, but between them vitamins A, C and D and calcium are still missing. If salad, fruit and vegetables are added, then vitamins A and C are brought into the diet. Milk and cheese add the missing calcium and vitamin D.

A healthy diet, therefore, is a mixture of foods which together provide sufficient nutrients. Notice that each of the foods above lacks some nutrient which the other foods make up between them. They each represent one of the group of foods shown in Figure 11.3A.

IT'S A FACT

Babies rely on milk to meet all their nutritional needs. Nutrients which are low in milk, such as iron, are already stored in large amounts in the baby's liver. However, older children and adults do not store enough of these nutrients, so milk alone does not provide a balanced diet for them.

Nutritionists have developed the idea of the 'basic four' food groups to help us choose a balanced diet. *Do you eat at least one helping of food from each group daily?*

You should also aim for variety. Daily helpings of the same food from each group may contribute to a balanced diet but not necessarily a healthy one. For example, the vitamin C content of fruits may range from next-to-nothing in raw pears to 150 mg per 100 g in stewed blackcurrants. Also, eating the same foods all the time is not only boring but may provide too much fat, sugar and salt. In excess these foods, together with too much alcohol, probably cause more disease in developed countries than any other single factor.

Eat a variety of foods from each group every day

milk and milk products

meat and alternatives

bread and cereals

fruits and vegetables

Figure 11.3A ● The four basic food groups

CHECKPOINT

The table compares the amount of energy and some nutrients that an adult human needs each day with the energy/nutrients in milk.

Energy/ foodstuff	Daily needs of adult	Content of 100 g of cow's milk
Energy	12 000 kJ	272 kJ
Protein	72 g	3.2 g
Calcium	0.5 g	0.10 g
Vitamin A	0.75 mg	0.06 mg
Vitamin C	30 mg	1.5 mg

Study the table and answer the following questions:

❶ On a diet of milk only, how much would an adult have to drink to satisfy daily energy needs?

❷ Milk is a 'balanced diet' for a baby. What is meant by a 'balanced diet'?

❸ Give one reason why milk is particularly important for the development of bones and teeth in babies.

❹ How much milk would an adult need to satisfy his/her daily need for calcium?

Alcohol

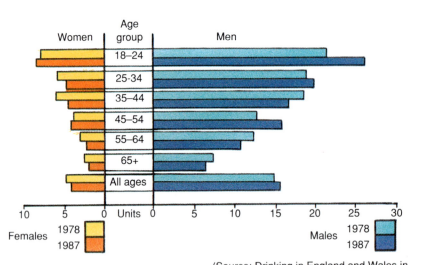

Figure 11.3B ● Alcohol consumption

(Source: Drinking in England and Wales in 1987, Office of Population Censuses and Surveys.)

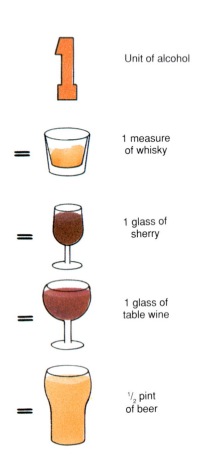

Figure 11.3C ● Units of alcohol

Beers, wines and spirits contain ethanol (alcohol in everyday language). Drinking alcohol is part of many people's social life. Figure 11.3B shows the amount of alcohol consumed each week by women and men of different ages. It also shows how drinking habits have changed. *Which age group drinks the most alcohol each week? How do drinking habits change with increasing age?* Units of alcohol are shown in Figure 11.3C.

Drinking too much alcohol is one cause of diseases such as cirrhosis of the liver, heart disease and damage to the nervous system. The liver metabolises alcohol, and the link between heavy drinking and cirrhosis is

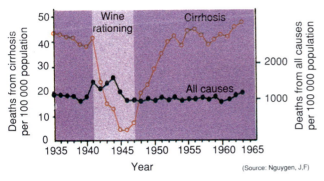

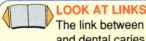

Figure 11.3D ● Deaths from all causes and deaths from cirrhosis in Paris, 1935–65

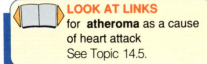

One reason women are more affected by alcohol than men is because of the water content of their bodies. In men 55–65% of the body weight is water; in women it is 45–55%. Alcohol is distributed by the body fluids, so in men it is more dilute.

Exchange of substances between mother and foetus occurs across the placenta (see page 281). Ethanol (alcohol) passes across the placenta very easily. A pregnant woman who regularly drinks beer, wine and/or spirits increases the risk of the foetus developing abnormally. Foetal growth is also reduced. **Foetal alcohol syndrome** may occur if the pregnant woman drinks heavily. At birth, the baby is smaller than average, may be mentally retarded and may suffer heart problems. Also the facial bones are incorrectly developed giving an abnormal look to the face.

LOOK AT LINKS
The link between sugar and dental caries is discussed in Topic 11.5.

LOOK AT LINKS
for **atheroma** as a cause of heart attack
See Topic 14.5.

well-established. Figure 11.3D shows the number of deaths from all causes and the number of deaths from cirrhosis in Paris between 1935 and 1965. Notice that deaths from cirrhosis fell by 80% when wine was rationed during World War II. The number of cirrhosis-related deaths rose rapidly to pre-war levels when wine rationing stopped. More recent figures from France suggest wine consumption has fallen. Estimates predict that a 50% decrease would cut cirrhosis by about 58%.

How much alcohol is too much? This depends on a person's sex, age, size and metabolic rate. For example, the 'safe' level of alcohol for a woman is only about two-thirds as much as for a man of the same weight. Also, different drinks contain different concentrations of alcohol (Figure 11.3C). Look at Figure 11.3B and compare the drinking habits of men and women.

It is difficult to give 'safe' limits for drinking alcohol because the level varies so much from person to person. The Royal College of Physicians suggests that a man should not drink more than the equivalent of four pints of beer a day. Other experts think this is too much and suggest two pints of beer a day is enough. All agree that drinking alcohol affects your behaviour and heavy drinking harms your health (Figure 11.3E). One of the first things to be affected by drinking alcohol is your driving ability. The message is clear – **DO NOT DRINK AND DRIVE**.

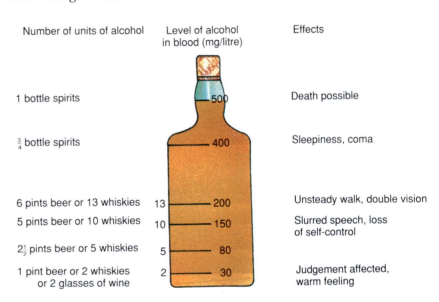

Figure 11.3E ● The effects of alcohol

Sugar and fat

The amounts of sugar and fat we eat also affect our risk of heart disease. If people eat too much sugar and fat they tend to put on weight. Overweight people have a higher risk of heart disease (Figure 11.3F).

Too much of the wrong sort of fatty food also increases the level of a substance called cholesterol in the blood. Cholesterol is found in nearly all body tissues. Large amounts of cholesterol are found in **atheroma** deposits, which block arteries and restrict the flow of blood through them. The more cholesterol there is in the blood, the greater the risk of heart attack (Figure 11.3G).

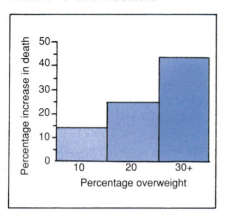

Figure 11.3F ● Increase in deaths from heart disease due to overweight

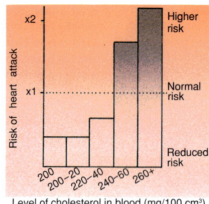

Figure 11.3G ● Cholestrol and the risk of heart disease

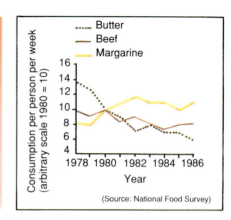

(Source: National Food Survey)

Figure 11.3H ● Consumption of saturated and unsaturated fat

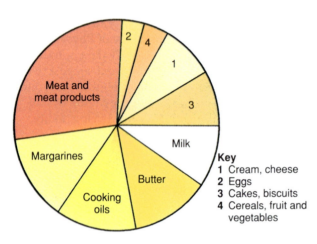

Key
1 Cream, cheese
2 Eggs
3 Cakes, biscuits
4 Cereals, fruit and vegetables

Figure 11.3I ● Sources of fat in the average UK diet

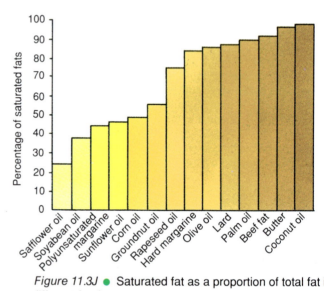

Figure 11.3J ● Saturated fat as a proportion of total fat in different foods

Eating food containing a lot of **saturated fats** seems to raise the level of cholesterol in the blood and therefore increase the risk of heart attack.

Look at Figures 11.3H, 11.3I and 11.3J. We need some fat in our diet for good health. It comes from different foods. Different fats contain different proportions of saturated and unsaturated fatty acids. *Which foods are high in saturated fats?* You should eat less of these. *Which foods can be substituted because they contain more unsaturated fats? What are the trends in consumption of saturated and unsaturated fats?*

The P/S ratio, which is the ratio of polyunsaturated fats to saturated fats in the diet, helps us calculate what proportion of saturated and unsaturated fats to eat. It can be worked out as follows

$$\frac{P}{S} = \frac{\text{Mass of polyunsaturated fat in diet}}{\text{Mass of saturated fat in diet}}$$

In 1985 the national average for P/S was 0.25. In 1984 a government report suggested a P/S value of 0.45 would reduce the risk of heart disease. That is, if 0.45 g of every gram of fat we ate was polyunsaturated.

So far the evidence seems straightforward: the higher your cholesterol level the greater your risk of developing heart disease. In fact the story is more complicated than this and more research is needed. However, it is much better that you act now to improve your diet than do nothing while waiting for more information.

LOOK AT LINKS
for **saturated fats**
See Topic 10.3.

LOOK AT LINKS
Fats are a store of energy and an important component of cell membranes
See Topics 9 and 10.

LOOK AT LINKS
for **sodium chloride**
See *KS: Chemistry*, Topic 13.1.

LOOK AT LINKS
for **osmosis**
See Topic 9.2.

LOOK AT LINKS
for **blood pressure**
See Topic 14.4.

IT'S A FACT

Your body loses salt when you sweat, which is why sweat tastes salty. People who work in hot places sweat a lot and may suffer from muscle cramps because of the salt they lose. Taking salt tablets helps replace the salt lost.

Salt

When we talk about salt in food we mean **sodium chloride**.

The muscles, nervous system and kidneys need salt to work properly. Salt also helps to maintain the correct **osmotic** balance between blood and tissues.

Too much salt in the diet can raise **blood pressure** and a person with blood pressure higher than normal is more likely to suffer from heart disease (see Figure 11.3K).

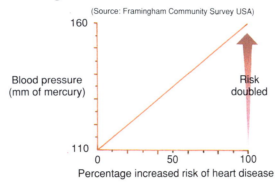

Figure 11.3K ● Blood pressure and the risk of heart disease

High blood pressure can also damage the kidneys and eyes, and increase the risk of an artery tearing open. When an artery tears which supplies blood to the brain it is called a cerebral haemorrhage or 'stroke'.

Very few young people have high blood pressure, but after the age of 35 it becomes more common. The reason is not clear, but different studies show that lifestyle can have an important effect. Eating less salt helps to keep blood pressure within normal levels.

Food additives

Food additives are substances that manufacturers put into food to make it tastier, improve its texture, make it look more attractive and prevent it from spoiling. Table 11.4 summarises the main types of food additive and what they do.

Table 11.4 ● Types of food additive and what they do

Types of food additive	What they do
Preservatives	Stop micro-organisms such as bacteria and fungi from spoiling food
Antioxidants	Prevent oxygen in the atmosphere oxidising fats and oils, turning them rancid
Emulsifiers	Keep oil and water in sauces mixed together
Drying agents	Prevent foods like flour from caking
Thickeners	Make foods like soups less 'runny'
Supplements	Nutrients added to help prevent deficiency diseases, e.g. vitamins A and D added to margarine
Colourings	Enhance natural colours to make foods look more attractive
Flavourings	Make foods tastier by bringing out their flavours
Moisteners (humectants)	Prevent food from drying out

When you look at the labels on food containers for information about energy values, you may notice the letter E printed with a number after it. This is an 'E Number', given to additives that are recognised as safe for use in food by the European Community. Each additive has its own E Number. Some examples are given in Table 11.5.

Table 11.5 ● The identity of some common E numbers

E Number	Additive	E Number	Additive
	Antioxidants		**Emulsifiers**
E300	I-Ascorbic acid	E400	Alginic acid
E320	Butylated hydroxyanisole (BHA)	E406	Agar
			Preservatives
	Colours	E210	Benzoic acid
E102	Tartrazine	E220	Sulphur dioxide
E110	Sunset yellow FCF		
E120	Cochineal		**Sweeteners**
E162	Beetroot red (betanin)	E421	Mannitol
		E420	Sorbitol

Vegetables in different proportion, cornflour, salt, yeast extract, flavour enhancers, sugar, emulsifiers (E471, E472), stabiliser and colour.

Figure 11.3L ● The ingredients of a vegetable soup mix. How many have side-effects?

LOOK AT LINKS
for **food poisoning**
See Topic 11.4.

Food additives are carefully tested to make sure they are safe to eat before they are given an E Number (see Figure 11.3L). However some approved additives can make some people ill. For example, some people are sensitive to the yellow colouring tartrazine (E102). If they eat food containing E102 they may have an asthmatic attack or develop a skin rash. More examples of the side-effects of some additives are listed in Table 11.6.

Doubts about the long-term effects of additives have led to some of them being withdrawn from use, even though they have been declared 'safe'. You must remember that additives help to stop food from spoiling and make it more convenient to use. However, public worries have led food manufacturers to cut down on their use, which in turn may create new problems. For example, bacteria and fungi multiply much more quickly in food without preservatives (in everyday language, the food goes 'bad'). Eating 'bad' food causes food poisoning.

Table 11.6 ● Additives with side effects

E Number		Effects on some people
E102	Tartrazine	Skin rashes, blurred vision, breathing problems, hyperactivity in children
E122	Carmoisine or Azorubine	Skin rashes, swellings
E150	Caramel	None proven although suspect for many years
E220	Sulphur dioxide	Irritation of gut
E320*	Butylated hydroxyanisole	Raises lipid and cholesterol levels in blood
E321*	Butylated hydroxytoluene	Skin rashes, behavioural effects, blood cell changes
E331	Sodium citrate	None known
E481	Sodium stearoyl 2-lactylate	None known

* Not allowed in baby foods

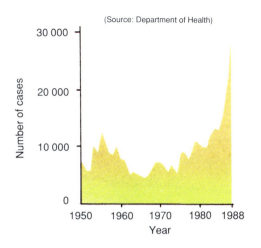

(Source: Department of Health)

Figure 11.3M ● The rise in reported cases of food poisoning

Look at Figure 11.3M. The rapid increase in the number of cases of food poisoning since 1980 coincides with the growing public pressure to cut down additives in food. *Do you think the two are linked?* (Scientists call this 'cause and effect'.) *Can you interpret the data in a different way?* For example, there has been an increase in the consumption of ready-made meals and takeaways. Also people go out for meals more often than they used to. *Do you think these factors influence the data in the graph?* Briefly summarise your opinion on food additives in the light of the evidence.

CHECKPOINT

❶ What is meant by the word 'preservative'?

❷ Why is a diet rich in plant fat probably more healthy than one rich in animal fat?

❸ Why do people who work in very hot countries sometimes take salt tablets?

❹ List some diseases caused by drinking too much alcohol.

❺ How many glasses of table wine contain the same amount of alcohol as $1\frac{1}{2}$ pints of beer?

❻ List some good effects and some bad effects of food additives.

Vegetarianism – fad or healthy alternative? ▪

Vegetarians do not eat meat: some because they think it cruel to kill animals for food; others for religious and cultural reasons. A growing number of people in Britain claim to be vegetarian, but what does it mean to be one? Table 11.7 compares the different types of vegetarian diet and the diets of carnivores and omnivores.

Table 11.7 ● Different types of diet. Lacto refers to milk; ovo to eggs.

Type of diet	Foods eaten				
	Beef and other 'red' meats	Pork, poultry, fish, seafood	Eggs	Milk, cheese and other milk products	Vegetables, fruits, grains, and grain products, legumes, nuts, seeds, oils, sugars
Omnivore	✓	✓	✓	✓	✓
Semivegetarian		✓*	✓	✓	✓
Lacto-ovovegetarian			✓	✓	✓
Lactovegetarian				✓	✓
Vegan					✓
Carnivore	✓	✓			

* May not include all types of food in this group

Many semivegetarians say that they 'feel better' for not eating 'red' meat, so practical reasons govern their choice of diet. Vegans choose their diet because of a principle, believing it not only cruel to kill animals for food but also to use them to produce dairy products.

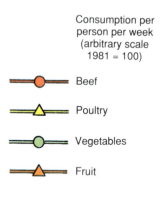
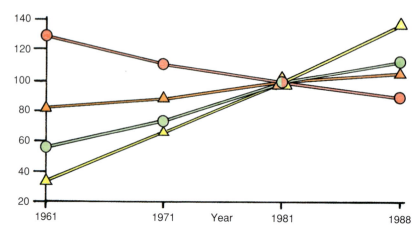

Figure 11.3N ● Changes in food consumption on England and Wales

Look at Figure 11.3N and notice how consumption of red meat has declined and consumption of white meat, vegetables and fruit has increased. People are more aware that food affects health and are changing their diets accordingly. *Do you think these changes are for the better?* Give reasons for your answer.

If foods are selected carefully then it is possible to obtain a complete range of nutrients from a vegetarian diet, especially if eggs, milk and cheese are included. In fact, properly balanced vegetarian diets have advantages over some non-vegetarian diets. They contain:

* more dietary fibre,
* less saturated fat and cholesterol,
* less high-energy food.

Vegetarians are less likely to be overweight than people who regularly eat meat. They are also less likely to develop heart disease, diabetes and some types of cancer. *Why should this be so?*

IT'S A FACT

The soya bean has overtaken wheat and maize as the most important cash crop in the USA. It contains a higher percentage of protein than beef does (see Figure 11.2F). Food manufacturers turn soya beans into 'meat' products or grind them into flour as a basis for savoury snacks.

CHECKPOINT

❶ Which nutrient found in meat could be in short supply in a vegetarian diet?

❷ How do eggs, milk and cheese help vegetarians to obtain a complete range of nutrients?

❸ Why do vegans have more difficulty than other groups in obtaining all the nutrients they need?

Thin or fat – what are the facts?

Magazines, newspapers, television and advertisements bombard us with body images. The message is: thin is beautiful, fat is ugly. This has not always been the case. In the late sixteenth century a much plumper body image was fashionable. Comparison between the past and the present shows how fashion trends shape our image of thinness and fatness.

Life insurance companies have calculated healthy weights for people of different heights (Figure 11.3O). Overweight people are more likely to be ill (Figure 11.3F) and are therefore a greater risk to insure.

Notice that Figure 11.3O shows a range of weights for each height. The range allows for differences in the size of the skeleton which forms the body frame. A small-framed person will tend to weigh in at the lower limit of the weight range for his/her particular height; a large-framed person of the same height will tend to be at the upper limit of the weight range.

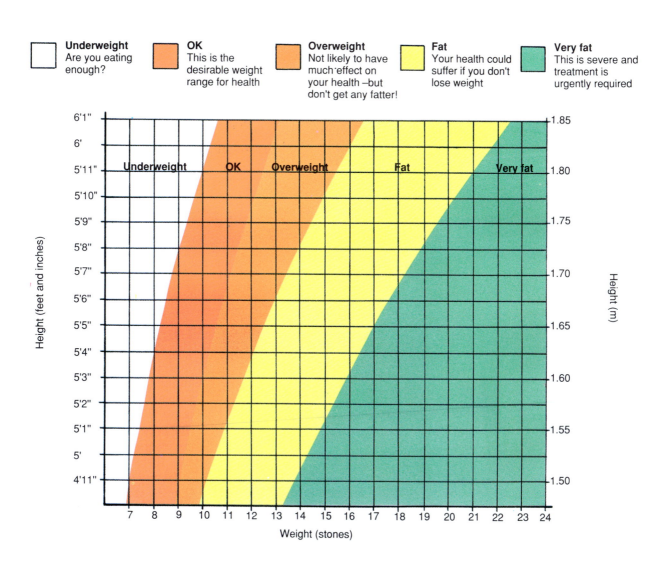

Underweight
Are you eating enough?

OK
This is the desirable weight range for health

Overweight
Not likely to have much effect on your health –but don't get any fatter!

Fat
Your health could suffer if you don't lose weight

Very fat
This is severe and treatment is urgently required

Figure 11.3O ● Healthy weights

LOOK AT LINKS
Growth is discussed in more detail in Topic 15.6.

TRY THIS

Assessing food advertisements. Next time you watch television, count how many advertisements you see in a given time and how many are for food.
- What percentage of advertisements are for food?
- What types of foods are being advertised?
- Use Tables 11.1 and 11.2 to help you to check the nutrients each advertised food contains.
- Which of the advertised foods would tend to make you put on weight? Give reasons for your answer.

Sports people and others with a muscular physique may weigh in as overweight for their height. Muscle is more dense than fat, so although the graph seems to be saying that they are overweight, they probably have less body fat than people of the same weight who do not take regular exercise.

Weight-for-height figures for children are more difficult to calculate than for adults. As you grow up the proportions of water, muscle and fat in your body alter as much as the lengths of your arms, legs and trunk. However, childhood obesity is a growing problem in the UK and other developed countries. As well as the increased risk to health, obese children may become targets for other children's teasing and ridicule. Although some obesity has medical causes, lack of exercise seems to be a major factor. In the USA there is a decline in levels of physical activity among children and obese children tend to be much less physically active than children of normal weight. Watching too much television could be part of the problem (Figure 11.3P).

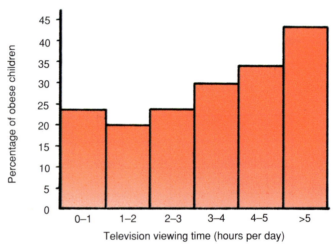

Figure 11.3P ● The relationship between obesity in children and the time spent watching television

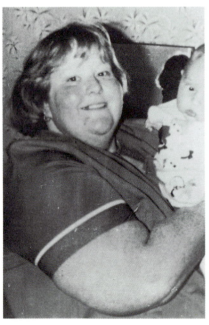

Figure 11.3Q ● Slimming (before and after)

Slimming

Congratulations Jackie Hendry for slimming from 184 kg to 64 kg in less than a year – a loss of 120 kg (Figure 11.3Q). How did you do it?

If a person eats more food than is necessary for his or her energy needs, then the excess is turned into fat and stored in fat cells under the skin and he or she puts on weight.

The only way to lose weight and slim is to make sure that the energy input (food) is less than the energy output (metabolism and physical activity). The options are:
- take more exercise, which increases energy output,
- eat less high-energy food, which decreases energy input.

The dietary requirements of people aged 75 or more are little different from the needs of younger adults. However, if they are less active (remember that moderate exercise within your capabilities is good for you at all ages – see p. 254), then the need for high-energy food is less. For some people adjusting the diet accordingly will help avoid an increase in weight. *How do you think the nutritional quality of meals for the elderly should be maintained?*

The first option is not very effective on its own. For example, a man trying to lose weight uses about 420 kJ/hour more taking a brisk walk than sitting down. If the walk makes him thirsty and he drinks a pint of beer at the end of it, he will take in more energy than he used up. The result: his weight increases.

Fortunately the second option is effective if carried out properly. The sensible approach to slimming includes:

- eating smaller amounts of food,
- eating fewer high-energy foods (Figure 11.3R),
- more physical activity.

A programme for slimming designed with these points in mind not only results in weight loss but also helps to maintain it. The aim is to alter gradually a person's exercise and eating habits. It is much easier to adjust to modest changes – smaller amounts of food, using stairs instead of lifts, for example – than to make sudden drastic changes. Weight is reduced quickly at the start of a weight control diet. Most of the popular diets promoted by the multi-million pound 'slimming' industry owe their success to sudden weight loss. However, few people stick to a diet that demands major upheavals in their eating habits and the weight soon goes back on.

Figure 11.3R ● Many low-energy foods have a high fibre content so eating lots of bread, fruit and vegetables can help you slim. Substituting artificial sweetners for sugar in tea and coffee and low-fat cottage cheese for full-fat cheese could also help you lose weight.

Anorexia

Some people take slimming too far, by following a strict low-energy diet to lose weight. They lose their appetite, eat little food and become dangerously thin. This disease is called anorexia nervosa. It is most common in teenage girls and young women from middle- to high-income families.

The effects of anorexia nervosa:

- Muscle tissue is used as a source of energy once the body's fat reserves are exhausted.
- Body temperature, metabolism and heart rate decrease.
- Depression sets in.
- Growth and sexual development in teenagers stop.
- Thoughts become dominated by food and eating.

Figure 11.3S ● The devastating effects of anorexia nervosa

People with anorexia nervosa do not recognise that they are in effect starving themselves (see Figure 11.3S). They often have a low opinion of themselves. Treatment focuses on building up their self-image. If these problems are overcome, then normal eating patterns and weight gain often follow. Many people recover from anorexia nervosa, and the earlier it is discovered, the more likely it is that treatment will be successful.

CHECKPOINT

❶ The table shows the percentage of British people in different age groups who are overweight.

Age group	Percentage overweight	
	Men	Women
20–24	22	23
25–29	29	20
30–39	40	25
40–49	52	38
50–59	49	47
60–65	54	50

(from: Report of Royal College of Physicians, 1983)

(a) Draw two bar charts to show the percentage of men and women overweight on the vertical axis and age group on the horizontal axis.
(b) How many times more overweight men are there in the 60–65 age group than in the 25–29 age group? Suggest reasons for the increase in weight.
(c) Why do you think that more men than women in all age groups except the first are overweight?
(d) Plan two menus – one that could lead to people becoming overweight, and one that overweight people could use to reach normal weight. Describe briefly the differences between the menus which bring out the gain and loss in weight. Do both of your menus give a balanced diet?
(e) 'Overweight people are more likely to suffer from ill health than normal weight people.' Briefly give reasons why you think this statement is true or false.

❷ 'You are what you eat.' Discuss the meaning of this popular saying by comparing a meal of hamburgers, egg and chips and steamed jam pudding with a meal of fish, salad and fruit.

11.4 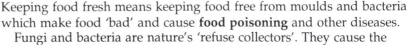 Keeping food fresh

Have you ever felt ill soon after eating a meal? If so, you may have experienced food poisoning. This section tells you how keeping food fresh helps to prevent the build up of microorganisms which cause food poisoning and other diseases.

LOOK AT LINKS
for more about **moulds** and **fungi**
See Topic 1.5.

LOOK AT LINKS
for more about **decay** and **decomposition**
See Topic 2.5.

LOOK AT LINKS
Food chains, food webs and **trophic pyramids** describe the feeding relationships of communities
See Topic 2.

LOOK AT LINKS
The bacteria which cause cholera and typhoid contaminate food and water
See Topic 4.3.

Keeping food fresh means keeping food free from moulds and bacteria which make food 'bad' and cause **food poisoning** and other diseases.

Fungi and bacteria are nature's 'refuse collectors'. They cause the decay and decomposition which clears away dead organic matter. Their activities release chemical elements from dead organisms into the environment making the elements available for absorption and the growth of new plants. Animals obtain the elements essential for their growth, development and healthy living through feeding on plants and/or other animals.

The problem

Food contains salts, which are inorganic; also food is dead organic matter. Fungi and bacteria can come into contact with our food (Figure 11.4A). Some of them cause diseases. They begin to decompose the food (Figure 11.4B) and release substances which make us unwell.

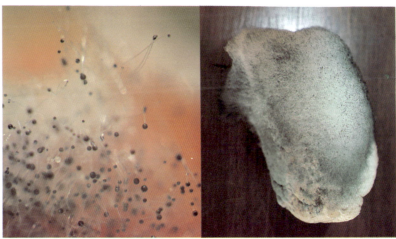

Figure 11.4A ● Moulds are a type of fluffy fungus – the black shiny capsules contain thousands of spores which are wafted away on air currents when the capsules burst. If the spores settle on food they grow into new moulds.

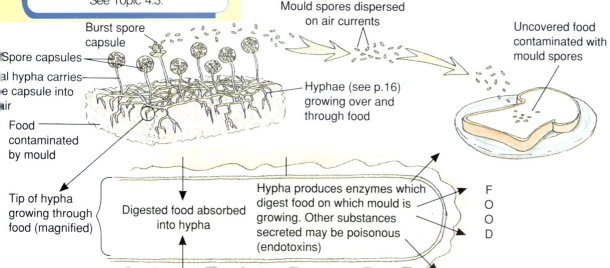

Mould spores dispersed on air currents

Burst spore capsule

Spore capsules

al hypha carries e capsule into ir

Food contaminated by mould

Hyphae (see p.16) growing over and through food

Uncovered food contaminated with mould spores

Tip of hypha growing through food (magnified)

Digested food absorbed into hypha

Hypha produces enzymes which digest food on which mould is growing. Other substances secreted may be poisonous (endotoxins)

F O O D

Figure 11.4B ● Mould covers the food on which it is growing

LOOK AT LINKS
for more about **intensive farming**
See Topic 6.1.

LOOK AT LINKS
for more about the **small intestine**
See Topic 11.6.

LOOK AT LINKS
for more about **diarrhoea** and **treatment by oral rehydration**
See Topic 4.3.

LOOK AT LINKS
Different species of bacteria cause **food poisoning**
See Topic 11.3.

LOOK AT LINKS
The number of cases of food poisoning is increasing
See Topic 11.3.

IT'S A FACT

Salmonella bacteria are spread to live chickens through chicken carcasses which are not properly sterilised but processed and added to chicken meal as a source of extra protein.

LOOK AT LINKS
for more about **antibiotics**
See Topic 20.5.

Case Study: Salmonella *food poisoning*

Look at Figure 11.4C. Most chickens are kept **intensively**. The crowding in poultry houses provides an ideal environment for spreading bacteria from bird to bird. *Salmonella enteriditis* is found in the gut of most chickens without causing ill effects. After slaughter *Salmonella* bacteria from the gut and soiled skin contaminate the carcasses before they are sent to the shops for sale. Recent evidence suggests that eggs from contaminated chickens are also infected.

Figure 11.4C ● (a) Intensively reared chickens – their droppings contain *Salmonella* bacteria

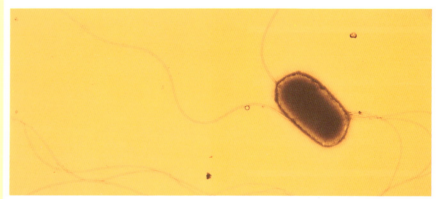

(b) *Salmonella enteriditis* is a rod-shaped bacterium

Someone who eats food contaminated with *Salmonella* soon develops the symptoms of food poisoning. The bacteria invade cells lining the small intestine. They multiply and produce a poison (**endotoxin**) which inflames the tissue and causes fever and acute pain. Symptoms also include vomiting and diarrhoea which results in an enormous loss of water. The body quickly dehydrates and the victim feels tired and unwell. Young children and elderly people are particularly vulnerable.

Cooking chicken thoroughly to at least 68 °C kills *Salmonella* bacteria and helps ensure it is 'safe' to eat. Extra care is needed when cooking chicken from frozen. The bird should be completely thawed, otherwise the centre of the carcass may not reach the 'safe' temperature.

Particular attention to personal hygiene is also important. Handling raw chicken and then preparing other food without first washing the hands increases the danger of contamination and *Salmonella* food poisoning. Improving the standards of hygiene at farms by 'mucking out' poultry houses more frequently also helps to control the disease.

Most victims of *Salmonella* food poisoning recover within a week. Antibiotics are not very helpful since an insufficient amount of the drug gets into the gut cells where the bacterium is causing the problem. Treatment aims to replace water, glucose and salts lost through the severe diarrhoea.

LOOK AT LINKS
Chlorine kills germs. For more about **chlorination of water supplies** and 'safe' **drinking water**
See Topic 5.2.

LOOK AT LINKS
for more about **food additives** and 'E' **numbers**
See Topic 11.3.

LOOK AT LINKS
for **osmosis**
See Topic 9.2.

SUMMARY

Bacteria and fungi contaminate food, spoiling it. Some bacteria and fungi make us unwell. Different methods of preserving food either kill the microorganisms or make them inactive.

Preserving food

Preventing food-borne diseases means keeping food fit for people to eat. 'Fresh' in this sense means preserving food in different ways.

- **Sterilisation** kills bacteria. Food is heated to a high temperature and then sealed in cans or other air-tight containers. Food is preserved for a long time but its flavour is affected.
- **Pasteurisation** of milk and cheese is a partial sterilisation. **Flash** heating milk to 72 °C for 15 seconds kills most bacteria but does not affect flavour. Some bacteria survive, so pasteurised milk should be kept in the refrigerator below 5 °C.
- **Refrigeration** below 5 °C stops bacteria from reproducing and slows their other activities. However, when the food warms up, the bacteria begin to reproduce once more and decomposition sets in. The food quickly spoils.
- **Freezing** between −18 °C and −24 °C stops all bacterial activity. As with refrigeration bacteria become active again as the food thaws.
- **Drying** is used to preserve vegetables, fruits and some meats. Bacteria deprived of water cannot reproduce. Preserving food by leaving it in the sun to dry is a method that has been in use for thousands of years.
- **Freeze drying** is used to preserve foods such as custard powder, coffee and soups. The food is first frozen and the ice is then drawn off in a vacuum. The dried food is stored in sealed containers.
- **Ohmic heating** cooks and sterilises food by passing an electric current through it. The food quickly heats up because of its resistance to the current.
- **Chemical preservatives** are substances which either stop the growth of bacteria or kill them. Important preservatives are sulphur dioxide, nitrites and nitrates. Each one is given an 'E' number.
- **Pickling** food in **vinegar** produces an acid environment which prevents bacterial growth. *What is the acid in vinegar?* Pickling in **brine** (a concentrated solution of sodium chloride) draws water from the food by osmosis. Bacteria are prevented from growing on the food. Also they lose water by osmosis and are killed. Pickled food has a distinctive taste.
- **Jam making** preserves food in a concentrated sugar solution. Again the food and bacteria lose water through osmosis. Prompt storage of treated food in clean, sterilised containers is necessary because some moulds can grow on jam.
- **Smoking** food over burning wood or peat deposits a thin coating of nitrites and nitrates on the food's surface. This kills bacteria and mould.
- **Irradiation** exposes the food to γ-**radiation** from a radioactive source. This kills bacteria and moulds and so prevents food spoilage. Irradiation does not affect enzymes in the food so that the ripening and texture of fruit, for example, is not affected. Food must not be put on sale for twenty-four hours after irradiation. The amount of radiation used is tightly controlled.

CHECKPOINT

❶ What is the link between fungi and bacteria causing decomposition and food poisoning?

❷ Why does rearing chickens intensively increase the risk of *Salmonella* food poisoning?

❸ List the different methods used to preserve food. Discuss **three** of the methods in more detail.

11.5  Teeth

Teeth – what is their structure and how are they adapted for different types of feeding? This section will tell you!

When animals take in (ingest) food we say they are feeding. Different animals have different structures for feeding. Teeth are the feeding structures in most vertebrates (except birds).

Tooth structure

LOOK AT LINKS
You will find more about different feeding structures in Topic 1.5.

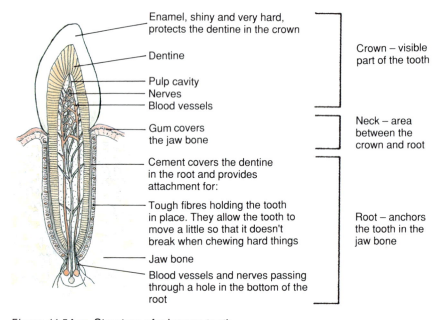

Figure 11.5A shows the internal structure of a human tooth. Notice that blood vessels and nerves enter the tooth through a hole at the bottom of the root. In carnivores and omnivores this hole becomes smaller when the tooth is fully grown, reducing the supply of blood. The tooth is then said to have a **closed root**; dentine is no longer produced, and the tooth stops growing. The roots of herbivores' teeth stay open, so dentine production continues and their teeth keep growing throughout life. These animals feed on rough material like grass. Continuous growth of the teeth makes sure that the abrasive food does not wear them away.

Figure 11.5A ● Structure of a human tooth

IT'S A FACT

Tooth enamel is the hardest substance in the body. It is made of crystals of calcium hydroxide phosphate, bound together by the protein **keratin**.

LOOK AT LINKS
What do teeth have in common with hair, nails, claws and feathers? They all contain **keratin**.
See Topic 10.4.

Types of teeth

The teeth of fish, amphibia and reptiles are usually cone-shaped. The teeth of mammals are different shapes and sizes (Figure 11.5B).

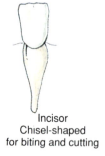

Incisor
Chisel-shaped for biting and cutting

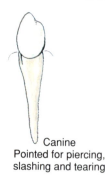

Canine
Pointed for piercing, slashing and tearing

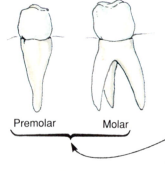

Premolar   Molar

Broad surface made uneven by 'bumps' called cusps. For crushing and grinding. A molar has three 'prongs' to its root; some premolars have two

Figure 11.5B ● The four basic types of human teeth

Figure 11.5C ● The arrangement of teeth in an adult human jaw. There are 32 teeth in total; eight on each side of the upper and lower jaw.

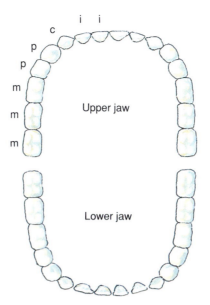

Key
i = incisor
c = canine
p = premolar
m = molar

The different types of teeth are positioned in the mouth according to their functions. Figure 11.5C shows the arrangement of teeth in an adult human jaw.

Humans have two sets of teeth. The first set of 'milk' (or deciduous) teeth form in the jaw before birth and begin to appear (or erupt) about three to six months after birth. There are 24 teeth in the first set, 20 of which are gradually replaced by the permanent teeth between the ages of six and twelve. The third molars (the 'wisdom' teeth) do not appear until the age of about 18 years, if at all.

The word **dentition** is used to describe the number and arrangement of teeth in an animal. Humans and other omnivores have all four basic types of teeth to deal with a mixed diet of plants and meat. The dentition of adult humans is described in a dental formula using the key letters in Figure 11.5C.

Number and kind of teeth on each side of upper jaw

$$i \frac{2}{2} \ c \frac{1}{1} \ p \frac{2}{2} \ m \frac{3}{3}$$

Number and kind of teeth on each side of lower jaw

Herbivores and carnivores have different dentition, to enable them to deal with their particular diets.

● Herbivore dentition

Sheep and cattle are herbivores. They eat tough plants and grasses. Instead of having incisors in the front of the upper jaw, sheep and cattle have a tough, horny pad which the incisors of the lower jaw bite against. There are no canines in the upper jaw and the canines in the lower jaw look like incisors. There is a space in both jaws in front of the premolars. This space is called the diastema (Figure 11.5D). The animal can push its long, muscular tongue through the diastema to sweep grasses into its mouth, where they are cut off by the action of the lower incisors against the pad in the upper jaw (Figure 11.5E).

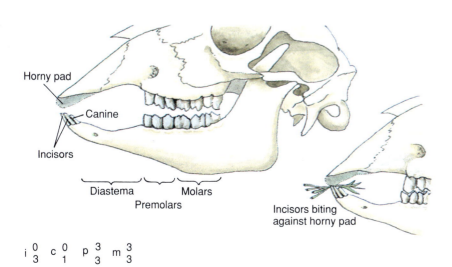

Horny pad
Canine
Incisors
Diastema
Premolars
Molars
Incisors biting against horny pad

$$i \frac{0}{3} \ c \frac{0}{1} \ p \frac{3}{3} \ m \frac{3}{3}$$

Figure 11.5D ● The dentition of a sheep and its dental formula

Figure 11.5E ● Herbivore feeding

The premolars and molars have layers of cement, enamel and dentine which wear away at different rates. This causes ridges of enamel to form, which make a good surface for grinding the food (Figure 11.5F). The joint of the jaw and skull can move from side to side as well as up and down, which also makes grinding food easier.

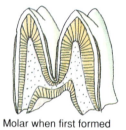

Molar when first formed

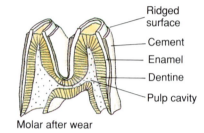

Ridged surface

Cement

Enamel

Dentine

Pulp cavity

Molar after wear

Figure 11.5F ● The crown of a sheep's tooth showing how a herbivores teeth wear into ridges

● Carnivore dentition

Carnivores' teeth are adapted for catching struggling prey and cutting through soft flesh and hard bones (Figure 11.5G).

Dogs and cats are carnivores. Their canines are long and well-developed for grasping and tearing, and they have powerful jaw muscles. Their incisors, premolars and molars are used for cutting. The last premolar on each side of the upper jaw and the first molar on each side of the lower jaw are very large and are called the carnassial teeth. They are especially suitable for cutting through flesh and bone (Figure 11.5H). The jaw joint only allows up and down movements.

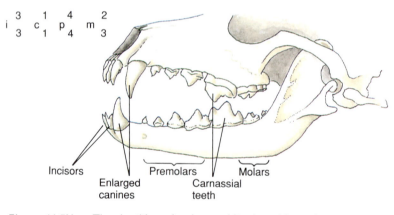

$$i \frac{3}{3} \quad c \frac{1}{1} \quad p \frac{4}{4} \quad m \frac{2}{3}$$

Incisors

Enlarged canines

Premolars

Carnassial teeth

Molars

Figure 11.5G ● Carnivore feeding

Figure 11.5H ● The dentition of a dog and its dental formula

CHECKPOINT

❶ Explain the words carnivore, herbivore and omnivore.

❷ Look at Figure 11.5A. Name the innermost part of the tooth? What does it contain?

❸ (a) Name the four basic types of teeth.
(b) Describe the functions of two of the basic types of teeth in humans.

❹ (a) How are the teeth of a sheep adapted for grinding food?
(a) How are the teeth of a dog adapted for grasping and cutting food?
(b) Compare Figures 11.5D and 11.5H and list the differences between sheep's teeth (herbivore) and dogs' teeth (carnivore).

Looking after your teeth

With proper care, a set of human teeth can last a lifetime. However, it is a sad fact that in the UK a child of twelve years has on average eight decayed teeth. What can we do to improve this?

Evidence shows that sugary foods in particular are bad for teeth. Figure 11.5I shows what happens. Reducing the amount of sugary food we eat helps to prevent tooth decay and gum disease. Look at Figure 11.5J. *Do you think cutting down on the amount of sweets you eat will help protect your teeth?* Give reasons for your answer.

Cleaning your teeth properly and regularly – at least after breakfast and last thing at night – is also very important. A disclosing tablet contains a dye, which colours plaque and shows you where extra cleaning is needed (Figure 11.5K).

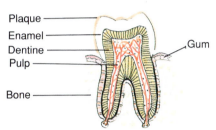

Plaque
Enamel
Dentine
Pulp
Gum

Bone

Bacteria in plaque break down sugar in the mouth, forming acids. NO PAIN

If teeth are not cleaned the acids attack and soften the enamel, which begins to decay (dental caries). The plaque thickens and irritates the gums. NO PAIN

If untreated the decay eats through the dentine layer. SEVERE PAIN

The decay eventually reaches the pulp cavity. If bacteria reaches the bone of the tooth an abscess forms. AGONY!

Figure 11.5I ● Tooth decay in progress

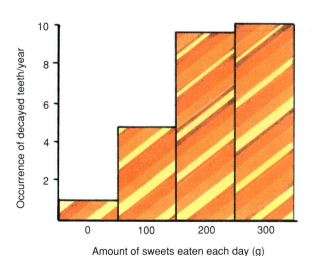

Figure 11.5J ● Tooth decay in children

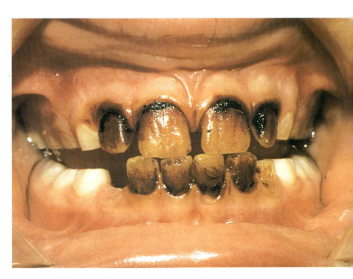

Figure 11.5K ● Dental plaque revealed using a disclosing tablet which stains plaque red

LOOK AT LINKS
for **fluoride** and **tooth decay**
See Topic 13.4.

You need a good toothbrush. It should have a fairly small head, dense bristles and a firm handle. You also need to use it correctly (see Figure 11.5L). Your dentist can give you more advice on cleaning your teeth. You will need a new toothbrush about every three months. After brushing, you can use a soft thread called dental floss to clean between your teeth where the brush cannot reach (Figure 11.5M).

Having your teeth checked every six months by a dentist helps to make sure that your teeth and gums stay healthy. Dentists like to concentrate on preventing tooth decay and gum disease.

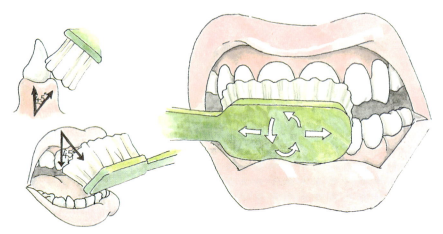

The angle between the bristles of the toothbrush and the neck of the tooth should be about 45°.

Pressing gently, but firmly, move the brush with short strokes from side to side and also with an up–down circular movement to help clean in the spaces.

Work round the mouth on only a few teeth at a time, brushing all the front, back and biting surfaces.

Figure 11.5L ● One way of cleaning teeth properly with a toothbrush

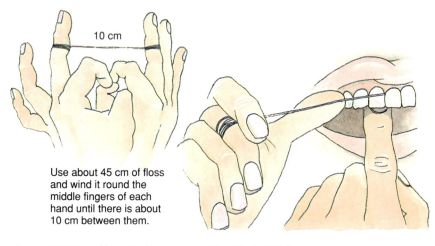

10 cm

Use about 45 cm of floss and wind it round the middle fingers of each hand until there is about 10 cm between them.

Starting with the upper teeth, use your fingers to guide the floss gently between two teeth until it reaches the gum line. Slowly slide it up and down the sides of both teeth. Repeat between all teeth, unwinding the floss from one hand to the other occasionally, for a fresh length.

Figure 11.5M ● How to clean your teeth with dental floss

CHECKPOINT

① (a) Make a list of sugary foods. Tick the four that you eat most often.
(b) How does eating sugary foods cause tooth decay and gum disease?

② (a) How often should you visit the dentist?
(b) What is dental floss?

③ Chlorine is added to drinking water to kill bacteria that would otherwise cause diseases like cholera and typhoid. Are you in favour of treating drinking water with chlorine? Give reasons for your answer.

11.6 Digestion and absorption

FIRST THOUGHTS

How are nutrients in the food you eat converted into substances your body can absorb and use? The answer is – by digestion. This section tells you all about it.

LOOK AT LINKS
for **carbohydrates** see Topic 10.2; for **lipids** see Topic 10.3; for **proteins** see Topic 10.4.

LOOK AT LINKS
for **enzymes**
See Topic 10.4.

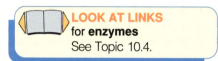
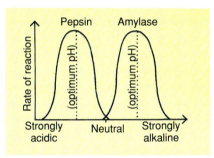

(a) The effect of pH on enzyme activity. Activity is greatest at the optimum pH for that particular enzyme. Strong acid/alkali denatures most enzymes

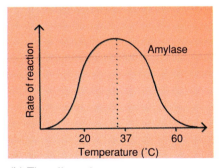

(b) The effect of temperature on enzyme activity. The activity of most enzymes at first increases with temperature reaching a maximum and then decreases. A high temperature denatures most enzymes

Figure 11.6A ● The effects of pH and temperature on the activity of the enzymes pepsin and amylase

Carbohydrates, **lipids** and **proteins** are the main nutrients in our food. They are complex molecules which the body cannot absorb directly. To be useful they must be broken down into their basic constituents, substances which the body can absorb.

Nutrient		Absorbed as...
Carbohydrate	→	Simple sugars
Protein	→	Amino acids
Fat	→	Fatty acids and glycerol

The chemical and mechanical processes of digestion convert these foods into absorbable substances in the intestine (gut). Vitamins and minerals in food are absorbed unchanged during digestion.

Digestive enzymes

The chemical processes of digestion depend on nearly one hundred different **enzymes**. Enzymes are catalysts that speed up the chemical reactions. Without them the nutrients in the food you eat would be broken down so slowly that you would starve to death.

The absence of even one enzyme can cause illness. For example, people who do not produce the enzyme lactase cannot digest the sugar lactose ('milk' sugar). They develop cramps and diarrhoea if they consume lactose in milk and milk products, because the sugar builds up in the gut.

Enzymes are grouped according to the reactions they catalyse (Table 11.8). The rate at which they work is influenced by various factors, especially pH and temperature. Figure 11.6A illustrates the effects of these factors. *What is meant by 'optimum pH' and 'optimum temperature'? Which nutrients do pepsin and amylase digest?* Look at Figure 11.6C to find out where pepsin and amylase are made in the gut.

Table 11.8 ● Enzymes that digest carbohydrates, proteins and lipids

Enzyme group	Example	Nutrient digested
Carbohydrases (catalyse the digestion of carbohydrates)	Amylase Lactase Sucrase Maltase	Starch Lactose Sucrose Maltose
Proteases (catalyse the digestion of proteins)	Pepsin Trypsin Chymotrypsin	Proteins Proteins Polypeptides
Lipases (catalyse the digestion of lipids)	Intestinal lipase	Fats

The gut and how it works

Figure 11.6B shows the human gut and its position in the body. It is a tube. At one end food is put into the mouth (**ingested**). At the other end the undigested remains of a meal are removed through the anus

(egested). In between mouth and anus the mechanical and chemical processes of **digestion** break down food into substances suitable for **absorption**. These processes are explained in Figure 11.6C. *Why do you think the action of renin is particularly important in babies? How do the actions of teeth and bile make it easier for enzymes to digest food?*

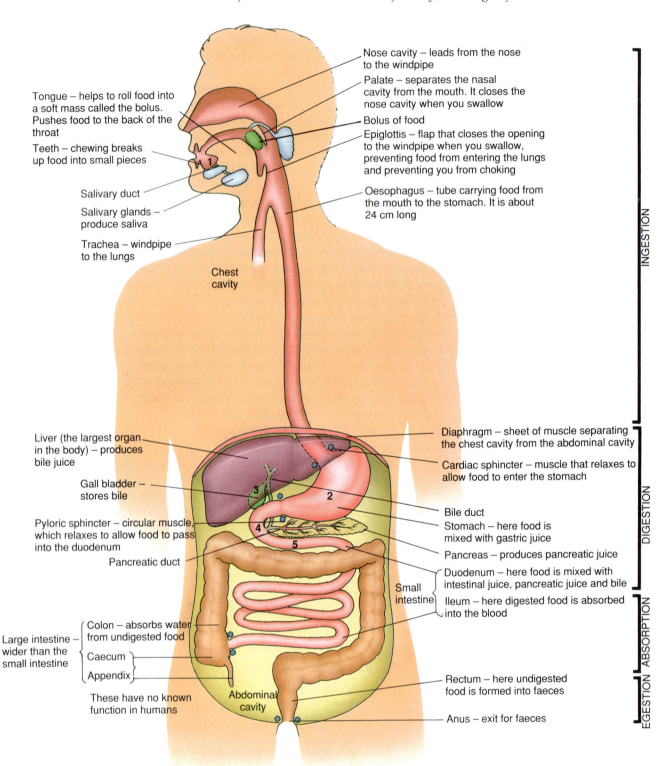

Nose cavity – leads from the nose to the windpipe

Palate – separates the nasal cavity from the mouth. It closes the nose cavity when you swallow

Bolus of food

Epiglottis – flap that closes the opening to the windpipe when you swallow, preventing food from entering the lungs and preventing you from choking

Oesophagus – tube carrying food from the mouth to the stomach. It is about 24 cm long

Tongue – helps to roll food into a soft mass called the bolus. Pushes food to the back of the throat

Teeth – chewing breaks up food into small pieces

Salivary duct

Salivary glands – produce saliva

Trachea – windpipe to the lungs

Chest cavity

INGESTION

Liver (the largest organ in the body) – produces bile juice

Gall bladder – stores bile

Pyloric sphincter – circular muscle, which relaxes to allow food to pass into the duodenum

Pancreatic duct

Large intestine – wider than the small intestine

Colon – absorbs water from undigested food

Caecum

Appendix

These have no known function in humans

Abdominal cavity

Diaphragm – sheet of muscle separating the chest cavity from the abdominal cavity

Cardiac sphincter – muscle that relaxes to allow food to enter the stomach

Bile duct

Stomach – here food is mixed with gastric juice

Pancreas – produces pancreatic juice

Duodenum – here food is mixed with intestinal juice, pancreatic juice and bile

Small intestine

Ileum – here digested food is absorbed into the blood

Rectum – here undigested food is formed into faeces

Anus – exit for faeces

DIGESTION

ABSORPTION

EGESTION

Figure 11.6B ● The human gut (The functions are explained in Figure 11.6C; the numbers ❶ to ❺ refer to the sections of Figure 11.6C)

● Chemical processes

What happens

❶ **Saliva** moistens the food. It contains the enzyme amylase, which begins the digestion of starch in the mouth.

❷ Pits in the stomach wall produce **gastric juice**, which contains hydrochloric acid and the enzymes pepsin and renin.

Hydrochloric acid:
• lowers the pH of the stomach contents
• kills bacteria on the food
• stops the action of the enzyme amylase in the saliva

Pepsin:
• starts the digestion of protein

Renin:
• clots milk so it stays long enough in the gut to be digested

❸ **Bile** is a green alkaline liquid produced by the liver and stored in the gall bladder. It neutralises acid from the stomach and breaks fats into small droplets which are easier for the enzymes to digest.

❹ **Pancreatic juice** contains four different enzymes which digest carbohydrate, fat and protein. It also contains sodium carbonate which helps to neutralise stomach acid.

❺ Glands in the wall of the duodenum produce **intestinal juice** which contains enzymes that complete the digestion of carbohydrates and fats.

Where it happens

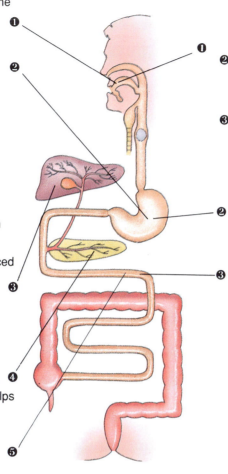

Figure 11.6C ● How we digest food

● Mechanical processes

What happens

❶ Chewing breaks the food up into small pieces which are easier to digest. Mucus, produced by membranes lining the mouth, and saliva make food slippery for easy swallowing.

❷ Muscles of the stomach wall rhythmically contract and relax, mixing food with the gastric juices into a liquid paste called chyme.

❸ Rhythmic contractions of the muscles in the small intestine mix chyme with bile, pancreatic juice and intestinal juice.

Digestive enzymes (amylase, pepsin etc) catalyse the breakdown of food by **hydrolysis**. Water splits molecules of food components into smaller molecules.

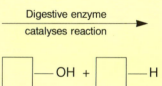

Hydrolysis breaks chemical bonds

Digestive enzyme catalyses reaction

Products – smaller molecules suitable for absorption

The liver
The liver is the chemical-processing factory of the body.

It stores:
• glycogen which can be hydrolysed to glucose and released back into the blood in response to the body's needs
• iron compounds from destroyed red blood cells.

It converts:
• excess of amino acids into urea, which is excreted in urine
• amino acids from one type into another (a process called **transamination**) in response to the body's needs.

It makes:
• bile
• plasma proteins.

Different parts of the gut perform different tasks in processing food as it passes through. The human gut is 7–9 m long. The longest part of it is composed of the small intestine and the large intestine. These lie folded and packed in the space of the abdominal cavity. The **liver** and **pancreas** are connected by ducts to the gut and play an important part in the digestion of food. They also have a major role in the metabolism of food substances once these have been absorbed into the body.

● Absorption of digested food

Figure 11.6D summarises what has happened to the food so far. *Which nutrients are not digested? Is fat digested in the stomach?*

The next stage – absorption – takes place mostly in the ileum, although alcohol and small amounts of simple sugars and water are absorbed by the lining of the stomach. Water is also absorbed by the colon.

Figure 11.6B shows that folding and coiling packs as much gut as possible into the restricted space of the abdominal cavity. The extra length means that food not only travels greater distances through the gut so that enzymes have more time to digest the food, but also that the surface area for absorption is increased. Closely packed finger-like projections called **villi** line the small intestine and increase its surface area further (Figure 11.6E).

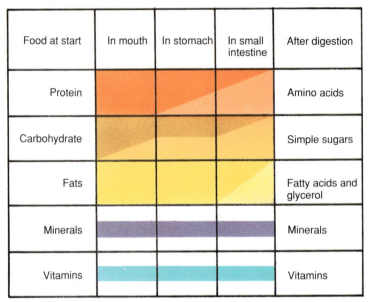

Food at start	In mouth	In stomach	In small intestine	After digestion
Protein				Amino acids
Carbohydrate				Simple sugars
Fats				Fatty acids and glycerol
Minerals				Minerals
Vitamins				Vitamins

Figure 11.6D ● The progress of digestion. Solid colours represent the nutrients in food before digestion

LOOK AT LINKS

Absorption involves **diffusion** and **active transport**. See Topic 9. Transport by the lymph system and by blood vessels is discussed in Topic 14.4.

Figure 11.6F shows inside a villus. Absorption occurs when nutrients pass through its cells and into the blood and lymph vessels. Most fatty acids and fat-soluble vitamins (see Table 11.2) pass into the lymph vessels. Water, glucose, glycerol, amino acids and other substances pass into the network of capillary blood vessels and are transported to the liver. Mucus helps to protect the gut wall from its own digestive enzymes. Microvilli project from the surface of each cell lining the villus. They increase the surface area for absorption about 20 times more compared with the villi alone! *Altogether, how large a surface area do villi and microvilli make for absorption?* (See Figure 11.6E.)

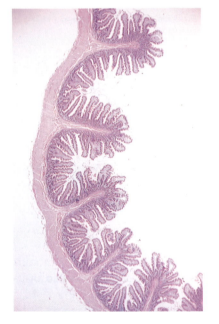

Figure 11.6E ● Section of the small intestine. Five million villi, each about 1 mm long, cover the inside of the small intestine. They provide approximately 10 m² of surface for absorption in your gut!

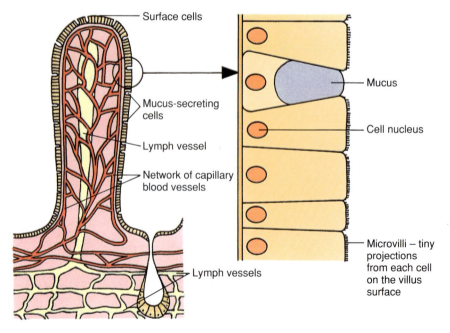

Figure 11.6F ● Inside a villus (part of its surface is shown at high magnification)

Undigested food passes out of the small intestine into the colon. Here, water poured onto the food during digestion is absorbed back into the body (Figure 11.6G). The food remnants dry out into a compact mass of faeces which is removed from the body through the anus – a process called **defaecation**.

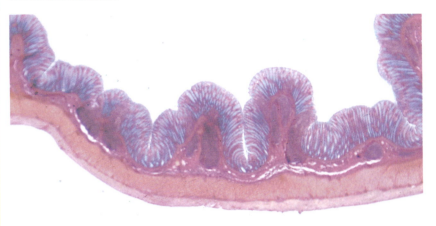

Figure 11.6G ● Section of large intestine. Compare with the section of small intestine shown in Figure 11.6E. *What differences can you see?*

Moving food through the gut

Repeated contraction and relaxation of **muscle** layers in the gut wall move food through the gut. This muscular action is called **peristalsis** (Figure 11.6H).

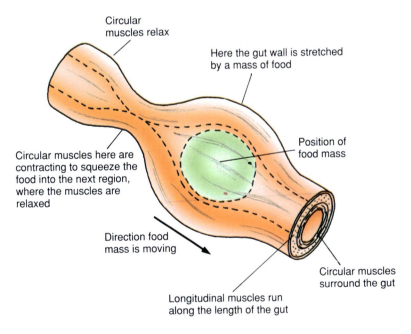

Figure 11.6H ● Peristalsis – the gut narrows when the circular muscles contract and shortens when the longitudinal muscles contract

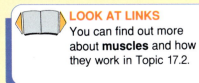

LOOK AT LINKS
You can find out more about **muscles** and how they work in Topic 17.2.

Another type of gut movement happens when the circular muscles alone contract. This muscular action is called **segmentation** (Figure 11.6I). Together peristalsis and segmentation move a meal through the gut in about 24 hours.

LOOK AT LINKS

for how an **X-ray** photograph is taken
See *KS: Physics*, Topic 14.2.

for **barium meals**
See *KS: Chemistry*, Topic 13.4.

TRY THIS

Imitating circular muscle
You need a lump of plasticine. Roll it into a cylinder about 2 cm in diameter. Lightly grip it in your hand – the diagram shows you how. Now gently squeeze the cylinder and notice what happens.

• How does your hand imitate the action of circular muscle?
• Briefly describe what happens to the plasticine when you squeeze it. What does this tell you about the movement of food through the gut?

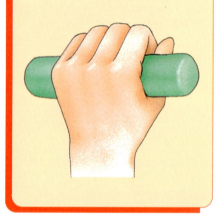

LOOK AT LINKS

for more about **omnivores**, **herbivores** and **carnivores**
See Topic 11.3.

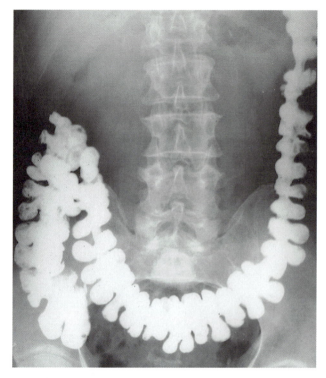

Figure 11.6I ● Segmentation. The X-ray picture shows that contraction of the circular muscle temporarily divides the gut into a series of segments

The gut suits the diet

Different guts are adapted to deal with different types of diet. Most humans, for example, are omnivores and the human gut is adapted to deal with a mixture of plant and animal foods. Rabbits and cows are herbivores, with guts adapted for plant foods, and cats are carnivores, with guts adapted for meat. Figure 11.6J compares them. List the differences between them. *How is each gut adapted to deal with a particular diet?*

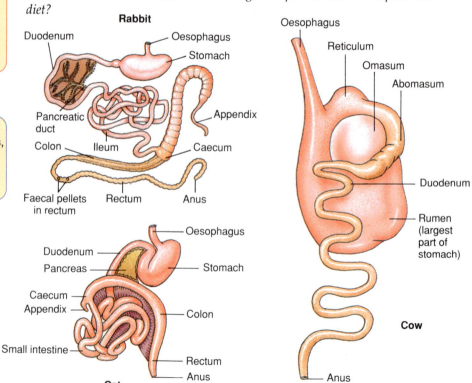

Figure 11.6J ● The guts of the cow, rabbit and cat

The blue whale is the largest animal that has ever lived. It is found in the North Atlantic, Pacific and Indian oceans. It feeds on huge quantities of shrimps filtered from the water, which it strains through a series of horny plates (called **baleen**) that grow down from the sides of the upper jaw. Other baleen whales, like the minke, fin and humpback, feed on shoals of fish as well as the minute protists (see page 14) that form the dense plankton layer at the water's surface.

Most animals do not possess enzymes to digest the cellulose in the walls of plant cells. However, some species of **bacteria** and **protozoa** do. These microorganisms live in different parts of the gut of an animal and digest the cellulose in food. In rabbits, for example, the caecum and appendix are the home of these microorganisms. In cows they live in the stomach, which consists of four parts. The rumen and reticulum receive the food first of all. The cellulose-digesting microorganisms live in the rumen. Here the food forms into balls of cud before returning to the mouth where it is thoroughly re-chewed (animals that chew cud are called ruminants). Swallowing takes the cud to the remaining two parts of the stomach, where protein digestion takes place.

Rabbits do not chew cud. Instead they eat the pellets of faeces produced from the food's first journey through the gut. The pellets are soft and contain a lot of undigested food which the cellulose-digesting microorganisms have another change of dealing with as they pass through the gut a second time. Faecal pellets produced after the food's second journey through the gut are hard.

Cats do not chew their food – they swallow it whole or in large chunks. The stomach is large so that it can store the meal while it is digested.

CHECKPOINT

❶ Mechanical digestion breaks up food, and chemical digestion converts its insoluble nutrients into soluble substances which the gut can absorb.

(a) Rewrite the following list under two headings – mechanical digestion and chemical digestion – placing each in the correct column.

action of teeth   action of bile
action of amylase   action of saliva
action of peristalsis

(b) Why do you think that mechanical digestion usually occurs before chemical digestion?

❷ Experiments to investigate the effects of temperature and pH on the activity of the protein-digesting enzyme pepsin produced the following results.

Temperature (° C)	Time taken for digestion to be completed (s)	
	pH5.8	pH8.5
20	275	
25	221	
30	169	
35	114	Digestion not completed during the experiment
40	137	
50	196	
60	Digestion not completed during the experiment	

Plot the results on graph paper and then comment on the results.

❸ Relative to the size of the rest of the body, frog tadpoles have a long, coiled gut and adult frogs have a short one. Briefly explain why you suspect frog tadpoles are herbivores and adult frogs are carnivores.

RESPIRATION

12.1 **What is respiration?**

Cells use the oxygen in air to release energy from sugars. The process produces carbon dioxide, which living things pass to the environment.

Figure 12.1A ● Breathing hard

Figure 12.1A shows a man and his best friend doing something that we all do – breathing: in their case, breathing hard!

Breathing describes inhaling (taking in) and exhaling (giving out) air. Look at Table 12.1. It shows that the proportions of gases in inhaled and exhaled air are different. Oxygen is used by the body and carbon dioxide is produced. Oxygen and carbon dioxide are exchanged between the inhaled air and the blood across the inner surface of the lungs. *What is oxygen used for in the body?*

Table 12.1 ● Differences between inhaled and exhaled air.

Gas	Inhaled air (%)	Exhaled air (%)
Nitrogen	78	78
Oxygen	21	16
Noble gases	1	1
Carbon dioxide	0.03	4
Water vapour	0	1

Cellular respiration

Digested food substances are **oxidised** in cells to release energy. These oxidation reactions are called **cellular respiration**.

Look at Table 12.1. *Why is there less oxygen in exhaled air than in inhaled air?* The reason is that some of the oxygen inhaled is used by the cells to oxidise food. Cellular respiration that uses oxygen is called **aerobic respiration**.

The aerobic respiration is **exothermic** – it gives out energy.

Glucose + Oxygen → Carbon dioxide + Water
$C_6H_{12}O_6(aq)$ + $6O_2(g)$ → $6CO_2(g)$ + $6H_2O(l)$

Energy released = 16.1 kJ/g glucose

In Figure 12.1A aerobic respiration in the cells of the man's leg muscles gives him a flying start. Soon, however, in spite of rapid breathing and strenuous pumping by the heart, oxygen cannot reach the muscles fast

LOOK AT LINKS
for **digestion**
See Topic 11.6.

LOOK AT LINKS
for **oxidation** and **exothermic reactions**
See *KS: Chemistry*, Topics 21.1 and 26.1.

LOOK AT LINKS
for the transport of oxygen in the blood
See Topic 14.2.

enough to supply their needs. The muscles then switch from aerobic respiration to **anaerobic respiration**, which does not use oxygen. Lactic acid is produced and collects in the muscles. The reactions are

Glucose $\rightarrow$ Lactic acid

$C_6H_{12}O_6(aq) \rightarrow 2CH_3CHOHCO_2H(aq)$

Energy released = 0.83 kJ/g glucose

The energy released is less than in aerobic respiration. An 'oxygen debt' builds up. After the muscles have respired anaerobically for a few minutes, the lactic acid they have built up stops them from working. The dog's leg muscles will also have been respiring anaerobically. Both the man and the dog will be unable to run further until the lactic acid is removed from the muscles. During the recovery period, which lasts several minutes, man and dog pant vigorously (Figure 12.1B). The rush of oxygen to the muscles promotes aerobic respiration. Some lactic acid is oxidised to carbon dioxide and water; the rest is converted into glucose. In this way the 'oxygen debt' is repaid.

Figure 12.1B ● Recovery period

Bacteria, yeasts and root cells of plants can also change from aerobic respiration to anaerobic respiration if they are short of oxygen. Certain bacteria live permanently without oxygen: in fact oxygen is poisonous to some of them.

Yeast cells use anaerobic respiration to convert glucose into ethanol and carbon dioxide, with the release of energy. The reaction, called **fermentation**, is used commercially for the production of ethanol (alcohol) by yeast.

Glucose $\rightarrow$ Ethanol + Carbon dioxide

$C_6H_{12}O_6(aq) \rightarrow 2C_2H_5OH(aq) + 2CO_2(g)$

Energy released = 1.17 kJ/g glucose

LOOK AT LINKS
for **fermentation**
See Topic 20.

The energy released is less than in aerobic respiration. Other differences between aerobic and anaerobic respiration are summarised in Table 12.2.

Why is there a difference in energy output? Aerobic respiration completely oxidises glucose to carbon dioxide and water and releases all the available energy from each glucose molecule. Anaerobic respiration converts glucose into ethanol or lactic acid. More energy can be obtained by oxidising lactic acid aerobically.

Cellular respiration is sometimes compared with the combustion of petrol in vehicle engines, but there is a vital difference. When fuel is burnt in an engine, energy is released suddenly in an explosive reaction. If cells were to release energy from food suddenly, the sharp rise in temperature would kill them. Cellular respiration is not a one-step chemical change, but a series of chemical changes that release energy from food gradually.

Table 12.2 ● A comparison of aerobic and anaerobic respiration

Process	Oxygen used or not used	Products of process	Energy released (kJ/g glucose)
Aerobic respiration	Used	Carbon dioxide and water	16.1
Anaerobic respiration in yeast	Not used	Ethanol and carbon dioxide	1.17
Anaerobic respiration in muscle cells	Not used	Lactic acid	0.83

Stages of cellular respiration

LOOK AT LINKS
for **mitochondria**
See Topic 9.1.

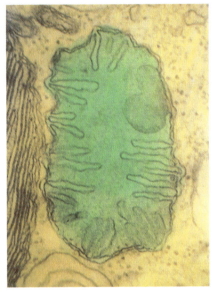

The first stage in both aerobic and anaerobic respiration is the conversion of glucose into pyruvic acid. This happens in the cell's cytoplasm. Then, if oxygen is present, pyruvic acid is oxidised in the **mitochondria** to carbon dioxide and water (Figure 12.1C).

Figure 12.1C ● Mitochondria (x 40 000)

If no oxygen is present, anaerobic respiration takes place and lactic acid is formed.

IT'S A FACT

Cells do not use the energy from the oxidation of food as soon as it is released. It is converted into the energy of the chemical bonds in a substance called adenosine triphosphate (ATP). ATP is found in the cells of nearly all living things, and is the link between the cell's energy-releasing activities and energy-using activities.

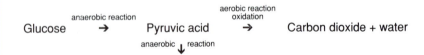

Glucose   anaerobic reaction →   Pyruvic acid   aerobic reaction / oxidation →   Carbon dioxide + water

anaerobic ↓ reaction

Lactic acid

Cellular respiration and photosynthesis

Cellular respiration and photosynthesis form a cycle. The products of one are the starting materials of the other (Figure 12.1D).

During the day photosynthesis produces more oxygen than is used in aerobic respiration. The surplus oxygen diffuses through the stomata out of the leaf. At night photosynthesis stops but plant cells still need oxygen for aerobic respiration. Oxygen therefore diffuses through the stomata into the leaf.

RESOURCE –ACTIVITY– PACK

SUMMARY

Cellular respiration is the name given to the processes in cells that release energy from food. Aerobic respiration in the presence of oxygen oxidises glucose to carbon dioxide and water with the release of energy. Anaerobic respiration in the absence of oxygen converts glucose into ethanol and carbon dioxide, or into lactic acid. Less energy is released in anaerobic respiration than in aerobic respiration.

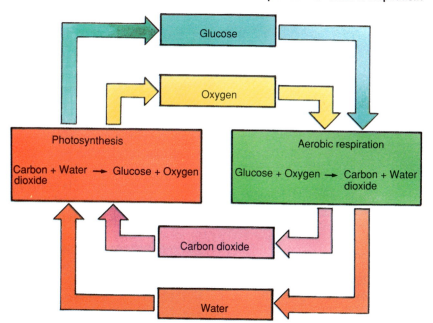

Figure 12.1D ● Respiration and photosynthesis form a cycle

CHECKPOINT

❶ Draw a table with two columns, one headed 'Photosynthesis', the other 'Aerobic respiration'. Study the descriptions below and decide which ones belong to which column.

(a) occurs only in plant cells
(b) occurs in all cells
(c) takes place only in daylight
(d) produces carbon dioxide
(e) takes place night and day
(f) produces food
(g) produces oxygen
(h) uses oxygen
(i) uses carbon dioxide
(j) releases energy
(k) uses water

❷ Complete the following paragraphs using the words provided. Each word may be used once, more than once, or not at all.

(a) aerobic   aerobically   anaerobically   with
     as much   releases   without   respire

_____ respiration requires oxygen. Anaerobic respiration takes place _____ oxygen. Some cells can _____ either aerobically or anaerobically. If oxygen is available these cells will respire _____.

(b) respire   repaid   oxygen   muscles
     oxygen debt   liver   lactic acid

During a sprint your leg _____ will need oxygen faster than it can be supplied by your blood. To compensate, the muscle cells _____ anaerobically. This process forms _____ which stops _____ from working. The oxygen needed to oxidise the lactic acid is called the _____.

❸ The flow diagram at the bottom of the page is a summary of the chemical changes that take place during cellular respiration.

(a) Which pathway, A, B or C, summarises aerobic respiration?
(b) What is the name of compound X?
(c) Why is there a difference in energy values between pathway A and
   (i) pathway B (ii) pathway C?

A
Glucose → Pyruvic acid → CO_2 + H_2O + 16.1 kJ/g glucose

B
Glucose → Pyruvic acid → CO_2 + X + 1.17 kJ/g glucose

C
Glucose → Pyruvic acid → Lactic acid + 0.83 kJ/g glucose

12.2 Surfaces for exchanging gases

FIRST THOUGHTS

One of the processes vital to living things is the exchange of gases across a surface. Read on to find out why this is so important.

LOOK AT LINKS
for **photosynthesis** and **cellular respiration**
See Topics 11.1 and 12.1.

LOOK AT LINKS
Diffusion of gases depends on the ratio of the area of the gas exchange surface to the volume of the body. For more on the relationship between surface area and volume
See Topic 9.5.

RESOURCE ACTIVITY PACK

LOOK AT LINKS
for *Hydra* and *Planaria*
See Topic 1.5.

Figure 12.2A shows some different organisms that live in water. The organisms are taking in and giving out air. In the tissues there is an exchange of gases because of:

● **photosynthesis** (in plants only)
● **cellular respiration** (in nearly all organisms)

These exchanges take place across the membranes of gas exchange surfaces. Notice the gills of the newt tadpole and the dragonfly nymph and the flat, thin leaves of the water weed. *Which gas (oxygen or carbon dioxide) does each process (a) use and (b) produce?*

Figure 12.2A ● (a) Newt tadpole   (b) Dragonfly nymph

In small animals like *Hydra* and *Planaria* gas exchange occurs across the body wall. Their surface areas are large enough to supply sufficient oxygen to the inside of the body and let carbon dioxide pass out. The body of the flatworm *Planaria* is flattened to a leaf-like shape which allows gases to diffuse easily through it. (Figure 12.2B).

Only small organisms can obtain enough oxygen by diffusion through the body wall. Larger animals have special organs for supplying oxygen to tissues in the deepest parts of the body and removing carbon dioxide from them.

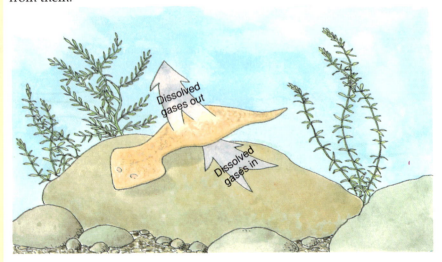

Figure 12.2B ● *Planaria* lives in water where gases are in solution. The gases diffuse into and out of the animal across its body wall

LOOK AT LINKS
for more about gas
exchange in leaves
See Topic 11.1.

Leaves are also flat. Figure 12.2C shows the small holes called stomata that pierce the undersurface. Oxygen and carbon dioxide diffuse into and out of the leaf through the stomata.

The cells on either side of the stomata are called guard cells. Each one is sausage-shaped and has many chloroplasts. Guard cells control the size of the stomata and therefore control the rate of diffusion.

Figure 12.2C ● Underside of a leaf. The oval holes are stomata. Guard cells on either side of a stoma are packed with chloroplasts, unlike the surrounding cells of the epidermis

Lungs

The fish in Figure 12.2D breathes air! It has simple sac-like structures, called lungs, through which it exchanges oxygen and carbon dioxide in air. It also has gills through which it exchanges oxygen and carbon dioxide in water. Breathing air is a safety device for this animal. It enables it to survive periods of drought when the swamps and rivers it lives in dry out temporarily.

IT'S A FACT

The ancestors of lungfish were among the first land-living vertebrates 300 million years ago. Breathing air through lungs allowed them to survive times of drought.

Figure 12.2D ● A South American lungfish

The human lungs and upper respiratory tract

Our lungs are large and honeycombed with passages. Gases are exchanged across the surfaces in the lungs. Figure 12.2E shows the position of the lungs in the human body. They lie inside the **thoracic cavity**. The **upper respiratory tract** is a tube from the nostrils and mouth to the lungs.

There are mechanisms to prevent food from entering the larynx from the pharynx. The act of swallowing closes the glottis and lifts up the whole larynx so that it is blocked by the base of the tongue. At the same time, the flap-like epiglottis moves backward and protects the opening of the glottis. If food does lodge in the larynx, it triggers off violent coughing, which usually clears the obstruction.

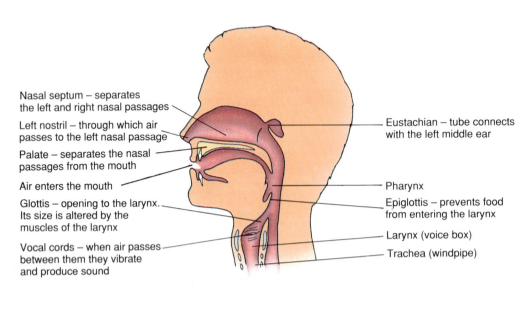

Nasal septum – separates the left and right nasal passages

Left nostril – through which air passes to the left nasal passage

Palate – separates the nasal passages from the mouth

Air enters the mouth

Glottis – opening to the larynx. Its size is altered by the muscles of the larynx

Vocal cords – when air passes between them they vibrate and produce sound

Eustachian – tube connects with the left middle ear

Pharynx

Epiglottis – prevents food from entering the larynx

Larynx (voice box)

Trachea (windpipe)

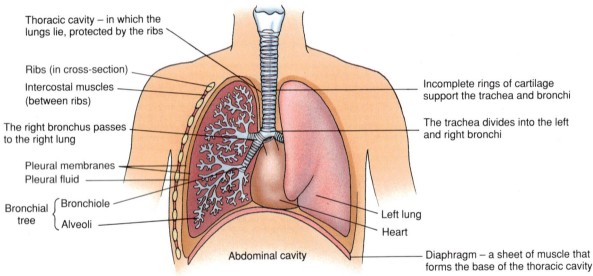

Thoracic cavity – in which the lungs lie, protected by the ribs

Ribs (in cross-section)

Intercostal muscles (between ribs)

The right bronchus passes to the right lung

Pleural membranes
Pleural fluid

Bronchial tree { Bronchiole
Alveoli

Abdominal cavity

Incomplete rings of cartilage support the trachea and bronchi

The trachea divides into the left and right bronchi

Left lung

Heart

Diaphragm – a sheet of muscle that forms the base of the thoracic cavity

Figure 12.2E ● The lungs (viewed from the front) and the upper respiratory tract (viewed from the left hand side) in the human body

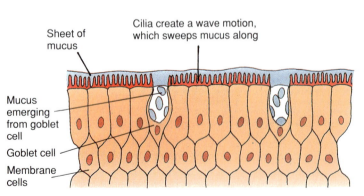

Sheet of mucus

Cilia create a wave motion, which sweeps mucus along

Mucus emerging from goblet cell

Goblet cell

Membrane cells

Figure 12.2F ● The membrane lining of the upper respiratory tract

● *The upper respiratory tract at work*

The upper respiratory tract is an air-conditioning system. It warms and filters inhaled air. The membrane which lines the upper respiratory tract is well supplied with blood, which warms the air to body temperature. Hairs in the nasal passage filter out large particles of dust. A sheet of mucus lines the upper respiratory tract and traps bacteria, viruses and dust particles (Figure 12.2F). The mucus comes from cells in the membrane lining called goblet cells. Cilia, rows of fine hairs, sway to and fro and sweep the mucus, and with it trapped bacteria, viruses and dust, into the pharynx. It is either swallowed, sneezed out or coughed up. The air which then passes to the lungs is cleaned and freed from germs. What happens when the body fails to remove all germs is described later in this topic.

LOOK AT LINKS
You can find out more about the transport of gases by the blood in Topic 14.2.

IT'S A FACT

There are millions of alveoli in a pair of human lungs. Together they give a surface area of approximately 90 m²!

● *The alveoli at work*

In the lung, each bronchus branches many times into small tubes called **bronchioles**. These form a network called the **bronchial tree** (see Figure 12.2E). The bronchioles divide and sub-divide into even smaller tubes which end in clusters of small sacks called **alveoli** (singular: alveolus).

The walls of the alveoli are very thin and surrounded by capillary blood vessels. Figure 12.2G shows gas exchange between the walls of the alveoli and the capillary blood vessels.

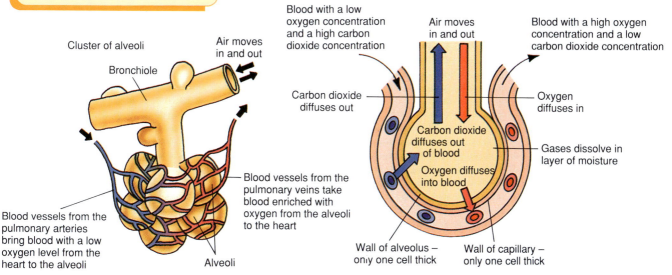

Section through alveolus showing gas exchange

Cluster of alveoli

Bronchiole

Air moves in and out

Blood with a low oxygen concentration and a high carbon dioxide concentration

Air moves in and out

Blood with a high oxygen concentration and a low carbon dioxide concentration

Carbon dioxide diffuses out

Oxygen diffuses in

Carbon dioxide diffuses out of blood

Oxygen diffuses into blood

Gases dissolve in layer of moisture

Blood vessels from the pulmonary veins take blood enriched with oxygen from the alveoli to the heart

Blood vessels from the pulmonary arteries bring blood with a low oxygen level from the heart to the alveoli

Alveoli

Wall of alveolus – only one cell thick

Wall of capillary – only one cell thick

Figure 12.2G ● The alveoli at work (notice that oxygen and carbon dioxide diffuse in solution between alveolus and capillary blood vessel

● *Breathing movements*

The cage formed by the ribs and diaphragm is elastic. As it moves the pressure in the lungs changes. It is the change in pressure that causes inhaling (breathing in) and exhaling (breathing out) (Figure 12.2H).

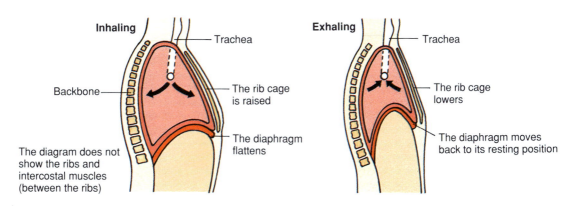

Inhaling

Trachea

Backbone

The rib cage is raised

The diaphragm flattens

The diagram does not show the ribs and intercostal muscles (between the ribs)

The volume of the thoracic cavity increases: therefore the pressure of the air inside it decreases. It becomes less than atmospheric pressure, and air is drawn into the lungs

Exhaling

Trachea

The rib cage lowers

The diaphragm moves back to its resting position

The volume of the thoracic cavity decreases: therefore the pressure of the air inside it increases. It becomes greater than atmospheric pressure and forces air out of the lungs into the trachea

Figure 12.2H ● Inhaling and exhaling

FIRST THOUGHTS

o
o
o

What is in the air we breathe? How does it affect our lungs? This section tells you.

LOOK AT LINKS
for **air pollution**
See Topic 7.

People who work in dusty conditions are likely to inhale very small particles which can damage the lung tissue. Particles of blue asbestos (in the past much used for insulation against fire) are particularly dangerous. They can lodge in the lungs and cause a type of cancer called **asbestosis**. There is now a ban on the use of blue asbestos.

SUMMARY

Gases are exchanged in solution by diffusion across gas exchange surfaces. These surfaces are thin and have a large surface area: adaptations for efficient diffusion.

Upper respiratory tract and lung diseases

The mechanism for sweeping dust, bacteria and viruses out of the upper respiratory tract and lungs has been described in Figure 12.2F. Coughing and sneezing remove the mucus with its load of dust and micro-organisms. Even so, any part of the upper respiratory tract and lungs can become infected by disease-causing microorganisms.
Infection of the:

- throat (pharynx) is called **pharyngitis**,
- voice-box (larynx) is called **laryngitis**,
- windpipe (trachea) is called **tracheitis**,
- bronchus and bronchioles is called **bronchitis**.

Infection of the lungs by a particular type of bacterium causes pneumonia. The patient becomes breathless because fluid collects in the alveoli, reducing the surface area available for the absorption of oxygen. Pneumonia is treated with antibiotic drugs such as penicillin.

The **pleural membranes** line the rib cage and cover the surface of the lung (Figure 12.2E). Between them the **pleural fluid** stops the lungs sticking to the chest wall. Sometimes bacteria infect the pleural membranes, making them rough and causing pain when they rub together. The infection is called pleurisy and it is treated with antibiotics.

Other common lung diseases like lung cancer and asthma are not caused by bacteria or viruses. Lung cancer can be caused by many different things such as smoking or can just occur naturally. Tumours form in the lung and if they are not discovered quickly the cancer can spread to other parts of the body. Asthma is often caused by allergies. If you have asthma sometimes it is difficult to breathe, you wheeze and your chest feels tight. The occurence of both lung cancer and asthma can be linked to the quality of the air that we breathe.

● *The air we breathe*

Motor vehicle exhaust fumes, smoke and dust from industry, cigarette smoke and all the other substances that human activities put into the air are part of our environment . Breathing this mixture affects our lungs.

Figure 12.2I shows the effect of fine coal dust on a coalminer's lungs. People who work in a dusty atmosphere can be affected in a similar way.

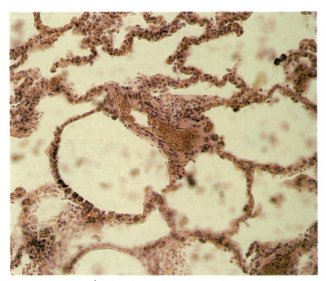

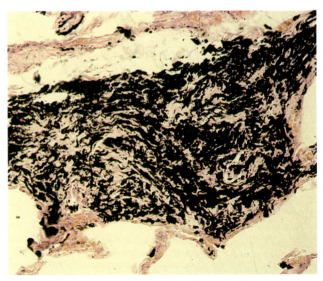

Figure 12.2I ● (a) Section of lung taken from a healthy person

(b) Section of lung taken from a coal miner with a lung disease called pneumoconiosis. Notice the deposits of coal dust in the lung tissue

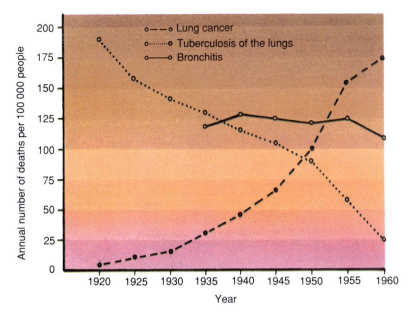

Figure 12.2J ● Deaths from lung disease in England and Wales from 1916-60. Deaths from lung cancer increased sharply when deaths from other forms of lung disease (in this case tuberculosis) were falling

Smoking is harmful: the evidence and attitudes

Research into the links between smoking and disease was prompted by the sort of data shown in Figure 12.2J.

Scientists soon suspected a correlation (relationship) between smoking cigarettes and **lung cancer** but could not prove the link completely. However, new data strengthened the case against cigarettes.

- When doctors saw the early evidence many of them gave up smoking cigarettes. Deaths from lung cancer among doctors went down compared with the population as a whole.

- Studies clearly established the relationship between the risk of dying from lung cancer and number of cigarettes smoked – the more cigarettes smoked, the greater the risk (Figure 12.2L).

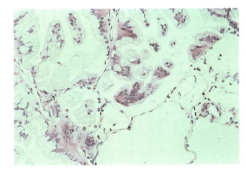

Figure 12.2K ● (a) Healthy lung tissue

(b) Cancerous lung tissue

People affected by **cystic fibrosis** produce too much mucus in the lungs. The mucus blocks the alveoli and bronchioles. Regular physiotherapy helps improve the person's breathing difficulties. Excess mucus also blocks the pancreatic duct (see p. 204). Reduced release of pancreatic juice with its digestive enzymes leads to inadequate digestion. The trapped enzymes may start to digest the pancreas itself.
Why can people affected in this way develop diabetes (see p. 313)? Cystic fibrosis is caused by a recessive allele.
Why are only people who are homozygous recessive for cystic fibrosis affected by the condition (see pp. 345, 346)?

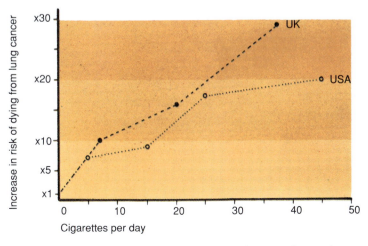

Figure 12.2L ● Death rates from lung cancer in men who smoke

● *Why are cigarettes dangerous?*

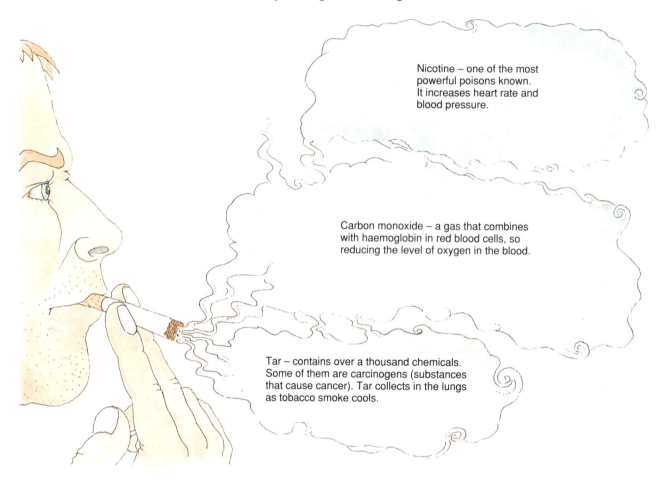

Nicotine – one of the most powerful poisons known. It increases heart rate and blood pressure.

Carbon monoxide – a gas that combines with haemoglobin in red blood cells, so reducing the level of oxygen in the blood.

Tar – contains over a thousand chemicals. Some of them are carcinogens (substances that cause cancer). Tar collects in the lungs as tobacco smoke cools.

Figure 12.2M ● Dangers to health from the chemicals in cigarette smoke

LOOK AT LINKS
The link between smoking and heart disease is covered in Topic 14.5.

A lighted cigarette produces a number of substances. Many of them are harmful (Figure 12.2M). Some irritate the membrane lining the upper respiratory tract. Others stop the cilia from beating (Figure 12.2F). Extra mucus (phlegm) forms in the trachea and bronchi, causing 'smoker's cough'; it is the only way to get rid of the build up of phlegm. Smoking also weakens the walls of the alveoli and repeated coughing can destroy some of them. This breakdown of the alveoli is called **emphysema** (Figure 12.2N). *Why does a person with emphysema become breathless and exhausted easily?*

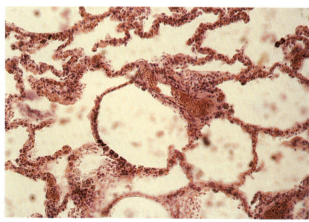

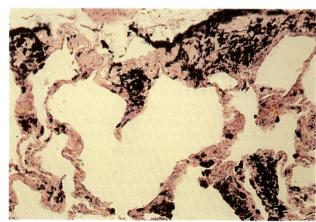

Figure 12.2N ● (a) Healthy alveoli

(b) Alveoli destroyed by emphysema

LOOK AT LINKS
for for development of the baby in the uterus
See Topic 15.5.

Pregnant women are always advised not to smoke. The chemicals in cigarette smoke enter the mother's bloodstream and reach the developing baby across the placenta.

Babies born to mothers who smoke are generally lighter than babies born to mothers who do not smoke and there is an increased risk of premature birth. It seems that chemicals in cigarette smoke prevent the baby from getting all the nourishment she or he needs from the mother. This is in addition to the other dangers listed in Figure 12.2M.

Children see parents, older brothers and sisters, pop stars and film stars smoking and decide that smoking is the 'grown up' thing to do, so they copy them. Other children copy them, and so on (Figure 12.2O). This is the way smoking usually starts and once 'hooked' it is difficult to give up because the nicotine in tobacco is habit-forming.

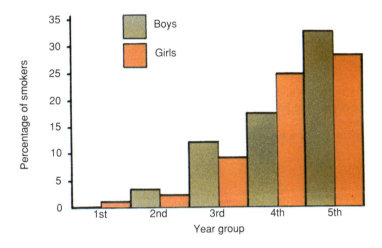

Figure 12.2O ● The smoking behaviour of girls and boys at secondary school. *Can you explain why girls who smoke outnumber boys who smoke in the fourth year?*

● **Living sensibly**

In 1971 The Royal College of Physicians said:

'Premature death and disabling illnesses caused by cigarette smoking have now reached epidemic proportions and present the most challenging of all opportunities for preventive medicine …

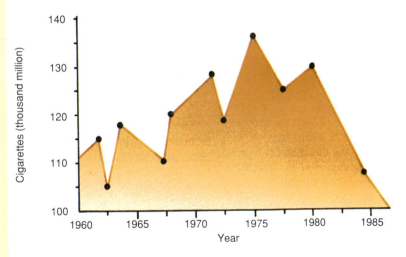

Figure 12.2P ● Drop in the sales of cigarettes in the UK. Health education and heavy taxes on tobacco have persuaded people to give up smoking

SUMMARY

Smoking cigarettes is a major cause of lung disease, especially lung cancer. It has taken scientists many years to establish the link. Fewer people smoke today than previously but many people continue to damage their health by smoking cigarettes.

Since then the campaign against cigarette smoking has been fought hard. Cigarette sales have gone up and down, but the overall trend is downwards (Figure 12.2P). Smoking is now banned in some public places, because non-smokers also suffer increased risks of ill health when they breathe in smoke from other people's cigarettes. This is called 'passive smoking'.

Between 1972 and 1984 cigarette smoking in Britain dropped by almost a third. In 1982, for the first time, there were fewer smokers than non-smokers. However, among women deaths from lung cancer continue to increase. Cigarette manufacturers are now putting their efforts into exports to the Third World! Cigarette smoking in these countries is increasing, and so too are deaths from lung cancer. Think about this and look at the evidence once again. The message is clear: **DO NOT START SMOKING**.

Gills

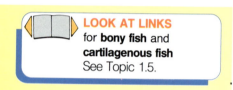

LOOK AT LINKS
for **bony fish** and **cartilagenous fish**
See Topic 1.5.

Figure 12.2Q looks inside the mouth of a fish. Notice the series of openings on either side. They are the **gill slits**. In **bony** fish, such as cod, herring and goldfish, each series of gill slits is covered by a flap of bony tissue called the **operculum** (plural: opercula). **Cartilagenous** fish such as shark and skate do not have opercula. The gill slits are easily seen as Figure 17.2D on page 318 shows.

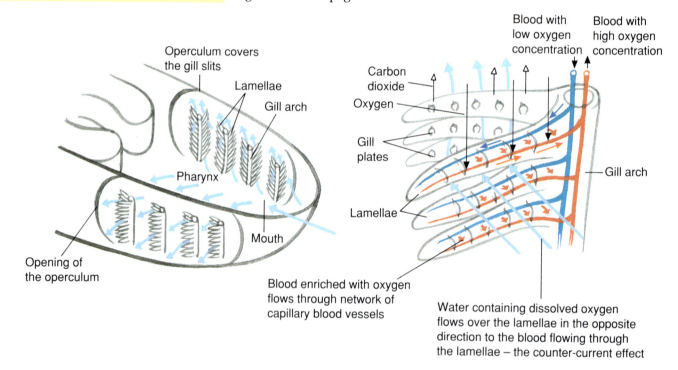

Figure 12.2Q ● Horizontal section through the head of a fish. The arrows indicate the direction of water current and blood flow. Part of a gill is shown in more detail

Also notice in Figure 12.2Q that a partition separates each gill slit from the next in line. Each partition is strengthened by a thin bar of bone called a **gill arch**. A gill consists of rows of leaf-like tissue (called **lamellae**) projecting from either side of the gill arch. Each lamella is folded into **gill plates**, greatly increasing the surface area of the gills overall. Blood vessels at the base of each gill branch into dense networks of capillaries in the lamellae.

The lamellae at work

As water passes over the lamellae, dissolved oxygen diffuses into the blood flowing through the capillary network. At the same time carbon dioxide diffuses out of the blood and is carried away in the water current. *Where in the water and in the blood flowing through the gills are oxygen and carbon dioxide in greatest concentration?*

Exchange of the gases is helped because the flow of blood through the lamellae and the flow of water over the lamellae are in opposite directions. *How does this **counter-current** effect make the diffusion of oxygen from water to the blood and diffusion of carbon dioxide from blood to the water more efficient?*

The respiratory current

The flow of water through the mouth and over the gills is called the **respiratory current**. Figure 12.2R shows what happens. The co-ordinated opening and shutting of the mouth and opercula results in a continuous flow of water over the gill surfaces. Water cannot enter the gut because muscles at its upper end contract, closing off the opening.

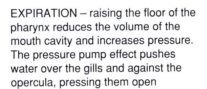

SUMMARY

Dissolved oxygen and carbon dioxide in solution are exchanged between water and the blood in the capillary blood vessels which run through the lamellae of the gills. A counter-current effect between water and blood enhances exchange of the gases.

EXPIRATION – raising the floor of the pharynx reduces the volume of the mouth cavity and increases pressure. The pressure pump effect pushes water over the gills and against the opercula, pressing them open

RESPIRATORY CURRENT

INSPIRATION – lowering the floor of the pharynx increases the volume of the mouth cavity and reduces pressure. Water flows in through the open mouth and over the gills

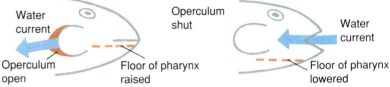

Water current

Operculum open

Floor of pharynx raised

Operculum shut

Water current

Floor of pharynx lowered

Figure 12.2R ● Respiratory current – inspiration and expiration are continuous processes

CHECKPOINT

❶ What is a gas exchange surface?

❷ Write a brief statement against each of the following items to show how efficient alveoli are as gas exchange surfaces:
a) large surface area
b) short diffusion distance
c) good blood supply
d) moist
e) in contact with air

❸ Complete the following paragraph using the words given. Each word may be used more than once or not at all.

uptake   respire   inhalation   oxygen   alveoli
energy   exchange   moist   exhalation

The _____ of oxygen and removal of carbon dioxide occur in the _____ of the lungs. These provide a large surface area (about 90 m²) for efficient gas _____. They are thin-walled, have an excellent blood supply, are _____ and kept well supplied with air by breathing. Air is taken into the lungs by _____ and removed by _____.

❹ Why do you think people should not smoke in the presence of a pregnant woman?

❺ Discuss how gills are adapted for the efficient exchange of gases between water and the blood of a fish.

❻ Describe the pressure changes taking place during inspiration and expiration of the respiratory current in fish.

TOPIC 13 EXCRETION

13.1 Getting rid of wastes

Metabolism produces waste substances, which must be removed from the body. How is it done? This section will tell you.

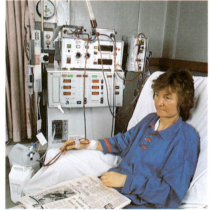

The person in Figure 13.1A is attached to a machine that has taken over the work of her kidneys. This is one way of treating people whose kidneys are not working (kidney failure).

The kidney machine removes waste substances from the patient's blood. Figure 13.1B explains how it works.

LOOK AT LINKS
Transplant rejection is an example of the immune reaction at work.
See Topic 14.3.

Figure 13.1A ● The patient must use the kidney machine two or three times a week. Each treatment takes about ten hours

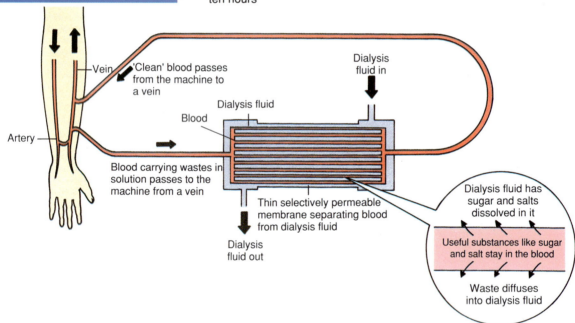

Vein — 'Clean' blood passes from the machine to a vein

Dialysis fluid in

Dialysis fluid

Blood

Artery

Blood carrying wastes in solution passes to the machine from a vein

Thin selectively permeable membrane separating blood from dialysis fluid

Dialysis fluid out

Dialysis fluid has sugar and salts dissolved in it

Useful substances like sugar and salt stay in the blood

Waste diffuses into dialysis fluid

Figure 13.1B ● A kidney machine at work. A thin semi-permeable membrane separates the patients blood from the dialysis fluid. Wastes from the blood diffuse into the dialysis fluid. The process is called dialysis. *Explain why wastes diffuse out of the blood but sugars and salts do not?*

Donor Card
I would like to help someone to live after my death
Keep the card with you at all times in a place where it will be found quickly

Figure 13.1C ● Donor card

A person with kidney failure may have a kidney transplant. A healthy kidney is taken from a person (the donor) who has just died (Figure 13.1C), or from someone living who wants to help the patient. Live donors are very often close relatives of the patient because their body tissues are similar, which reduces the chances of the patient's body rejecting the transplant. We have two kidneys, but it is possible to live a healthy life with only one.

LOOK AT LINKS
for **metabolism**
See Topic 10.2.

Where do wastes come from?

Thousands of chemical reactions take place inside a living cell. These reactions are called **metabolism**.

Compounds are broken down and new ones are made. The waste products formed could be harmful to the body if they were allowed to accumulate (Table 13.1). Removal of these waste substances is part of the process of excretion.

Table 13.1 ● Waste substances produced by metabolism

Waste substances	How waste is made	Where waste is made	Where waste is excreted
Carbon dioxide	Cellular respiration	All cells	From lungs or other gas exchange surfaces
Urea and other compounds containing nitrogen	Deamination of amino acids	Liver cells	From the kidney
Bile pigments	Breakdown of haemoglobin	Liver cells	In bile which passes into the duodenum
Oxygen	Photosynthesis	Inside the chloroplasts of plant cells	From leaf cells through the stomata

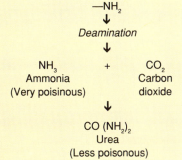

IT'S A FACT

Urea forms when **amino acids** break down in liver cells. The flow chart shows what happens to the amino group ($-NH_2$). The process is called **deamination**.

$$-NH_2$$
$$\downarrow$$
Deamination
$$\downarrow$$

NH_3 + CO_2
Ammonia   Carbon
(Very poisonous)   dioxide

$$\downarrow$$

$CO(NH_2)_2$
Urea
(Less poisonous)

Water and mineral salts are needed for cells to work properly. However, water and salts in excess of the body's needs must be excreted. In humans and other mammals the kidney is the main excretory organ.

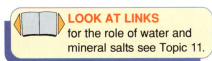

LOOK AT LINKS
for **amino acids**
See Topic 10.4.

LOOK AT LINKS
for the role of water and mineral salts see Topic 11.

The kidneys at work

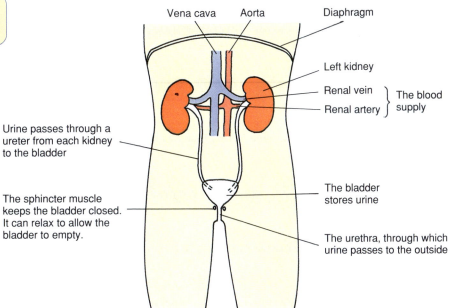

Figure 13.1D ● The position of the kidneys in the human body

Vena cava   Aorta   Diaphragm

Left kidney

Renal vein ⎫ The blood
Renal artery ⎭ supply

Urine passes through a ureter from each kidney to the bladder

The sphincter muscle keeps the bladder closed. It can relax to allow the bladder to empty.

The bladder stores urine

The urethra, through which urine passes to the outside

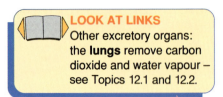

LOOK AT LINKS
Other excretory organs:
the **lungs** remove carbon
dioxide and water vapour –
see Topics 12.1 and 12.2.

Figure 13.1D shows you where the kidneys are in the human body. Each one consists of about one million tiny tubules called **nephrons**. The nephron is the working unit of the kidney (Figure 13.1E). Blood is brought to the kidneys by the renal arteries. It contains wastes (mostly urea) which the nephrons remove from the blood along with glucose, salts and other substances in solution. Some of these are useful and the nephron reabsorbs them into the blood. In this way the composition of blood is kept constant (**homeostasis**).

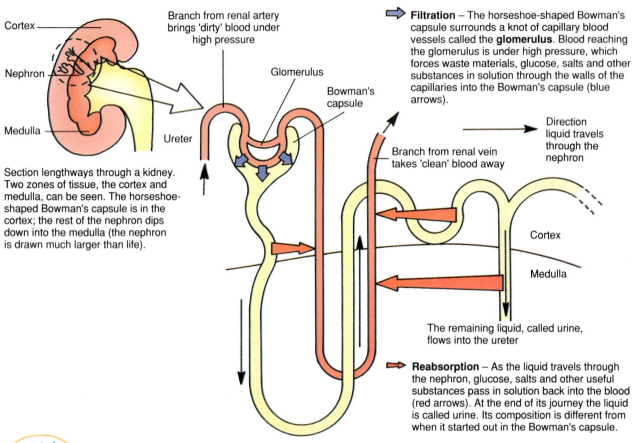

Cortex

Nephron

Medulla

Ureter

Section lengthways through a kidney. Two zones of tissue, the cortex and medulla, can be seen. The horseshoe-shaped Bowman's capsule is in the cortex; the rest of the nephron dips down into the medulla (the nephron is drawn much larger than life).

Branch from renal artery brings 'dirty' blood under high pressure

Glomerulus

Bowman's capsule

Branch from renal vein takes 'clean' blood away

Filtration – The horseshoe-shaped Bowman's capsule surrounds a knot of capillary blood vessels called the **glomerulus**. Blood reaching the glomerulus is under high pressure, which forces waste materials, glucose, salts and other substances in solution through the walls of the capillaries into the Bowman's capsule (blue arrows).

Direction liquid travels through the nephron

Cortex

Medulla

The remaining liquid, called urine, flows into the ureter

Reabsorption – As the liquid travels through the nephron, glucose, salts and other useful substances pass in solution back into the blood (red arrows). At the end of its journey the liquid is called urine. Its composition is different from when it started out in the Bowman's capsule.

Figure 13.1E ● A kidney nephron

IT'S A FACT

The organs of excretion in insects are called malpighian tubules. They are slender tubes bathed in the blood which fills the inside of the insect's body. Each tube is closed at one end and opens at the other into the intestine. Wastes in the blood are removed by the malpighian tubes and converted into uric acid, which is excreted through the anus.

Look at Table 13.2. Notice the composition of blood in the glomerulus, before treatment by the nephron, and the composition of the liquid (called urine) at the end of its journey through the nephron.

Table 13.2 ● Composition of the blood in the glomerulus and the urine of a healthy adult

	Blood in the glomerulus	Urine
	(percentage by mass)	
Water	91.7	96.5
Proteins	7.5	0
Urea	0.03	2
Ammonia	trace	0.05
Sodium ions (Na^+)	0.3	0.6
Potassium ions (K^+)	0.02	0.15
Chloride ions (Cl^-)	0.36	0.6
Glucose	0.1	0

LOOK AT LINKS
for **homeostasis**
See Topic 16.5.

These questions refer to Table 13.2.

1 What is the mass of water in 100 cm³ of urine?

2 Why do you think there are no proteins in urine?

3 What percentage of glucose does the nephron reabsorb into the blood?

4 Name the main waste product formed from the deamination of amino acids.

13.2 Excretion in plants

Plants do not have specialised structures like nephrons for the excretion of waste. Oxygen, which is a waste product of photosynthesis, and carbon dioxide, which is a waste product of respiration, diffuse out through the leaves. In land plants excess water which is a waste product of respiration is removed by evaporation. If more water is taken up by the roots than can be removed by evaporation, then the excess oozes from the leaves.

IT'S A FACT

In tropical rainforests where humidity is high, some plants have special structures round the edges of their leaves to get rid of excess water. Why is excess water not removed from rainforest plants by evaporation?

Plants dispose of wastes by combining them with inorganic salts to form insoluble crystals. The wastes are stored in this state in different parts of the plant (Figure 13.2A).

SUMMARY

The kidney is the chief organ of excretion in humans and other mammals. It is made up of many tubules called nephrons. Each nephron removes wastes produced by metabolism and substances in excess of the body's needs. Plants excrete wastes by storing them in harmless forms in different parts of the plant body. Wastes stored in leaves are lost when the leaves fall in autumn.

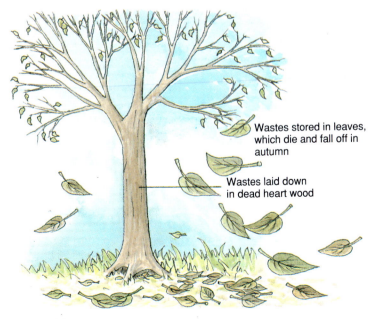

Wastes stored in leaves, which die and fall off in autumn

Wastes laid down in dead heart wood

Figure 13.2A ● Plant waste disposal

CHECKPOINT

❶ The diagram shows the position of the kidneys in the body and their connections with blood vessels and the bladder.
(a) Name the parts labelled A – G.
(b) Briefly explain why the concentration of urea in A is much less than in B.
(c) What is the role of E?
(d) Where is urine stored before it is released to the outside?

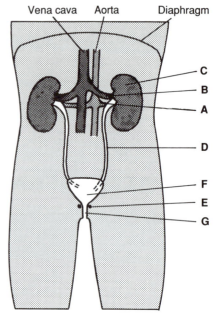

❷ Copy the table below. It lists parts of the kidney and its connections to blood vessels and the bladder. Tick each substance present in the part listed in the left hand column (assume the person is healthy).

	Protein	Glucose	Urea	Water	Salt
Blood in the renal artery					
Blood in glomerulus					
Fluid that passes into the Bowman's capsule					
Urine in the bladder					
Blood in the renal vein					

❸ State three ways in which blood leaving the kidney is different from blood entering the kidney.

❹ Match the lettered statements with the appropriate numbered parts.

A takes urine from the kidney to the bladder
B prevents urine leaving the bladder
C removes urea from the blood
D carries urine out of the body
E carries blood to the kidney
F stores urine
G carries blood away from the kidney

1 sphincter
2 renal artery
3 renal vein
4 kidney
5 urethra
6 ureter
7 bladder

❺ Complete the following paragraph about the kidney, using the words provided. Each word may be used once, more than once, or not at all.

Bowman's capsule   ureter   cortex   one million   medulla
renal vein   bladder   glomerulus

A kidney cut lengthways has two regions, an outer _____ and an inner _____. Under a microscope many tiny nephrons are visible: there are about _____ in each kidney. Each nephron has a cup-shaped capsule called the _____ at one end. This surrounds a small bunch of capillaries called the _____. Leading away from each capsule is a narrow tubule which twists and turns and eventually joins the _____, which takes urine from the kidney to the _____.

❻ What is dialysis and why do some people need it?

TOPIC 14

TRANSPORT IN LIVING THINGS

14.1 Transport in plants

FIRST THOUGHTS

How do adequate supplies of water and food reach all parts of a plant? A transport system is the answer! In this section we shall discuss the systems which transport water, minerals and food in a plant.

Water and food move into and out of plant cells by **osmosis**, **diffusion** and **active transport**. Flowering plants require special tissues for the transport of these materials to where they are needed. The tissues are xylem and phloem (see Figure 14.1A). Xylem transports water and phloem transports food. Xylem and phloem tissue form vascular bundles which is why flowering plants, along with conifers and ferns, are sometimes referred to as vascular plants.

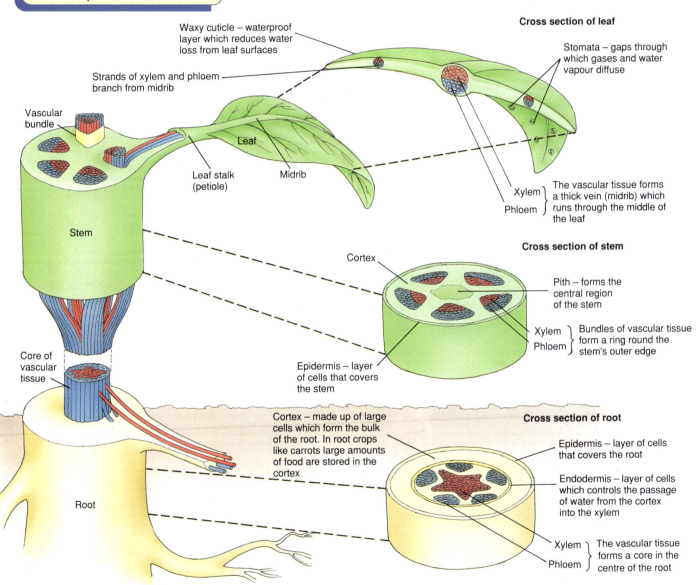

Cross section of leaf

Waxy cuticle – waterproof layer which reduces water loss from leaf surfaces

Stomata – gaps through which gases and water vapour diffuse

Strands of xylem and phloem branch from midrib

Vascular bundle

Leaf

Leaf stalk (petiole)

Midrib

Xylem
Phloem } The vascular tissue forms a thick vein (midrib) which runs through the middle of the leaf

Stem

Cross section of stem

Cortex

Pith – forms the central region of the stem

Xylem
Phloem } Bundles of vascular tissue form a ring round the stem's outer edge

Core of vascular tissue

Epidermis – layer of cells that covers the stem

Cortex – made up of large cells which form the bulk of the root. In root crops like carrots large amounts of food are stored in the cortex

Cross section of root

Epidermis – layer of cells that covers the root

Endodermis – layer of cells which controls the passage of water from the cortex into the xylem

Root

Xylem
Phloem } The vascular tissue forms a core in the centre of the root

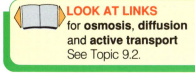

LOOK AT LINKS
for **osmosis**, **diffusion** and **active transport**
See Topic 9.2.

Figure 14.1A ● The arrangement of xylem and phloem (vascular tissue) in a flowering non-woody plant. (Notice that the central core of vascular tissue in the root changes to a ring of bundles in the stem at ground level.)

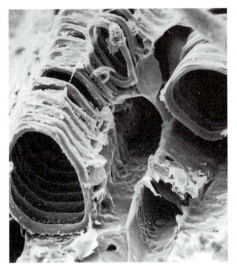

Figure 14.1B ● Xylem in a stem – notice the rings of lignin, a substance which strengthens and waterproofs the cell walls of each vessel

RESOURCE
–ACTIVITY–
PACK

Xylem transports water

Mature xylem cells are dead. They are cylindrical and join end-to-end to form strands of xylem tissue. The cross-walls separating adjoining cells break down so that each xylem strand becomes a long, hollow vessel through which water can pass freely (see Figure 14.1B). Xylem vessels run from the tips of roots, through the stem and out into every leaf as Figure 14.1A shows.

Water enters the plant through the root hairs by osmosis. It moves across the root cortex by osmosis and diffusion into the xylem vessels. Once in the xylem vessels, water forms unbroken columns from the roots, through the stem and into the leaves. Water evaporates from the leaves – mainly through the tiny pores in the underside of the leaf called stomata (see Figure 14.1C). The process is called **transpiration**.

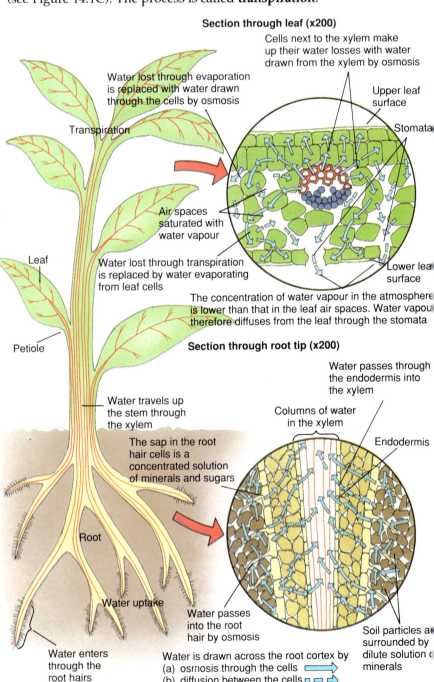

Section through leaf (x200)

Cells next to the xylem make up their water losses with water drawn from the xylem by osmosis

Water lost through evaporation is replaced with water drawn through the cells by osmosis

Upper leaf surface

Transpiration

Stomata

Air spaces saturated with water vapour

Leaf

Water lost through transpiration is replaced by water evaporating from leaf cells

Lower leaf surface

The concentration of water vapour in the atmosphere is lower than that in the leaf air spaces. Water vapour therefore diffuses from the leaf through the stomata

Petiole

Section through root tip (x200)

Water passes through the endodermis into the xylem

Columns of water in the xylem

Water travels up the stem through the xylem

The sap in the root hair cells is a concentrated solution of minerals and sugars

Endodermis

Root

Water uptake

Water passes into the root hair by osmosis

Soil particles are surrounded by dilute solution of minerals

Water enters through the root hairs

Water is drawn across the root cortex by
(a) osmosis through the cells
(b) diffusion between the cells

Figure 14.1C ● The transpiration stream

LOOK AT LINKS
for turgor
See Topic 9.2.

Cacti are adapted to survive in hot dry environments. Water loss is reduced by:

- a thick cuticle covering the surfaces
- leaves which are often reduced to spines
- deep pits in the plant's surface surrounding the stomata
- a shiny surface which reflects light and heat
- very thick stems storing water and shallow wide-spread roots.

Explain how each of the adaptations listed help cacti conserve water. Find out about the unusual method cacti have of obtaining carbon dioxide for photosynthesis.

As water transpires, more is drawn from the xylem. This movement of water exerts 'suction' on the water filling the xylem vessels of the stem. As the water is 'sucked' upwards through the xylem of the stem, more water is supplied to the bottom of the xylem by the roots. There is therefore a continuous moving column of water from the roots to the leaves. This is called the transpiration stream (see Figure 14.1C).

The force that drives the transpiration stream is equivalent to as much as thirty atmospheres pressure – sufficient to move water to the top of the tallest tree.

● *Factors affecting transpiration*

If the loss of water through transpiration is not made up by intake of water from the soil, then the stomata close. The closing of the stomata reduces transpiration. However, if the plant still does not get enough water, then its cells beginto lose **turgor** and the plant wilts.

Other factors in the environment also affect the rates of transpiration and water intake. For example, plants transpire more quickly (and therefore absorb more water from the soil) when it is light than when it is dark. Light stimulates the stomata to open wide. This means that water vapour can transpire more easily. Figure 14.1D shows the effects of other environmental factors on the rate of transpiration.

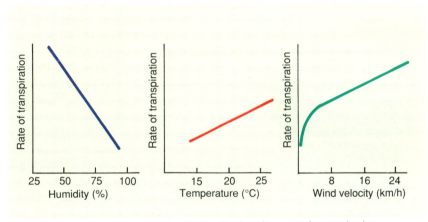

Figure 14.1D ● Environmental factors affecting the rate of transpiration

Phloem transports food

Phloem tissue consists of different types of cell. **Sieve cells** and **companion cells** are the most important. Figure 14.1E shows that sieve cells are cylindrical and joined at their end walls to form strands of tissue called sieve tubes.

Mature sieve cells do not have nuclei. However, they are alive: each cell contains cytoplasm. The end walls, called sieve plates, are pierced with holes (hence their name). A companion cell lies next to each sieve cell. Companion cells and sieve cells work together to

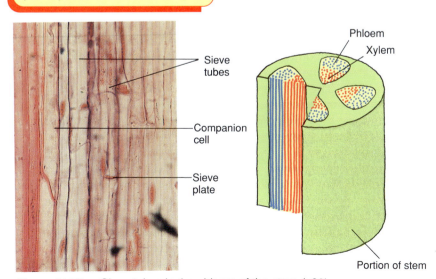

Figure 14.1E ● Sieve tubes in the phloem of the stem (x20)

transport sugars to where they are needed in the plant. The movement of sugars and other substances from one region to another through the sieve cells is called **translocation**.

Which surface of a leaf transpires the most water? Cobalt chloride paper is blue when dry and pink when wet. Take a suitable plant and sandwich blue cobalt chloride papers against the upper and lower surfaces of some of its leaves. Record the time taken for each cobalt chloride paper to change colour. Explain your results.

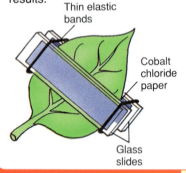

Thin elastic bands

Cobalt chloride paper

Glass slides

Control of the size of the stoma depends on the two sausage-shaped guard cells which surround the opening. Unlike the other cells of the lower leaf surface, guard cells contain chloroplasts. During the day, photosynthesis increases the concentration of sugar in the guard cells compared with neighbouring cells. There is a net flow of water by osmosis (see p. 143) into the guard cells, which become turgid. At night time, photosynthesis stops and the concentration of sugar in the guard cells drops. There is a net outflow of water by osmosis and the guard cells lose turgor.

Guard cells open the stoma when turgid and close it when they lose turgor. Opening and closing depends on the hoops of cellulose fibres which encircle each guard cell like a bandage round the finger. When water enters the guard cells, the hoops prevent outward expansion, so the cells expand lengthways. Because the guard cells are attached at each end, lengthways expansion forces them to bow out, opening the stoma.

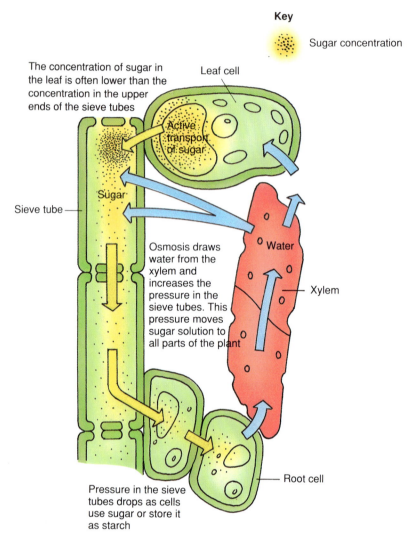

Key

Sugar concentration

The concentration of sugar in the leaf is often lower than the concentration in the upper ends of the sieve tubes

Leaf cell

Active transport of sugar

Sieve tube

Sugar

Osmosis draws water from the xylem and increases the pressure in the sieve tubes. This pressure moves sugar solution to all parts of the plant

Water

Xylem

Root cell

Pressure in the sieve tubes drops as cells use sugar or store it as starch

Figure 14.1F ● Translocation depends on the differences in concentration of sugar in different parts of the plant. Diffusion transports sugar from where it is in high concentration to where it is in low concentration.

Figure 14.1F shows how translocation takes place. Sugar passes from the leaf cells to the sieve cells by active transport. Once in the sieve tubes, the sugar, along with other substances, is carried to where it is needed. Sugar that passes to the roots can move out of the sieve tubes and into the cells of the cortex. The sugar is converted into starch and stored.

Scientists understood 300 years ago that xylem transports water in plants. *How do we know that phloem transports food?* The evidence is more recent. Two of the experiments that provided the answer are summarised below:

● **Tree ringing experiments**   A ring of bark is cut from the stem of a woody plant. The strip of bark removed is just thick enough to contain phloem tissue from the vascular bundles. The xylem remains intact. After a few days there is a bulge of growth above the ring and no further growth below the ring. *Why do you think there is extra growth above the ring? Why does this result suggest that food is transported in the phloem? What will happen to the plant eventually?* (See Figure 14.1G overleaf.)

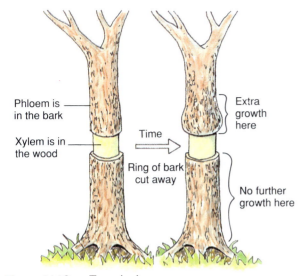

Phloem is in the bark

Xylem is in the wood

Time

Ring of bark cut away

Extra growth here

No further growth here

Figure 14.1G ● Tree ringing

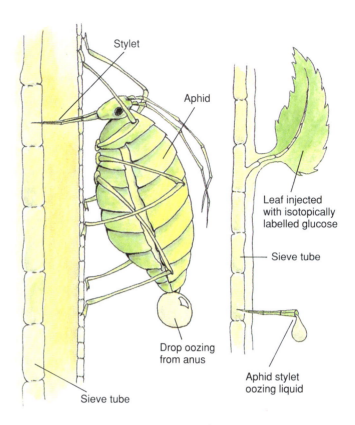

Stylet

Aphid

Drop oozing from anus

Sieve tube

Leaf injected with isotopically labelled glucose

Sieve tube

Aphid stylet oozing liquid

Figure 14.1H ● Aphids and isotopic tracers

IT'S A FACT

Do you know why a car parked under a lime tree is soon covered with a sticky substance? Aphids feeding on the tree sprinkle the juice which passes through them onto the car beneath.

● **Aphids and isotopic tracers** The aphid can be used to study the transport of sugar in the phloem sieve tubes. To feed, the aphid inserts its tube-like mouthparts (called a stylet) through the surface of a stem into a sieve tube. The pressure of the liquid in the sieve tube force-feeds the aphid to the point where drops ooze from its anus. When an aphid is cut from its stylet, liquid will continue to ooze from the stylet for several days. Isotopically labelled glucose is injected into a leaf and samples of liquid are collected regularly from a newly placed stylet. *What test would you do to find out if the liquid on which the aphid was feeding contains sugar? If sugar is transported in the phloem what change would you expect in the samples of liquid obtained from the stylet? How would you detect the change?* (see Figure 14.1H.)

CHECKPOINT

❶ With the help of diagrams describe the differences in structure between xylem and phloem.

❷ Distinguish between transpiration and translocation.

❸ Describe the passage of a water molecule from the soil through a plant and out into the atmosphere.

❹ Why does a plant lose more water on a dry, warm, windy day than when it is cool, still and wet?

❺ By what route does sugar made in the leaves move to the roots?

❻ Atmospheric pressure can support a column of water 11 metres high. How tall could a tree grow and water still reach its topmost branches?

14.2 **Blood and its functions**

Blood transports gases, food and other vital materials to the tissues of the body. It protects the body from disease by producing antibodies and forming clots. This topic explains how the blood in the human body performs these functions and many others beside.

IT'S A FACT

Anybody who is healthy, weighs over 50 kg and is between the ages of 18 and 65 can give blood. There are about five litres of blood in the body of an adult human. A donor may give up to half a litre of blood at one time. That is one tenth of the body's total blood content.

When a person is seriously injured or undergoes a major operation they often lose a lot of blood. The blood they lose is replaced by **blood transfusion** with blood from blood donors. A needle is inserted into a vein in the arm of the donor and blood flows through the needle and into a sterilised bottle via a tube. The blood is mixed with sodium citrate to stop it from clotting and stored at 5 °C in regional blood banks. When a hospital needs blood it should be available from its regional blood bank.

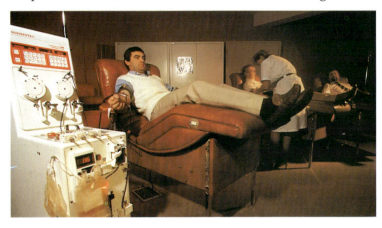

Figure 14.2A ● Giving blood

What is blood made of?

Figure 14.2B shows what happens when a sample of blood is spun for a time in a centrifuge. The blood separates into a straw coloured liquid called **plasma** and a dark red–brown mass of blood cells.

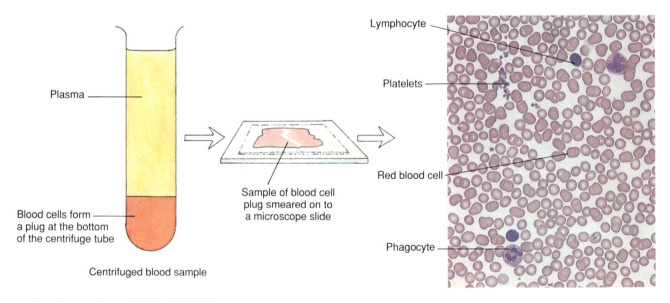

Plasma

Blood cells form a plug at the bottom of the centrifuge tube

Centrifuged blood sample

Sample of blood cell plug smeared on to a microscope slide

Lymphocyte

Platelets

Red blood cell

Phagocyte

LOOK AT LINKS
for **centrifuging**
See *KS: Chemistry*, Topic 11.5.

Figure 14.2B ● Human blood sample spun in a centrifuge and examined under a microscope. (Notice the characteristic shapes of the nuclei of phagocytes and lymphocytes.)

IT'S A
FACT

One millilitre of blood contains
3000 white blood cells and
5 000 000 red blood cells.

Sickle-cell anaemia is caused by a
mutation in the genes controlling
the synthesis of haemoglobin. The
amino acid valine replaces
glutamic acid. The substitution
results in a drastic reduction in the
oxygen-carrying capacity of the
haemoglobin.

LOOK AT LINKS
for **cellular respiration**
See Topic 12.1.

The plasma transports heat released by metabolism in the liver,
muscles and fat to other parts of the body. The plasma consists of:
* Water – 90% by volume.
* Blood proteins – these include antibodies that help to protect the body
  from disease and fibrinogen, one of the proteins that helps blood to
  clot.
* Foods, vitamins and minerals (see Topic 11).
* Urea (see Topic 13).
* Hormones – substances which help to co-ordinate different body
  functions (see Topic 16).

The red plug of blood cells at the bottom of the test-tube consists of:
* Red blood cells – the cells contain the red pigment haemoglobin that
  gives blood its colour. Red blood cells do not have nuclei.
* White blood cells – two types, lymphocytes and phagocytes. The
  nucleus of each type of white blood cell has a characteristic shape (see
  Figure 14.2B).
* Platelets – these look like fragments of red cells.

Red cells transport oxygen and carbon dioxide

Red cells are made in the marrow of the limb bones, ribs and vertebrae.
Old red cells are destroyed in the liver. Red cells contain the protein
haemoglobin. Haemoglobin readily combines with oxygen in tissues
where the concentration of oxygen is high to form **oxyhaemoglobin**.
Oxyhaemoglobin breaks down to release oxygen in tissues where the
concentration of oxygen is low.

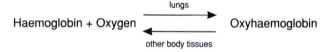

Haemoglobin + Oxygen ⇄ Oxyhaemoglobin

lungs

other body tissues

Blood which contains a lot of oxyhaemoglobin is called **oxygenated
blood** and is bright red in colour (see Figure 14.2C). Blood with little
oxyhaemoglobin is called **deoxygenated blood** and looks a deep
red–purple.

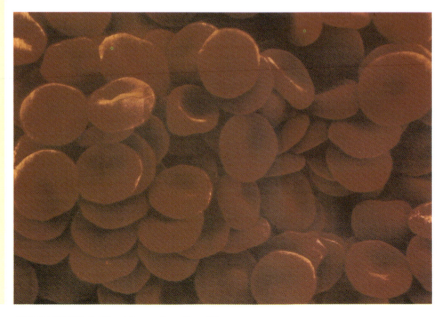

Figure 14.2C ● Oxygen makes the difference

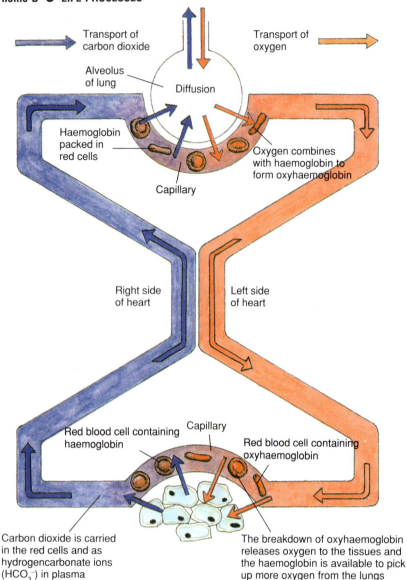

Transport of carbon dioxide

Transport of oxygen

Alveolus of lung

Diffusion

Haemoglobin packed in red cells

Oxygen combines with haemoglobin to form oxyhaemoglobin

Capillary

Right side of heart

Left side of heart

Red blood cell containing haemoglobin

Capillary

Red blood cell containing oxyhaemoglobin

Carbon dioxide is carried in the red cells and as hydrogencarbonate ions (HCO$_3^-$) in plasma

The breakdown of oxyhaemoglobin releases oxygen to the tissues and the haemoglobin is available to pick up more oxygen from the lungs

Figure 14.2D ● Transport of oxygen and carbon dioxide between the lungs and tissues of mammals

Cellular respiration releases carbon dioxide which diffuses into the blood. Some of it forms hydrogencarbonate ions (HCO$_3^-$) in the plasma. Haemoglobin carries the rest of the carbon dioxide in the red cells. When blood reaches the lungs, carbon dioxide diffuses from the plasma and red cells into the alveoli and is exhaled (see Figure 14.2D).

White cells protect the body

Some types of virus and bacteria enter (infect) the blood and tissues and cause disease. The two types of white blood cell, lymphocytes and phagocytes, protect the body by working quickly to destroy viruses and bacteria. Lymphocytes and phagocytes also destroy any other cells or substances which the body does not recognise as its own. Materials 'foreign' to the body are called **antigens**. When antigens come into contact with lymphocytes they stimulate the lymphocytes to produce proteins called **antibodies** which begin the process of destruction. The phagocytes finish the job. Figure 14.2E shows what happens.

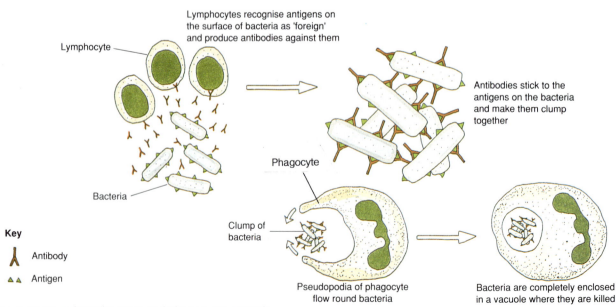

Lymphocytes recognise antigens on the surface of bacteria as 'foreign' and produce antibodies against them

Lymphocyte

Antibodies stick to the antigens on the bacteria and make them clump together

Bacteria

Phagocyte

Clump of bacteria

Key

Antibody

Antigen

Pseudopodia of phagocyte flow round bacteria

Bacteria are completely enclosed in a vacuole where they are killed

Figure 14.2E ● Lymphocytes and phagocytes at work

LOOK AT LINKS
Where white blood cells
are made in the body is
discussed in Topic 14.4.

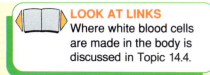

● Antibodies are specific

Antibodies produced against a particular antigen will attack only that antigen. The antibody is said to be **specific** to that antigen. This means that antibodies produced against typhoid bacteria will not attack pneumonia bacteria. It seems that each of us can produce tens of millions of different antibodies to deal with all the antigens we are ever likely to meet in a life time.

● Active immunity

The action of lymphocytes and phagocytes against invading microorganisms is called an **immune reaction**. The antibodies produced may stay in the body for some time ready to attack the same microorganisms when next they invade the body. Even if the antibodies do not stay in the body for long, they are soon made again because the first-time battle between microorganisms and lymphocytes primes the lymphocytes to recognise the same microorganisms next time. This means that re-infection is dealt with by immune reactions which are even faster and more effective than the first reaction. In other words you become resistant. This is why you rarely catch diseases like chicken pox and measles more than once.

Immunisation

Immunisation promotes active immunity to disease-causing microorganisms. It involves the doctor or nurse giving you an injection or asking you to swallow some substance. The substance injected or swallowed is called a vaccine and the process of being immunised is called immunisation or vaccination.

Vaccines are made from one of the following:
* Dead microorganisms, e.g. whooping-cough vaccine is made from dead bacteria.
* A weakened form of microorganism which is harmless. Vaccines made like this are called attenuated vaccines, e.g. the vaccine against tuberculosis and Sabin oral vaccine (the vaccine against poliomyelitis) are both attenuated vaccines.
* A substance from the microorganism which does not cause the disease, e.g. diphtheria vaccine.

LOOK AT LINKS
Genetic engineering is
used to produce new
vaccines
See Topic 20.

What effect does a vaccine have? Antigens from the dead or attenuated microorganisms in the vaccine stimulate the lymphocytes to produce antibodies. So, when the same active, harmful microorganisms invade the body, the antibodies made in response to the vaccine destroy them.

The active immunity produced by vaccines can protect a person from disease for a long time, although several more vaccinations called boosters may be needed after the first one. Boosters keep up the level of antibodies and so maintain a person's immunity.

Children in the UK are immunised against six diseases which used to cause many deaths. They are diphtheria, tetanus, whooping cough, poliomyelitis, tuberculosis and German measles. These diseases are now rare but those that spread through person-to-person contact would soon increase if the number of people vaccinated against them fell to levels where infections easily spread among unprotected individuals.

● Whooping cough and German measles

Whooping cough is a highly infectious disease caused by the bacterium *Bordetella pertussis*. The patient suffers from a wracking cough with characteristic 'whoops' which may last for 2–3 months. Whooping cough may make the patient vulnerable to bronchopneumonia and also increases the risk of brain damage. These complications and deaths from the disease are most common in babies under six months old.

Whooping cough vaccine is a suspension of killed *Bordetella pertussis*. It is usually given with diphtheria and tetanus vaccines in a triple vaccine.

Vaccination against whooping cough started in 1957. Before then about 100 000 cases of the disease were reported in the UK each year. By 1973 more than 80% of the population had been vaccinated and the number of annual cases had fallen to approximately 2400. However, there was a scare over the safety of the vaccine and vaccinations fell to around 30% in 1975. Epidemics of whooping cough followed in 1977–79 and 1981–83. Methods of producing the vaccine were improved. A publicity campaign pointed out the advantages of vaccination and helped to restore public confidence so that vaccination rose to 67% in 1986. This was enough to halt further epidemics.

Side-effects after injection of whooping cough vaccine occur in about one in 100 000 injections. Symptoms include fever and headache. In a very few cases there can be permanent brain damage. The fever soon goes away but the slight possibility of brain damage causes anxiety amongst parents. For this reason a double vaccine is available that contains diphtheria and tetanus vaccines but not the vaccine for whooping cough. Parents who choose the double vaccine have to decide whether the chances of their child catching whooping cough are less than the chances of their child suffering brain damage through having the vaccine.

If large numbers of children are not vaccinated against whooping cough then the level of protection for the population as a whole falls and outbreaks of the disease increase. The 'fors' and 'againsts' for whooping cough vaccination show how difficult it is to balance individual well-being and freedom of choice against what is good for everybody. *Where do you think the balance lies?* The information in Table 14.1 will help you to make a choice.

Table 14.1 ● Comparison of complications and side-effects with whooping cough and triple vaccine

Result of vaccination or illness	Vaccination with triple vaccine (incidence per 100 000 vaccinations)	Whooping cough (incidence per 100 000 cases)
Death	0.2	4000
Permanent brain damage	0.6	2000
Inflammation of the brain	3.0	4000
Convulsions	90	8000

The case for vaccination against the virus which causes German measles (rubella) is more straightforward, especially for girls. The vaccine does not have the same level of risk compared with whooping cough vaccine.

German measles in children is not serious. The body, arms, legs and face may be covered with pink spots and at worst there may be a fever and the lymph glands may swell up. These symptoms clear up after two or three days. However, it is much more serious if a woman catches German measles during the first four months of pregnancy. The virus can cross the placenta and affect the developing baby. The baby may be born dead or blind or deaf or with a damaged nervous system. Up to 90% of babies whose mothers catch the disease when pregnant are affected. This is why it is important for girls to be vaccinated against German measles before they have babies.

IT'S A FACT

Sub-unit vaccines are designed to stimulate immune reactions without the unwanted side-effects of whole vaccines. They are made from particular proteins or small parts of disease-causing microorganisms.

● Passive Immunity

Not all vaccines contain antigens which stimulate the body to produce antibodies. Instead, antibodies can come ready-made from other animals. For example, anti-tetanus vaccine contains anti-tetanus antibodies produced by horses. The bacterium which causes tetanus lives in the soil and multiplies very rapidly in places where there is little air, such as in a deep wound. The bacterium produces a lethal poison which acts so quickly that the body's lymphocytes do not have time to make antibodies against it. This is why if you have a deep dirty cut you should be injected with vaccine containing anti-tetanus antibodies. These can act immediately to stop the disease from developing. Immunity which comes from antibodies made in another animal is called **passive immunity**.

Although it is short-lived, passive immunity is important for babies. While they are breast-feeding, babies receive antibodies from their mother's milk which protect them from disease-causing microorganisms. By the time this protection wears off, the baby is able to make its own antibodies.

ABO Blood Grouping
(program)

Find out how blood tests are carried out, using the program. The program takes you through the taking of blood samples, testing, transfusions, how blood forms clots, and the genetics of the ABO system.

Blood groups

Although red blood cells all look alike under the microscope, they may carry different antigens called antigen A and antigen B on the cell surface. Plasma contains antibodies which attack foreign red cell antigens but does not contain antibodies which would attack a person's own red cell antigens. The possible combinations of antibody and antigen are shown in Table 14.2. These determine which blood group a person belongs to.

Table 14.2 ● Blood groups

Antigen on red cells	Antibody in plasma	Blood group	Percentage of UK population with blood group
A	Anti-B	A	40
B	Anti-A	B	10
A and B	Neither	AB	5
Neither	Anti-A and anti-B	O	45

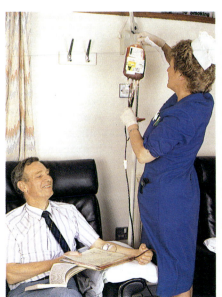

Figure 14.2F shows a patient having a transfusion. Blood at the right temperature is fed at the correct rate from the bag through the tube into a vein in the arm.

Before a person receives a blood transfusion it is important to know that the donor's blood group is compatible with that of the patient. If it is not then the donor's red blood cells clump in the patient's blood vessels and cause serious harm. Table 14.3 shows which blood groups are compatible. If a

Figure 14.2F ● A patient receiving a blood transfusion

donor's blood causes the patients red blood cells to clump, their blood groups are said to be incompatible.

Table 14.3 ● Blood transfusion: compatibility between donor and patient

Group	Donate to	Receive from
A	A and AB	A and O
B	B and AB	B and O
AB	AB	All groups
O	All groups	O

Group O people are called universal donors. *Why can they give blood to anybody?* Group AB people are universal recipients. *Why can they receive blood from anybody?*

Usually whole blood (plasma and cells) is given but sometimes plasma only is used to restore blood volume, especially in cases of serious burns or blood loss. It takes several weeks for the patient to make new blood cells to replace the ones lost.

Platelets help to stop bleeding

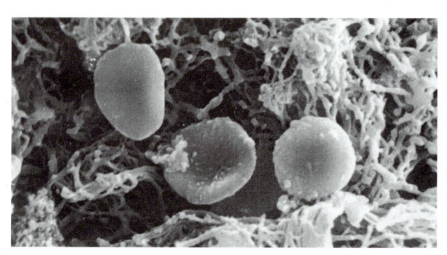

When you are cut a series of complex events begins which eventually stops the bleeding. A clot begins to form when platelets are damaged by the rough surface caused by a cut or torn tissue. They release a substance which results in the formation of an insoluble protein called fibrin . Fibrin forms a mesh of fibres across the wound that traps red cells, forming a clot (see Figure 14.2G).

Figure 14.2G ● A blood clot: red cells trapped by a mesh of fibrin fibres

Disorders of the blood

● Leukaemia

Leukaemia is a condition in which large numbers of immature white blood cells are produced and released into the blood stream. Overproduction of abnormal white cells results in the formation of too few red cells and other blood cells.

The microscopic examination of blood smears and bone marrow smears is an important tool in the diagnosis of leukaemia. The number of white cells and their appearance are helpful indicators to the presence of the disease.

Leukaemia can be treated with drugs and by radiotherapy. For some types of leukaemia the prospects of improvement and recovery in a patient are poor, but other types respond very well to treatment. For example, in about 90% of cases involving children with certain kinds of leukaemia, treatment stops the disease developing further and improves their chance of a healthy life. More than 50% of patients are cured.

IT'S A FACT

The drug vincristine has helped to improve the treatment of leukaemia. The drug is extracted from the rosy periwinkle plant which grows in the rain forests of Madagascar.

LOOK AT LINKS
The inheritance of haemophilia and other genetic diseases is described in Topic 19.3.

● Haemophilia

Some people lose a lot of blood if they injure themselves because their blood does not clot properly. This disease is called haemophilia. In the most common form of the disease the blood lacks a substance called **factor VIII** which is one of the substances involved in the production of fibrin. Haemophiliacs are treated with injections of factor VIII. Haemophilia is a **genetic disease** that runs in families. Some of the royal families of Europe carry the genes for haemophilia.

● AIDS

AIDS (**A**cquired **I**mmune **D**eficiency **S**yndrome) is caused by a virus called the **H**uman **I**mmunodeficiency **V**irus (HIV) (see Figure 14.2H). The virus attacks the lymphocytes which play an important part in the body's defence against disease. This means that the body of a person with HIV is far less well protected than normal. It is not usually HIV itself which causes the suffering of AIDS patients: they may die from another infection, often a type of pneumonia, which develops after HIV has destroyed part of the body's defences. We call AIDS a **syndrome** as there are a variety of symptoms which an HIV-infected person may have. Many of these symptoms are the result of other infections.

Unlike many other viral infections, controlling AIDS presents a

The test for **AIDS** depends on detecting antibodies to **HIV** in blood samples. However, diseases such as tuberculosis and multiple sclerosis trigger the production of antibodies which can give a **false-positive** result with the 'Aids test'. Also drug abusers, haemophiliacs who require multiple blood transfusions and people exposed to a variety of infections all at once may give a positive result because the immune system is weakened, even though they are not infected with HIV. These uncertainties lend support to the scientists who think that HIV is not the true cause of AIDS.

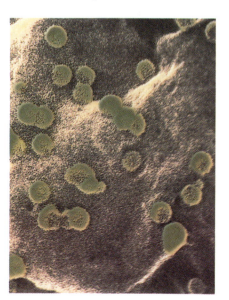

particular difficulty. There is a long time interval (from a few months to several years) between the virus getting into a person's body and symptoms developing. During this period the person is infectious and able to pass on the virus to other people without knowing it.

Figure 14.2H ● Human immunodeficiency virus (HIV)

CHECKPOINT

❶ List the different types of blood cell.

❷ What is meant by 'oxygenated blood' and 'deoxygenated blood'?

❸ What is an antigen?

❹ Why are diseases like chicken pox and measles rarely caught more than once?

❺ What is the difference between a vaccine and a vaccination?

❻ What is an attenuated vaccine?

❼ How do booster vaccinations help to maintain a person's immunity to disease?

❽ What is the triple vaccination?

⑨ Briefly explain why it is important for women to be vaccinated against German measles before they have babies.

⑩ What is the difference between active immunity and passive immunity?

⑪ Briefly explain how the presence or absence of antigen A and antigen B determines a person's blood group.

⑫ What is blood serum?

⑬ Before a blood transfusion is given, why is it important to know that the donor's and patient's blood are compatible?

14.3 Understanding immunology

LOOK AT LINKS
Immunology is outlined in section 14.2. Re-reading pages 236 to 239 will remind you of important words and essential background information.

Immunology is the science concerned with the processes and mechanisms that establish specific immunity in the body. These processes and mechanisms make up the body's **immune response** which is summarised in Figure 14.3A. Notice that there are two phases: a **primary response** when the antigen first invades the body, and a **secondary response** should the same antigen invade the body again. The secondary response depends upon a specific **immunological memory**.

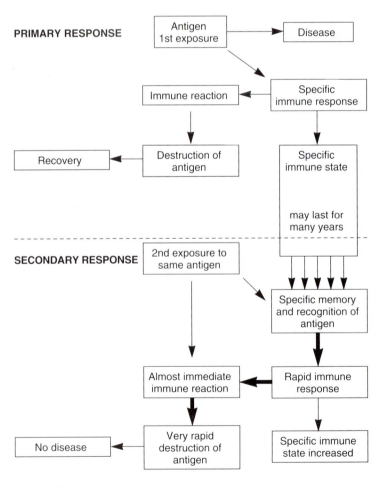

Figure 14.3A ● The body's immune response

Action of lymphocytes and phagocytes

Lymphocytes are the white cells in the blood that recognise and react to antigens. They originate in the bone marrow, which is a yellow fatty material that fills the hollow centre of the bone shaft.

There are two categories of lymphocyte: **B-cell** lymphocytes and **T-cell** lymphocytes. Contact with an antigen stimulates B-lymphocytes and T-lymphocytes to attack it. The B-lymphocytes divide and produce **antibodies** specific to the antigen that triggers the response. The antibodies circulate in the blood and are some of the **plasma proteins** listed on p. 235. Figure 14.2E on p. 236 shows B-lymphocytes at work.

T-cell lymphocytes do not produce antibodies. They bind with an antigen and divide to form a variety of cells that have different functions. Some of these cells, called **T-helper** cells, control the production of antibodies by B-lymphocytes. Others, called **T-cytotoxic** cells, destroy cells which are infected with a virus.

Phagocytes are white cells in the blood that behave rather like a 'sanitation squad'. Figure 14.2E on p. 236 shows that some types of phagocyte engulf and destroy bacteria which have been attacked by antibodies. Others migrate through the tissues of the body to deal with microorganisms which gain entry through cuts, scratches or other openings in the skin. This causes an **inflammatory** action with swelling, redness and heat at the site of infection. Still others damage large parasites like round worm (see Figure 3.7A on p. 61) which are too large to be engulfed.

Lymphocytes and phagocytes work together to keep most of us healthy for most of our lives. They quickly destroy viruses, bacteria or any other cells or substances which infect the blood or tissues and which the body does not recognise as its own.

Immunological memory

The body takes a few days to produce antibodies against a first-time infection (primary response). However, second-time-round the reaction is much quicker (secondary response). B-lymphocyte and T-lymphocyte memory cells left over from the first-time division of lymphocytes (called **clonal expansion**) are activated. They quickly divide (a second clonal expansion) on re-exposure to the same antigen.

Memory cells are specific for a particular antigen. It is due to the action of memory cells that we do not usually catch mumps or chicken pox more than once in a life-time. The rapid response of immunological memory destroys the viruses before they make us ill.

Tissue transplants and rejection

A person who suffers serious burns may need a **skin graft** to prevent infection and loss of water from the exposed areas. If the skin is taken from another part of the victim's body, then the 'new' skin grows and joins up with the skin surrounding the burn. After a time the affected area is as good as new. However, if skin is taken from another person (except the victim's identical twin), tissue rejection sets in after a few days and the 'new' skin dies. Microscopic examination shows the presence of large numbers of T-lymphocytes and phagocytes in the failed graft. These cells seem to be responsible for the rejection. *Why do other transplanted tissues and organs suffer the same fate unless steps are taken to prevent rejection?*

IT'S A FACT

Antibodies are proteins of a type called **immunoglobins** (abbreviated to **Ig**). In some people dust and pollen are antigens that stimulate the lymphocytes to produce a type of immunoglobin called **IgE**. This **allergic** response results in the symptoms of **hay fever**.

LOOK AT LINKS
for **interferon**
See Topic 20.5.

LOOK AT LINKS
for **HIV**
See Topic 14.2.

IT'S A FACT

Human Immunodeficiency Virus (HIV) attacks T-helper cells.
Why do you think HIV attacking T-helper cells reduces the efficiency of the body's response to infections?

LOOK AT LINKS
The division of lymphocytes is called clonal expansion because the daughter cells (see page 147) are genetically identical to their parents and each other (see p. 256).

IT'S A FACT

The most frequent types of transplant are:
- skin – treatment for burns
- kidney – alternative treatment to dialysis (see page 224) in cases of kidney failure
- heart – treatment for heart failure
- bone marrow – treatment for children with incurable blood diseases
- liver – treatment for liver failure.

Like viruses and bacteria, mammal cell membranes carry an enormous variety of antigens. In humans these antigens are called **human lymphocyte antigens** abbreviated to **HLA**. Even closely related people rarely have identical types of HLA. If an organ from one person (e.g. the heart) is transplanted to another person, the recipient's T-lymphocytes are activated if the donor's HLA type is different from the recipient's. T-lymphocytes attack the transplanted organ and cause **transplant rejection**.

Preventing rejection

Different methods reduce the chances of rejection:

- **Tissue typing** using **monoclonal antibodies** identifies the different HLA in the donor and recipient. Surgeons only undertake transplantation if the HLA are very similar or match. *Why does tissue typing reduce the chances of rejection?*
- **Immunosuppressive drugs** are used to prevent an immune response even if the HLA of donor and recipient do not match. **Cyclosporine** from the fungus *Trichoderma polysporum* is less poisonous than other immunosuppressive drugs to other tissues of the body. Given at the time of transplant, it prevents the recipient's T-cell lymphocytes (remember they are culprits in tissue rejection) from acting against the antigens in the transplanted tissue. Because cyclosporine and other immunosuppressive drugs reduce the effectiveness of the immune system, the recipient is more susceptible to infections. If infection occurs, then the drugs are stopped. However, there is a danger that immune reactions causing rejection of the transplant may then develop.

LOOK AT LINKS
for **monoclonal antibodies**
See Topic 20.5.

SUMMARY

The body produces specific antibodies against the millions of different antigens it meets in a life-time. This immune response is remembered by memory cells should an antigen invade the body again. The action of T-lymphocytes and phagocytes is probably responsible for the rejection of tissue transplants.

LOOK AT LINKS
for **smallpox**
See Topic 20.5.

LOOK AT LINKS
for **immunisation** and **vaccines**
See Topic 14.2.

IT'S A FACT

'Vaccination' comes from the Latin word *vacca* meaning a cow. The word 'immunisation' is sometimes used instead of 'vaccination'.

Who's behind the science

Edward Jenner (1749–1823) was a British country doctor living in Gloucestershire. He did not understand immunology as we do. However, he was a good scientist who tested ideas formed from everyday experience. He learnt from local farmers that milkmaids who caught the mild disease **cowpox** from handling cows, rarely caught the much more serious and often fatal disease **smallpox**. During an outbreak of smallpox in the neighbourhood, Jenner deliberately infected several of his patients with cowpox. The patients soon developed cowpox but were not affected by smallpox.

Jenner took the experiment a dangerous step further. He infected a boy who had just recovered from cowpox with pus from the spots of someone suffering from smallpox. The boy did not develop smallpox. His survival added weight to earlier ideas that giving a person a mild dose of a disease protects against more serious forms of the disease.

Jenner published his results in 1798 and the work established **vaccination** (immunisation) as a powerful weapon in the fight against disease. At first people were suspicious and it took time for the technique to be accepted. Now smallpox has been eliminated from all parts of the world and **vaccines** protect millions of people from a variety of diseases.

CHECKPOINT

❶ Distinguish between the following pairs:
antibody	antigen
primary immune response	secondary immune response
lymphocyte	phagocyte

❷ What happens when lymphocytes recognise antigens?

3 Distinguish between the activities of B-cell lymphocytes and T-cell lymphocytes during an immune response.

4 The diagram shows an immune complex formed when an antibody and antigen stick together. Find out why immune complexes only form between antibodies and the antigen for which the antibodies are specific.

5 What is inflammatory reaction?

6 Briefly summarise the basis of immunological memory.

7 Discuss the biology behind the precautions taken to reduce the chances of rejection of tissue transplants.

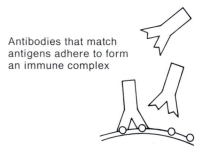

Antibodies that match antigens adhere to form an immune complex

---

14.4 The blood system

> *Why do we need a blood system?* Small animals like flatworms do not have one. Osmosis, diffusion and active transport carry gases, minerals and other materials to where they are needed in the flatworm's body. However, large animals like humans require a system specialised to carry materials from one part of the body to another. This is the role of the blood system.

How does the heart work? What is the role of arteries, veins and capillaries? Find out the answer to these questions in this section.

Blood vessels

The blood system consists of a network of tubes called blood vessels through which the heart pumps blood. The major blood vessels are the **arteries** and the **veins**. Table 14.4 compares arteries and veins.

Table 14.4 ● Arteries and veins compared. The smooth lining to the vessels helps bloodflow and prevents clots from forming.

Arteries	Veins
Thick outer wall / Narrow diameter / Thick layer of muscles and elastic fibres	Fairly thin outer wall / Large diameter / Thin layer of muscles and elastic fibres
Carry blood away from the heart to organs and tissues	Return blood to the heart from organs and tissues (except hepatic portal vein)
Blood at high pressure	Blood at low pressure. Body muscles squeeze the veins to help push the blood to the heart
Have a pulse because the vessel walls expand and relax as blood spurts from heart	Do not have a pulse since blood flows smoothly
Have thick walls to withstand pressure of blood	Have thin wall and large diameter reducing resistance to the flow of blood returning to the heart

Blood in the veins is at a much lower pressure than in the arteries. One-way valves inside the veins prevent blood from flowing backwards (see Figure 14.4A). The force of the heart beat keeps blood flowing away from the heart through the arteries, so there is no need for valves inside them.

The arteries and veins in the human body form two circuits: the lung circuit and the head and body circuit (see Figure 14.4B). The veins of the head and body bring deoxygenated blood to the heart. The heart pumps deoxygenated blood through the pulmonary arteries to the lungs, where it is oxygenated. The oxygenated blood returns to the heart through the pulmonary veins completing the lung circuit. The heart then pumps the oxygenated blood through the arteries of the head and body circuit.

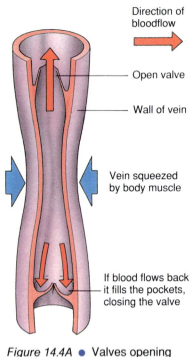

Figure 14.4A ● Valves opening and closing inside a vein

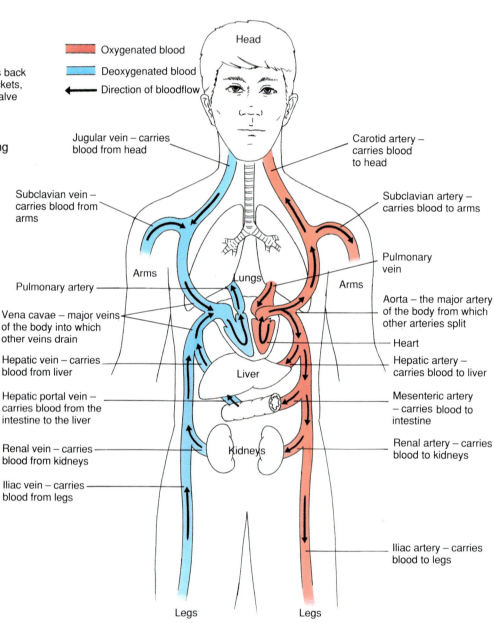

Figure 14.4B ● The lung circuit and the head and body circuit. Oxygen passes from tissue to blood and carbon dioxide passes from blood to tissue in the lungs. Oxygen passes from blood to tissue and carbon dioxide passes from tissue to blood in all other organs and tissues.

● *Capillaries*

Small blood vessels branch from the main arteries and veins. The vessels branching from arteries are called **arterioles**, those branching from veins are called **venules**. Arterioles and venules branch further to form capillaries. Capillaries join arterioles to venules and so link arteries and veins. Capillaries form dense networks, called beds, in the tissues of the body. This means that no cell is very far away from a capillary. (See Figure 14.4C.)

Capillaries are tiny vessels only 0.001 mm in diameter with walls one cell thick. The blood in capillaries supplies the nearby cells with the food, oxygen and other materials they need. The blood also carries away urea, carbon dioxide and other wastes produced by the cells' metabolism.

The pumping of the heart brings blood at high pressure to the arteriole end of the capillary network. The pressure forces plasma through the thin capillary walls. The liquid is now called tissue fluid and carries the food and oxygen to the cells which are not in direct contact with a capillary. (See Figure 14.4D.)

There is just enough room in the smallest capillaries for red cells to pass in single file. A lot of plasma is forced through the one-cell-thick walls as the red cells squeeze through the capillaries. The pressure drops as blood passes through the capillaries to the venule end of the bed. Tissue fluid can then seep back into the capillaries along with dissolved urea and carbon dioxide. Most tissue fluid returns to the blood by this route. The remaining small amount of fluid drains into the lymph vessels.

Figure 14.4C ● A capillary network

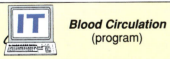
Blood Circulation
(program)

You can build up a complete picture of the blood circulation system using this program. Look at the models of heart structure. The program shows the way that blood flows through the heart.

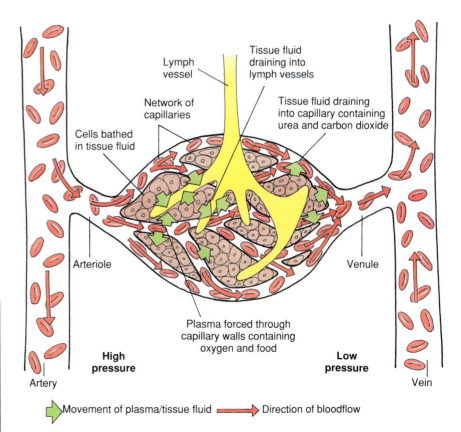

Figure 14.4D ● Capillaries at work

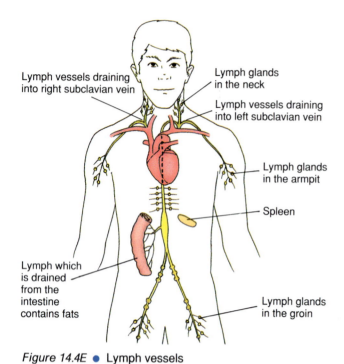

Lymph vessels draining into right subclavian vein

Lymph glands in the neck

Lymph vessels draining into left subclavian vein

Lymph glands in the armpit

Spleen

Lymph which is drained from the intestine contains fats

Lymph glands in the groin

Figure 14.4E ● Lymph vessels

The lymph vessels

The lymph vessels return fluids to the blood system that would otherwise collect in the tissues. The smallest lymph vessels are the size of capillary blood vessels; the largest are the size of veins. Capillary lymph vessels are blind-ended tubes which form a network in the body's tissues. Tissue fluid diffuses through their walls and slowly passes into the larger lymph vessels. Figure 14.4E shows how the lymph vessels are arranged in the human body. The fluid, now called lymph, is moved by the contraction of the body's muscles during normal daily activity. Valves in the lymph vessels, similar to those found in veins, prevent the backflow of fluid.

The spleen is also part of the lymph vessel system. Lymphocytes are made in the spleen and in the lymph glands. The spleen also collects damaged and old red blood cells, breaking them down and releasing the haemoglobin in them. Iron in the haemoglobin is removed and re-used by bone marrow to make new haemoglobin.

If the circulation of lymph is upset by disease or poor diet, then fluid may gather in the tissues and cause swelling. This is called **oedema**.

The human heart

The heart is a pump which propels blood through the arteries and veins. It is made of a type of muscle called **cardiac muscle** which contracts and relaxes rhythmically for a lifetime. The more efficient the heart is, the more efficient are the exchanges of food, oxygen, carbon dioxide and other dissolved materials between the blood and the tissues of the body.

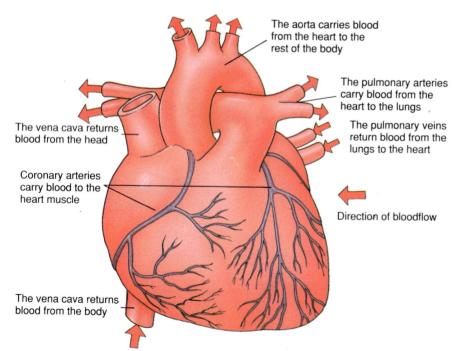

The aorta carries blood from the heart to the rest of the body

The pulmonary arteries carry blood from the heart to the lungs

The pulmonary veins return blood from the lungs to the heart

The vena cava returns blood from the head

Coronary arteries carry blood to the heart muscle

The vena cava returns blood from the body

Direction of bloodflow

Figure 14.4F ● View of the heart from the front

● Heart structure

The heart lies in the chest cavity surrounded by a membrane called the **pericardium**. It is protected by the rib cage. Figure 14.4F and G show the heart and the blood vessels leading to and from it.

The heart is full of blood and it seems odd that heart muscle needs an additional blood supply. However, like other tissues, heart muscle needs a steady supply of the food and oxygen dissolved in blood. The walls of heart muscle are so thick that these materials cannot diffuse quickly enough from inside the heart to all of the heart muscle. So, some of the heart muscle is supplied with blood by the coronary arteries.

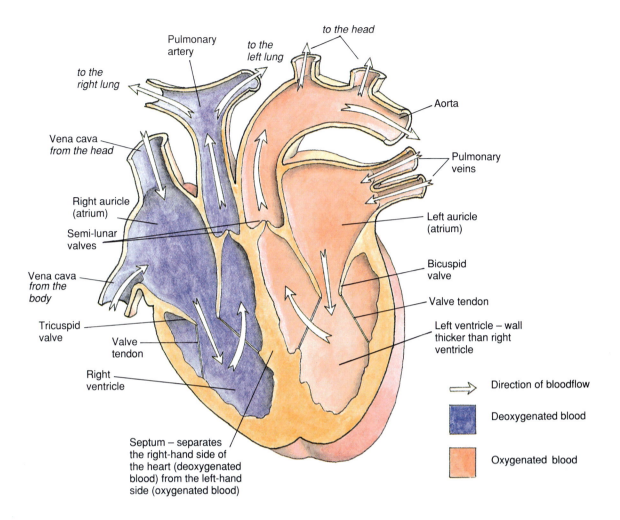

Figure 14.4G ● A cross-section of the heart viewed from the front

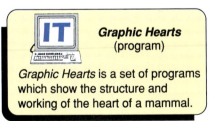

Graphic Hearts
(program)

Graphic Hearts is a set of programs which show the structure and working of the heart of a mammal.

≡ **SCIENCE AT WORK** ≡

Cancer cells that have broken loose from a cancer growth are often caught in the lymph glands. Lymph glands near to cancer growths under treatment are therefore usually removed by surgery.

● *Flow of blood through the heart*

When all the heart muscles relax (called diastole):

• deoxygenated blood from the head and body enters the right atrium through each vena cava,

• oxygenated blood from the lungs enters the left atrium through the pulmonary veins.

The atria fill with blood and then contract (called auricular systole). The increase in pressure opens the tricuspid and bicuspid valves and forces blood into the ventricles. When full, the ventricles contract (called ventricular systole). The increase in pressure closes the tricuspid and bicuspid valves and forces blood past the semi-lunar valves which guard the openings of the pulmonary artery and aorta. The right ventricle pumps blood into the pulmonary artery on its way to the lungs; the left ventricle pumps blood into the aorta which takes it round the rest of the body. The semi-lunar valves close when the ventricles relax.

In effect the heart is a double pump, each ventricle pumping blood along a different route through the body. Look back at Figure 14.4B. The distance travelled by blood through the head and body circuit is greater than the distance it travels through the lung circuit. This difference explains why the wall of the left ventricle is thicker than the wall of the right ventricle. Its contractions must be more powerful to pump blood the greater distance.

● The heartbeat

Diastole and systole produce an unmistakable two-tone sound which is easily heard through an instrument such as a stethoscope. This sound is the heartbeat. In a healthy adult the heart beats on average 72 times a minute but this can vary between 60 beats per minute and 80. Exercise makes the heart beat faster, bringing more blood to the muscles.

The beating of the heart is controlled by the pacemaker, which is a group of special cells in the right atrium. Occasionally the natural pacemaker goes wrong. The heart rate slows down causing drowsiness and shortage of breath. When the heart's natural pacemaker does not function properly an electronic pacemaker can be fitted which helps to keep the heart beating at the proper rate.

● The pulse

Each heart beat sets up a ripple of pressure which passes along the arteries. The ripple can be felt as a 'pulse' as the artery's muscular wall expands and relaxes.

Feeling a patient's pulse can help doctors and nurses to tell if the heart is beating properly. The neck pulse and wrist pulse are the most useful pulses (see Figure 14.4H). The pulse is felt with the finger tips pressed lightly over the artery (the thumb is not used since this has its own pulse). The number of beats in a minute are counted using the second hand of a watch.

Use a pulse sensor to measure your pulse rate. Compare the readings when you are relaxed and breathing easily with when you have exercised hard. You may be able to make the sensor beep in time to your heart beat.

RESOURCE
ACTIVITY
PACK

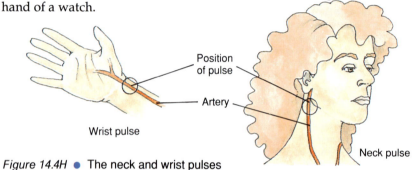

Position of pulse

Artery

Wrist pulse

Neck pulse

Figure 14.4H ● The neck and wrist pulses

Blood pressure

Figure 14.4I ● Measuring blood pressure with a sphygmomanometer

Blood flows along the blood vessels because it is under pressure. We call this **blood pressure**. It is measured with an instrument called a **sphygmomanometer** (see Figure 14.4I). An inflatable armband is wrapped around the patient's upper arm. The armband is inflated until it stops the blood from flowing along the main artery of the arm. The air is then let out of the armband very slowly while the doctor listens for the return of the pulse with a stethoscope placed below the armband. At the point when the blood moves back into the artery below the armband and the pulse returns, the pressure in the armband just equals the pressure of the blood.

LOOK AT LINKS
Blood pressure is measured in millimetres of mercury (mm Hg) rather than the SI unit of pressure, the pascal (Pa). A column of mercury 100 mm high exerts a pressure of 13 300 Pa at its base, so 1 mm Hg = 133 Pa. For more details, see *KS: Physics*, Topic 16.

Two readings are taken: the pressure of blood when the heart contracts (called systolic blood pressure) and the pressure of the blood when the heart relaxes (called diastolic blood pressure). The normal systolic pressure of a young adult is about 120 mm Hg; normal diastolic pressure is about 75 mm Hg.

Constant high blood pressure is harmful. It makes the heart work harder. Eventually the overworked heart may fail altogether. High blood pressure can also damage the kidneys and eyes, and increase the risk of an artery tearing open. A cerebral haemorrhage, or stroke, occurs when one of the arteries that supplies blood to the brain ruptures.

CHECKPOINT

❶ Briefly explain the significance of the difference between arteries and veins.

❷ How does blood, rich in digested food absorbed from the intestine, reach the liver?

❸ Explain the function of the valves in the veins and in the lymph vessels.

❹ What is oedema?

❺ Briefly describe the function of lymph glands.

❻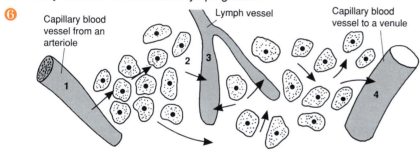

The diagram shows the movement of tissue fluid, plasma and lymph between capillary blood vessels, tissue cells and lymph vessels.
(a) Choose the word which correctly names the fluid in each of the locations numbered 1–4.
    water   lymph   salt solution   plasma   tissue fluid
(b) Explain why fluid leaves the capillary blood vessel at 1 but enters the capillary blood vessel at 4.

❼ Describe the route that blood takes from the time it enters the heart to the time it reaches the lungs.

❽ What are diastole and systole?

❾ What does 'heart rate' mean?

❿ What is meant by 'blood pressure'?

⓫ The diagram shows a section through the human heart viewed from the front.
(a) Name the parts of the heart labelled A–H.
(b) Which of the following sequences of letters describes the path taken by blood through the heart?
    G B A E H C D F   F D C H E A B G
    E A B G H C D F   H C D F G B A E
(c) Which two parts of the diagram show that chambers A and C are relaxed?
(d) Why is the wall of chamber D thicker than the wall of chamber B?

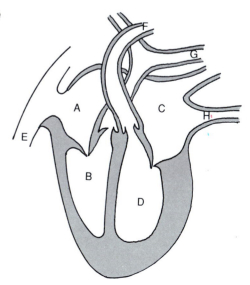

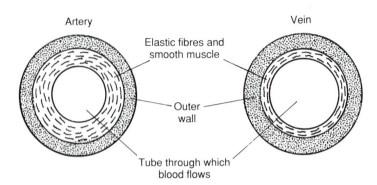

Artery

Vein

Elastic fibres and
smooth muscle

Outer
wall

Tube through which
blood flows

The diagram shows sections through an artery and a vein.
(a) Give two differences in structure between an artery and a vein.
(b) Briefly explain how the differences you have given in (a) are linked to the way in which arteries and veins work.
(c) Complete the following:
An artery carries blood from the _____ at _____ pressure.
A vein carries blood to the _____ at _____ pressure.
Arteries and veins are linked by tiny blood vessels called capillaries which are about _____ mm in diameter and have walls _____ _____ thick. Capillaries form at the end of small blood vessels branching from the arteries and veins. The vessels branching from the arteries are called _____; those branching from the veins are called _____ .

14.5 Understanding heart disease

Heart disease is responsible for more than a quarter of all deaths in the UK. More people in the world's developed countries die from heart disease than from any other single cause. Why?

The smooth lining of healthy blood vessels allows blood to flow easily through them. However, the lining can be damaged and roughened by a fatty deposit called **atheroma**. The build-up of atheroma makes blood vessels narrower and cuts down the flow of blood (see Figure 14.5A). This increases the risk of blood clots forming. A blood clot can block a blood vessel. The clot is called a **thrombus** and a blockage is called a **thrombosis**.

Atheroma in the coronary arteries is one cause of heart disease. The first signs of trouble may be cramp-like chest pain brought on by quick walking, anger, excitement or any other activity or emotion that makes the heart work harder than usual. The pain is called **angina**.

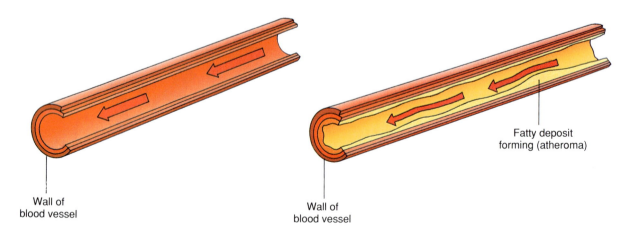

Wall of
blood vessel

Wall of
blood vessel

Fatty deposit
forming (atheroma)

Figure 14.5A ● Atheroma building up inside a blood vessel

People live with some types of angina for years but other types get worse and may later result in a heart attack (called a coronary thrombosis).

A heart attack happens when the blood supply to the heart is interrupted. This usually causes a gripping pain in the person's chest. The pain often spreads to the neck, jaw and arms, and the victim may also sweat and feel faint and sick.

The affected part of the heart is damaged and sometimes a heat attack is so severe that the heart stops beating altogether. This is called **cardiac arrest**. A victim of cardiac arrest will die unless the heart starts beating again within a few minutes.

Deaths from heart disease

Diseases of the heart and blood vessels kill more people in the UK than any other single cause (see Figures 14.5B and 14.5C). More than a quarter of all deaths are due to heart disease alone. This problem is not confined to the UK. Heart disease is the biggest single killer of middle-aged men in many of the world's developed countries.

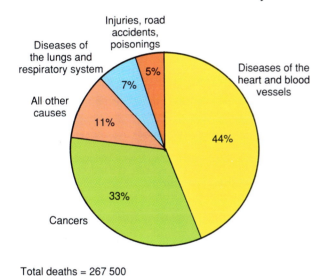

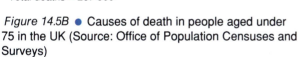

Total deaths = 267 500

Figure 14.5B ● Causes of death in people aged under 75 in the UK (Source: Office of Population Censuses and Surveys)

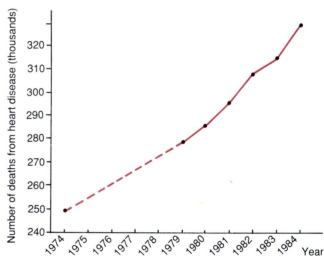

Figure 14.5C ● Deaths in the UK from heart disease between 1974 and 1984 (Source: Health and Personal Social Services statistics HMSO, 1986)

Risk factors

Research has identified some of the causes of heart disease by comparing groups of people who have high rates of heart disease with groups that have low rates. When a difference between the groups is found that might help to explain why some people are more likely to develop heart disease than others, it is called a **risk factor**.

Three risk factors are unavoidable:

- **Sex** – men are more likely to die of heart disease than women.
- **Age** – the risk of heart disease increases with a person's age.
- **Inherited genes** – the tendency to die from heart disease can run in families.

Being male, old or having a family history of heart disease does not mean that people in these categories will necessarily die from it. If the risk factors are known, people can live sensibly to increase their chances of reaching old age. Living sensibly means avoiding the other, controllable risk factors such as smoking and stress (see Figure 14.5D).

LOOK AT LINKS
for **inheritance**
See Topic 19.

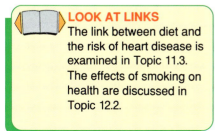

LOOK AT LINKS
The link between diet and the risk of heart disease is examined in Topic 11.3. The effects of smoking on health are discussed in Topic 12.2.

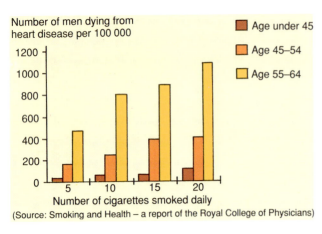

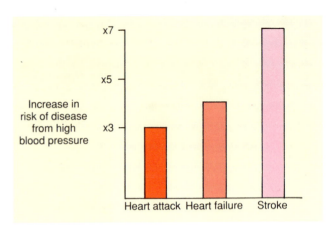

Figure 14.5D ● The evidence linking the risk of heart disease with avoidable risk factors
(a) Smoking and the risk of heart disease   (b) High blood pressure and the risk of heart disease

Exercise and ...

Blood pressure
- Blood pressure is lower in physically fit people
- Regular exercise can help control high blood pressure and reduce the risk of heart disease

Regular exercise increases heart fitness but is there evidence that exercise directly reduces the risk of heart disease? Many different studies strongly support the idea that regular exercise protects against heart disease and some of the evidence is summarised in Figure 14.5E. *What conclusions can you make from the data?*

Overweight
- Overweight is often due to lack of exercise and is linked with high blood pressure
- Regular exercise increases fitness, reduces body fat, lowers blood pressure and reduces the risk of heart disease

Stress
- Stress increases blood pressure and the risk of heart disease as well as making people irritable and depressed
- Regular exercise reduces the risk of heart disease and gives an increased feeling of well being

Coronary arteries
- Regular exercise reduces the build up of atheroma and makes coronary arteries wider reducing the risk of heart disease
- Exercise can promote the growth of new blood vessels after a heart attack

London Transport Study
- The study found that bus drivers had twice as many fatal heart attacks as the conductors, who climbed the stairs and walked the aisles of the bus all day long

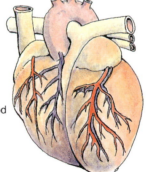

Figure 14.5E ● Exercise and heart disease

 **LOOK AT LINKS**
for **oxygen debt**
See Topic 12.1.

SUMMARY

Heart disease kills more people in the UK than any other single cause. Risk factors increase the likelihood of heart disease. Some risk factors are unavoidable; others can be controlled by diet and exercise. Exercise increases heart fitness.

The healthy heart at rest beats an average of 60–80 beats per minute (bpm). This is the **heart rate**. The volume of blood pumped from the heart each minute (**cardiac output**) depends on the heart rate and the volume of blood pumped out with each beat (**stroke volume**). Heart rate, cardiac output and stroke volume measure the heart's effectiveness and fitness.

A fit heart pumps more blood to the body's tissues and organs than does an unfit heart. At rest it has 25% more output than an unfit heart, rising to 50% more during vigorous exercise. This meets the increased demand for oxygen from the muscles more efficiently. Stroke volume is also greater and a fit heart beats more slowly. The fit heart therefore does not have to strain to pump blood around the body.

Vigorous exercise increases the rate and depth of breathing (panting). The response is triggered by the rising acidity of the blood due to the accumulation of carbon dioxide and increasing **oxygen debt** in the muscles. More air is drawn into the lungs and the rush of oxygen to the muscles promotes oxidation of the accumulated lactic acid. Panting also quickly gets rid of excess carbon dioxide. Blood acidity decreases and breathing movements slow to the resting rate.

CHECKPOINT

❶ The diagram shows three places A, B and C where blood vessels supplying blood to the heart could become blocked causing a heart attack.
   (a) Name the blood vessels supplying the blood to the heart.
   (b) What is the blockage of the blood vessel called?
   (c) Which of A, B or C would cause the most serious heart attack? Give reasons for your answer.

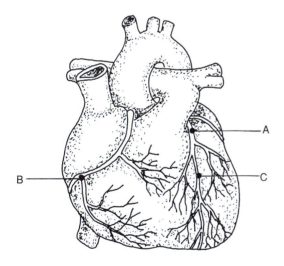

❷ The diagram shows lengthways sections of a healthy blood vessel and a diseased blood vessel.

Healthy blood vessel   Diseased blood vessel

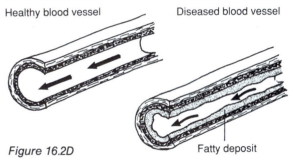

Figure 16.2D   Fatty deposit

   (a) What is the fatty deposit called?
   (b) The artery narrows where the fatty deposit is forming. Briefly explain how the blood vessel could become completely blocked.
   (c) What is a heart attack?
   (d) Imagine you are walking along a busy street with someone who suddenly has a heart attack. The person remains conscious but needs your help. Explain what you should do.

❸ Briefly explain how a man with a family history of heart disease can cut down the risk of having a heart attack.

❹ 'Exercise is good for you' – so we are told. Using the data available, describe the effect of exercise on heart fitness.

TOPIC 15 REPRODUCTION

15.1 Outlines of reproduction

FIRST THOUGHTS

Reproduction more than anything else characterises living things. It continues the thread of life, unbroken from generation to generation. This section gives you an overview of the processes of reproduction.

All living things eventually die. They may be killed by other organisms or die of old age. Before dying, some individuals produce new individuals of their own kind. This is **reproduction** (see Figure 15.1A).

There are two basic types of reproduction:

- **Asexual reproduction** – one parent gives rise to new individuals which are genetically identical to it. The genetically identical individuals produced are called **clones**.
- **Sexual reproduction** – two parents give rise to new individuals. Offspring acquire genes from each parent and are therefore genetically different from one another and from their parents.

Figure 15.1A ● Reproduction (a) A female scorpion carrying her young

(b) A tree surrounded by its seedlings

Asexual reproduction

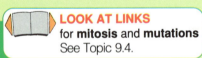

LOOK AT LINKS
for **mitosis** and **mutations**
See Topic 9.4.

Figure 15.1B illustrates different ways in which asexual reproduction takes place. The offspring are identical to the parent because the cells divide by **mitosis** to give new daughter cells. During mitosis DNA replicates itself. Mistakes in DNA replication are called **mutations**, but they are rare.

Figure 15.1B ● Ways of reproducing asexually
(a) *Paramecium* dividing by fission – the parent divides into equal parts
(b) *Hydra* budding – outgrowths of the parent's body (buds) separate from the parent. Each one becomes a separate individual.
(c) Parts of the parent body grow into new individuals. For example parts of stems can sprout roots and grow into new plants.
(d) Regeneration – the parent body breaks into pieces, each piece can grow into a new individual
(e) A female aphid produces eggs which develop into new individuals without fertilisation

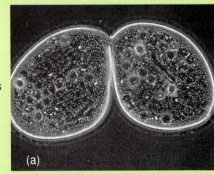

(a)

Sexual reproduction

Figure 15.1C ● Maleness and femaleness in *Spirogyra* – a new strand of *Spirogyra* develops from the fused nuclei.

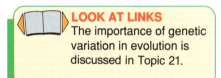
LOOK AT LINKS
for *Spirogyra*
See Topic 1.5.

LOOK AT LINKS
The importance of genetic variation in evolution is discussed in Topic 21.

Figure 15.1C shows strands of the green alga *Spirogyra* lying side-by-side. A cell of one of the strands is joined by a tube to a cell of the other strand. The content of one of the joined cells passes through the tube and the nucleus fuses with the nucleus of the other cell. The genetic material in each nucleus combines with the genetic material of its opposite number. The combination of genetic material is different from the combination in each of the parent cells. This is an important feature of sexual reproduction; the production of much more **genetic variation** is possible than by mutation alone. Figure 15.1C illustrates another important feature of sexual reproduction – maleness and femaleness. The male cell of *Spirogyra* is the empty one. Its contents have moved toward the female cell.

(c)

(d)

(e)

● Sex cells

There are two types of sex cell: male sex cells called sperms and female sex cells called eggs. Sperms and eggs are called **gametes**.

Table 15.1 ● Comparing sperms and eggs

Sperms	Eggs
Small	Large
Move towards the egg	Do not move much
Have no food store	Have food store

The sperms of many types of organism are very similar (see Figure 15.1D). Human sperm is fairly typical (see Figure 15.1E). In fact, the sperm is little more than a mobile nucleus designed to bring genetic material from the male parent to the egg of the female parent.

Animal eggs are bigger than sperms because they contain yolk which is a food store. Some eggs have more yolk than others. This is why the eggs of different animals are different in size.

Seaweed

Mosses and liverworts

Sea-urchin

Rat

Chicken

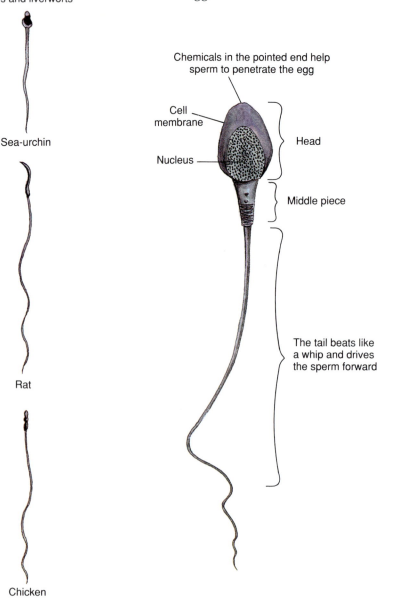

Chemicals in the pointed end help sperm to penetrate the egg

Cell membrane

Nucleus

Head

Middle piece

The tail beats like a whip and drives the sperm forward

● How many chromosomes?

Cells have a fixed number of chromosomes. For example, a human skin cell has 46 chromosomes and divides by mitosis to give daughter cells each with 46 chromosomes. Cells which, through mitosis, receive the full number of chromosomes from their parent cells and in turn hand them on to their daughter cells are described as **diploid** (or **2n**).

However, not all parent cells divide by mitosis to produce diploid daughter cells. When the cells that give rise to sex cells (sperms and eggs) divide, the diploid number of chromosomes in each nucleus is halved. For example, human sex cells each have 23 chromosomes compared with the normal diploid number of 46 chromosomes found in each body cell. The process which halves the diploid number of chromosomes is called **meiosis**. The half number – 23 chromosomes in the case of each human sperm and egg – is described as **haploid** (or **n**). Figure 15.1F shows the different stages of meiosis.

Figure 15.1D ● Plant and animal sperm   Figure 15.1E ● Human sperm

Parent cell

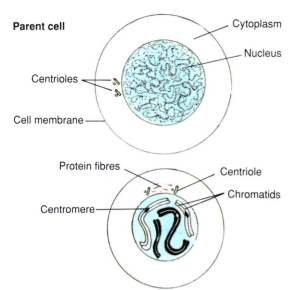

Cytoplasm

Nucleus

Centrioles

Cell membrane

Protein fibres

Centriole

Chromatids

Centromere

Stage 1 Chromosomes divide into pairs of identical chromatids joined to one another by the centromere. Centrioles move to opposite ends of the cell. Protein fibres form around each of them, just as in mitosis.

Stage 2 Chromosomes pair up. The two chromosomes of a pair are called **homologous chromosomes**. Notice the exchange of a segment of one chromosome with the corresponding segment of its homologous chromosome.

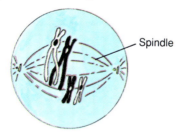

Spindle

Stage 3 The nuclear membrane disappears. A spindle forms between the centrioles just as in mitosis. Homologous pairs of chromosomes arrange themselves on the equator.

Stage 4 Homlogous pairs of chromosomes separate. Each chromosome (consisting of two chromatids) of the pair moves to opposite ends of the cell.

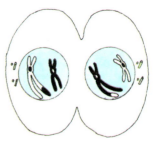

Stage 5 The chromosomes gather into two bunches. The cell begins to divide and a new nuclear membrane forms around each bunch.

Stage 6 The cell divides.

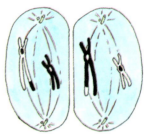

Stage 7 The nuclear membranes disappear and new spindles form at right angles to the first. The chromosomes (still as pairs of chromatids) arrange themselves on the equator.

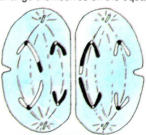

Stage 8 The centromeres divide. The chromatids separate and bunch at opposite ends of each cell. The chromatids are now the new chromosomes. Each cell begins to divide.

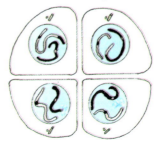

Stage 9 A nuclear membrane forms around each bunch of chromosomes and the cells divide.

Figure 15.1F ● The stages of meiosis. (Only four chromosomes are shown in detail.)

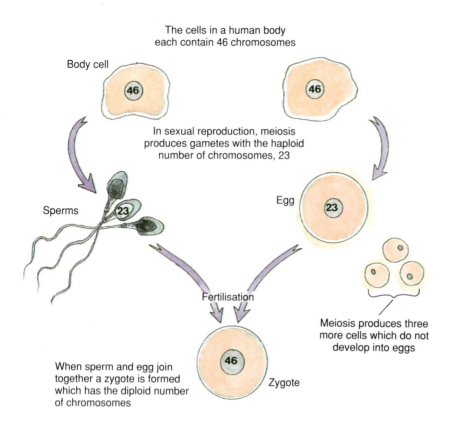

The cells in a human body each contain 46 chromosomes

Body cell

In sexual reproduction, meiosis produces gametes with the haploid number of chromosomes, 23

Sperms

Egg

Fertilisation

Meiosis produces three more cells which do not develop into eggs

When sperm and egg join together a zygote is formed which has the diploid number of chromosomes

Zygote

Why are sex cells haploid? During fertilisation (when sperm and egg join together) the sperm nucleus and the egg nucleus fuse. The chromosomes from each nucleus combine to produce new combinations of genes. Half come from the haploid sperm and half come from the haploid egg. The fertilised egg (called the zygote) is therefore diploid and develops into a new individual which inherits characteristics from both parents, not from just one as in asexual reproduction (see Figure 15.1G). *Can you think what would happen genetically in each generation of new individuals if sperms and eggs were diploid and not haploid?*

Figure 15.1G ● Fertilisation – restoring the diploid state

The carp is a fish which lives on the bottom of muddy rivers, lakes and ponds. In May and July a female carp lays up to 750 000 eggs, which the male fertilises. Only two fertilised eggs need survive to develop into mature adults for the population to remain stable.

● *Patterns of fertilisation*

Sperms swim. Swimming brings sperms to eggs. Sperms and eggs must therefore be in a liquid for fertilisation to take place. Most aquatic organisms release sperms and eggs into the surrounding water where fertilisation takes place. This is called **external fertilisation** because it takes place outside the parent's body (see Figure 15.1H).

Although moss plants grow on land, even they need water to complete their life cycle. Without it moss sperms cannot swim to moss eggs to fertilise them (see Figure 15.1I). This is why mosses and other such plants grow in damp places.

Figure 15.1H ● Frogs mating. The male sheds sperms over the eggs which the female lays in the water.

Figure 15.1I ● Moss sperms are released in the film of water covering the plant. They swim towards the eggs.

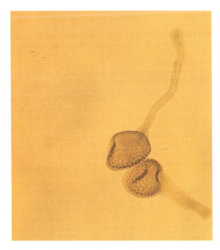

Figure 15.1J ● Sprouting pollen tubes

Without water sperms and eggs perish. *How do organisms that spend their lives on dry land overcome the problem?* Seed-producing plants can live in much drier places than moss plants. Their sperm is protected inside drought-resistant pollen grains. These are usually carried by animals or blown by the wind to the female part of the plant.

Each pollen grain sprouts a pollen tube which grows through the female tissues towards the eggs inside (see Figure 15.1J). Sperms pass down the pollen tube and one of them fertilises the egg. This is an example of **internal fertilisation**.

Internal fertilisation is one of the adaptations that insects, reptiles, birds and mammals have adopted for life on dry land. The male places sperm directly into the body of the female – a process that usually needs some kind of coupling, called **copulation**. Special organs help to transfer the sperm from male to female. In many insects, for example, sperm is transferred from male to female protected in a tiny package called the spermatophore (see Figure 15.1K).

In mammals the male has an organ called the penis which penetrates the opening to the reproductive system of the female (see Figure 15.1L). Sperms pass through the penis into her reproductive system where fertilisation takes place. The sperms swim towards the eggs in a liquid which is produced by the male reproductive system specially for the purpose and which is transferred to the female with the sperms.

Figure 15.1K ● Bush crickets transferring a spermatophore

Figure 15.1L ● Copulating zebras

CHECKPOINT

1. What is asexual reproduction? Briefly describe the different ways it takes place.

2. What are 'gametes'?

3. Explain, in simple terms, how sperms swim.

4. Why are eggs larger than sperms?

5. What is meant by the term 'fertilisation' and what is formed as a result of it?

6. What is 'meiosis'?

7. Briefly explain the meaning of 'diploid' and 'haploid'.

8. The nuclei of muscle cells in a species of lobster each contain 250 chromosomes. How many chromosomes would be found in the nucleus of each of the lobster's sperm?

9. 'The number of chromosomes in the nuclei of cells is kept constant from one generation to the next through meiosis and sexual reproduction.'
   Comment on this statement with reference to a named species.

15.2 Plant reproduction: flowers

How are flowers designed for sexual reproduction? What is the difference between pollination and fertilisation? This section gives you the answers.

Flowers are shoots which are specialised for sexual reproduction. Although flowers come in all shapes and sizes, they are all made up of similar parts. The sepals, petals, stamens and carpels of the buttercup are shown in Figure 15.2A.

Figure 15.2A ● The meadow buttercup

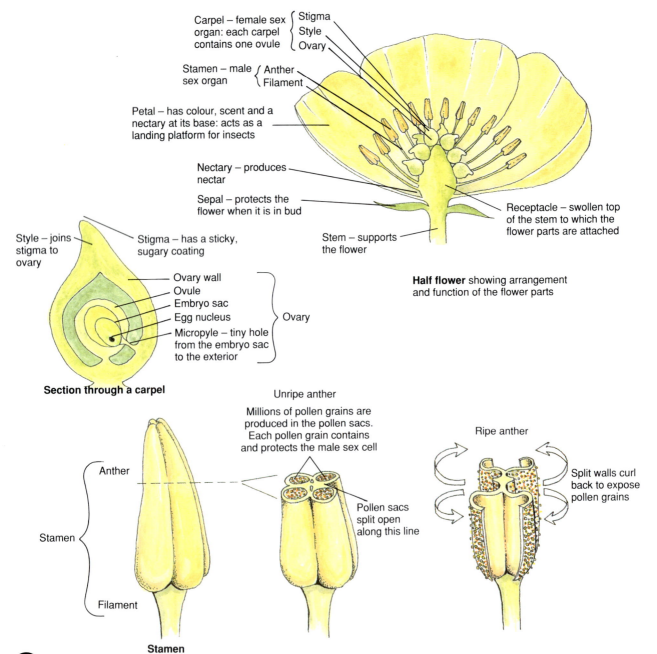

Carpel – female sex organ: each carpel contains one ovule {Stigma, Style, Ovary}

Stamen – male sex organ {Anther, Filament}

Petal – has colour, scent and a nectary at its base: acts as a landing platform for insects

Nectary – produces nectar

Sepal – protects the flower when it is in bud

Receptacle – swollen top of the stem to which the flower parts are attached

Stem – supports the flower

Half flower showing arrangement and function of the flower parts

Style – joins stigma to ovary

Stigma – has a sticky, sugary coating

Ovary wall
Ovule
Embryo sac
Egg nucleus
Micropyle – tiny hole from the embryo sac to the exterior

} Ovary

Section through a carpel

Unripe anther

Millions of pollen grains are produced in the pollen sacs. Each pollen grain contains and protects the male sex cell

Pollen sacs split open along this line

Ripe anther

Split walls curl back to expose pollen grains

Anther

Stamen

Filament

Stamen

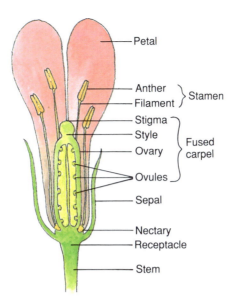

Petal
Anther } Stamen
Filament
Stigma
Style } Fused
Ovary carpel
Ovules
Sepal
Nectary
Receptacle
Stem

Types of flower

Flowers like the buttercup are simple flowers. Their parts are free and separate from each other and they have many stamens and carpels. The parts of other types of flower may be arranged differently with some parts joined together or have fewer parts than the buttercup has. Figure 15.2B illustrates examples of different arrangements.

Figure 15.2B ● Arrangements of flower parts
(a) Half-flower of the wallflower. The one large carpel is formed from a number of carpels fused together. You can tell that the carpel is fused because it contains a number of ovules.

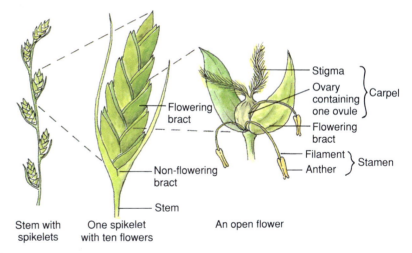

Stigma
Ovary containing one ovule } Carpel
Flowering bract
Flowering bract
Filament } Stamen
Anther
Non-flowering bract
Stem

Stem with spikelets   One spikelet with ten flowers   An open flower

(b) Rye grass flowers have no sepals or petals

Pollination

The pollen grain must pass from the anther to the stigma before the male sex cell inside it can fertilise the female sex cell in the ovule. This transfer of pollen is called pollination.

● Wind pollination

Wind-pollinated flowers are adapted in ways that make sure that pollen is widely scattered. Some have long, slender anthers which hang clear of the sepals and petals, which otherwise might get in the way of the pollen being blown away. The flowers of different grasses (for example rye grass in Figure 15.2B) do not have sepals or petals at all. The dangling male catkins of the hazel tree produce a cloud of pollen which is blown away well before the leaves, which could get in the way, are fully open (see Figure 15.2C).

Figure 15.2C ● The anthers of hazel flowers produce large amounts of pollen. Notice that the leaves are in tight buds.

Figure 15.2D ● Hibiscus flowers showing strongly marked honey guides

● *Insect pollination*

Insects carry pollen from the anthers of one flower on to the stigmas of another flower. Insect-pollinated flowers, therefore, are adapted to attract insects to them. They are often large, brightly coloured and sweetly scented. They also produce a sweet liquid called nectar. Insects visit flowers to feed on pollen and nectar, attracted by the colour and scent. Marks on the petals called honey guides lead the insect to the nectar (see Figure 15.2D). As insects feed at the flower, their bodies become covered in pollen which is carried to the next flower the insect visits (see Figure 15.2E).

Figure 15.2E ● Pollen sticking to the body and legs of a honey bee on a dandelion. Pollen is carried by the bee in pollen sacs, one on each of the back legs: one pollen sac can be seen. The curled stigmas of the dandelion florets can also be seen.

IT'S A FACT

Pollination need not always lead to fertilisation. Chemicals in the stigma stop pollen from a different species growing pollen tubes. In the case of self-pollination the pollen tube may not grow as fast, if at all, as that of pollen from another flower.

The differences between wind-pollinated and insect-pollinated flowers are summarised in Table 15.2.

Table 15.2 ● Comparing insect-pollinated and wind-pollinated flowers

Part of flower	Insect-pollinated		Wind-pollinated	
Petals	Brightly coloured Usually scented Most have nectaries	Large flowers which attract insects	If present, green or dull colour No scent No nectaries	Small flowers
Anthers	Positioned where insects are likely to brush against them		Hang loosely on long thin filaments so that they shake easily in the wind	
Stigma	Positioned where insects can brush against them Sticky and flat or lobe-shaped		Long, branching and feathery to make a large area for catching wind-blown pollen grains	
Pollen	Small amounts produced Large grains Rough or sticky surface which catches on the insects' bodies		Large amounts produced Small, light grains with smooth surfaces – easily carried on the wind	

● *Self-pollination and cross-pollination*

The wild arum is a trap for small insects. Insects that feed on dung and rotting flesh are attracted by its horrible smell. They fall down the slippery tube-like bract to the nectar at the bottom, brushing pollen on to the stigmas of the carpels as they fall. The insects cannot escape because of the downward pointing hairs. The anthers ripen within hours of pollination taking place, the hairs wither and the insect crawls out. As they brush past the anthers they are covered with fresh pollen which is transferred to another arum when the insect is trapped again.

LOOK AT LINKS
What are the advantages of cross-pollination compared with self-pollination? It increases genetic variation
See Topic 21.

Figure 15.2F ●
(a) Self-pollination. Pollen is transferred from the anthers to the stigma(s) of the same flower or the anthers of one flower to the stigma(s) of another flower on the same plant

(b) Cross-pollination. Pollen is transferred from the anthers of a flower on one plant to the stigma(s) of a flower on a different plant

Figure 15.2F shows the difference between self-pollination and cross-pollination and how self-pollination can take place. Different adaptations make cross-pollination more likely than self-pollination. For example, self-pollination is less likely to occur if anthers and stigmas on the same flower or different flowers on the same plant are ripe at different times. The two possibilities are that:

- The anthers ripen and pollen falls before the stigma is ripe.
- The stigma is ready to receive pollen before the anthers are ripe.

However, pollination would not occur at all if all of the anthers (or stigmas) in flowers of a particular species were ripe at a time when all of the stigmas (or anthers) were not. Success depends on some anthers and some stigmas in different flowers being ripe at the same time.

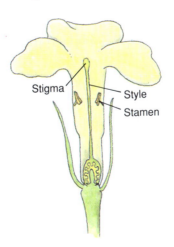

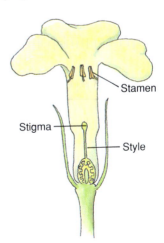

Figure 15.2G ●
(a) Half-flower of the pin-eye primrose with long style and stigma above stamens

(b) Half-flower of the thrum-eye primrose with short style and stigma below stamens

The primrose shown in Figure 15.2G illustrates another method for making cross-pollination more likely. When a bee pushes into a thrum-eyed flower in search of nectar, its head is dusted with pollen. If the next flower it visits is pin-eyed some pollen will brush onto the stigma. When a bee visits a pin-eyed flower, pollen sticks to its mouthparts at the place where it will touch the stigma of a thrum-eyed flower. Also, thrum-eyed flowers have large pollen grains and stigmas with small pits, pin-eyed flowers have small pollen grains and stigmas with large pits. Since large pollen grains fit best into large pits and small grains into small pits, cross-pollination is more likely.

The only way to make sure of cross-pollination is for a plant to have either all male or all female flowers. Few plants are like this but examples are the poplar, ash and willow trees.

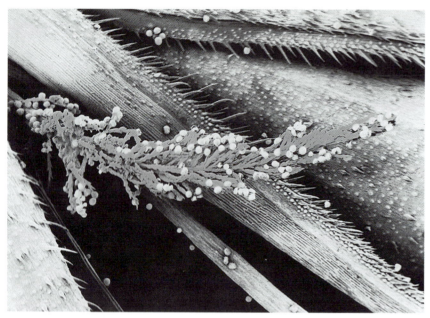

Figure 15.2H ● Pollen grains on the surface of a stigma

Fertilisation

Pollination brings pollen grains to the stigma (see Figure 15.2H). A male sex cell is inside each pollen grain. *How does it reach the egg cell in the ovule inside the ovary?* The sugar coating the stigma's surface helps pollen grains to stick to it. If conditions are right pollen grains begin to grow pollen tubes (see Figure 15.2I). The pollen tube grows down through the style to the micropyle. The tube nucleus dies and the two male nuclei pass down the pollen tube and into the embryo sac. There, one nucleus fuses with the egg nucleus. This fusion of male and female nuclei is **fertilisation**. (see Figure 15.2J.)

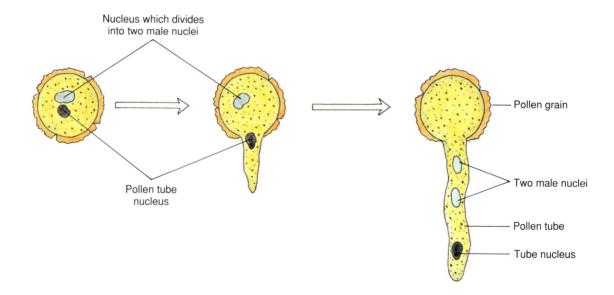

Figure 15.2I ● Growth of the pollen tube – notice the different nuclei

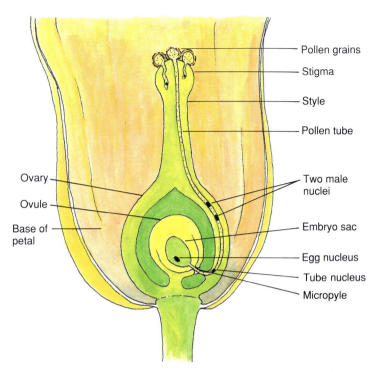

Figure 15.2J ● The male nucleus fertilises the egg nucleus

Although more than one pollen grain may grow a pollen tube, it is a male nucleus of the first pollen tube to reach the egg nucleus which fertilises it. In an ovary which has more than one ovule each egg nucleus fuses with only one male nucleus.

The fertilised egg divides and develops into the embryo which will become the new plant. The other male nucleus from the pollen tube fuses with two more nuclei in the embryo sac, developing into a special tissue which forms a food store for the embryo to use when it grows.

CHECKPOINT

❶ Look at Figure 15.2A

(a) Which parts are the male and female sex organs?
(b) The male sex organs produce millions of tiny pollen grains. What is inside each pollen grain?
(c) Which part of the female sex organ contains the ovule?
(d) What does the ovule contain?

❷ What is pollination?

❸ Pollination usually happens by wind or by insects. Choose one wind-pollinated flower and one insect-pollinated flower from Figures 15.2A–D and study the pictures carefully. How is each flower you have chosen adapted for the way it is pollinated?

❹ Briefly describe different ways flowers are adapted to make cross-pollination more likely than self-pollination.

❺ Does pollination take place before or after fertilisation?

❻ What is the name of the tiny hole through which the pollen tube goes to the ovule?

❼ What happens when the pollen tube reaches the ovule?

15.3 Plant reproduction: seeds and fruits

This section discusses seeds and fruits and the ways in which seeds are dispersed. As you read it think about:
• how seeds are formed,
• what fruits are,
• types of fruit,
• why dispersal of seeds is important,
• how fruits help the dispersal of seeds.

A seed is a fertilised ovule. The embryo plant and its store of food inside are covered by a tough seedcoat called the **testa**.

When the embryo is almost fully developed the tissues round it lose water, leaving the seed hard and dry. The seed can stay like this for a long time until conditions are right for it to grow.

Food may be stored in a thick, fleshy, wing-like structure called the **cotyledon**. Figure 15.3A shows that the seed of the broad bean has two cotyledons. Seeds like this are called **dicotyledonous** seeds. Figure 15.3A also shows which parts of the embryo will grow into which parts of a new plant.

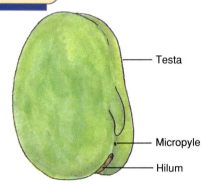

— Testa

— Micropyle

— Hilum

(a) The outside of the whole seed

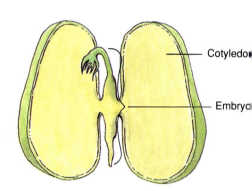

Cotyledon

Embryo

(b) The two cotyledons slightly separated showing the embryo plant between them

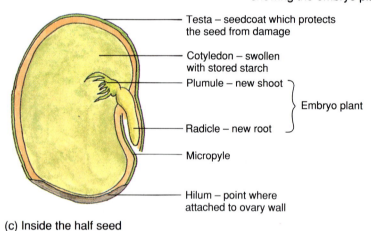

Testa – seedcoat which protects the seed from damage

Cotyledon – swollen with stored starch

Plumule – new shoot ⎫
⎬ Embryo plant
Radicle – new root ⎭

Micropyle

Hilum – point where attached to ovary wall

(c) Inside the half seed

Figure 15.3A ● The broad bean seed

Grasses produce seeds which store food in only one cotyledon. They are called **monocotyledonous** seeds.

After fertilisation it is usually the ovary which develops into the fruit. The wall of the ovary is then called the **pericarp**. As the fruit develops the pericarp becomes either dry and hard (examples are acorn, dandelion, wallflower, sycamore and poppy) or juicy and fleshy (examples are plum, tomato, blackberry, rosehip, honeysuckle and holly). Juicy fruits are more correctly called **succulent fruits** (see Figure 15.3B).

The fruit contains the seed or seeds. The number of seeds in a fruit depends on how many ovules there were in the ovary to begin with and how many were fertilised.

Figure 15.3B ● The parts of the flower that develop into parts of the fruit
(a) Plum – a succulent fruit with one seed inside the stone

(b) Acorn – a dry fruit

In some plants, parts of the flower other than the ovary develop into the fruit. These are called **false fruits**. In many false fruits the receptacle grows to form the fruit (see Figure 15.3C).

Figure 15.3C ● The apple is an example of a false fruit. The receptacle forms the fleshy fruit around the pericarp which contains the seeds.

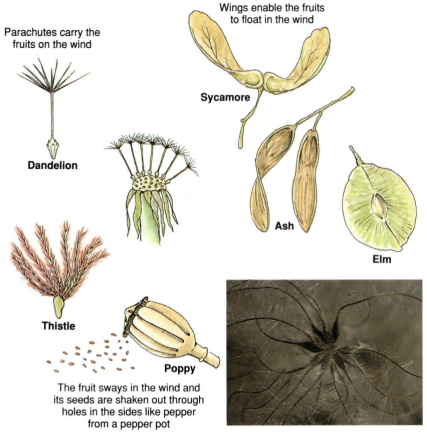

Parachutes carry the
fruits on the wind

Dandelion

Thistle

Wings enable the fruits
to float in the wind

Sycamore

Ash

Elm

Poppy

The fruit sways in the wind and
its seeds are shaken out through
holes in the sides like pepper
from a pepper pot

Dispersal of fruits and seeds

When a fruit is ripe it breaks away from the parent plant and its seeds are scattered. The scattering is called **dispersal**. It is important that seeds are dispersed far and wide so that the plants which grow from them will not be overcrowded.

The two main ways of dispersal are by wind and by animals. The pericarps of different fruits develop in different ways for different methods of dispersal (see Figure 15.3D).

Figure 15.3D ● Methods of fruit and seed dispersal
(a) Wind-dispersed fruits and seeds. 'Parachute', 'wing' and 'pepper-pot' mechanisms all help dispersal. The photograph shows the fruits of Old Man's Beard – the hairs make a large surface area which catches the wind

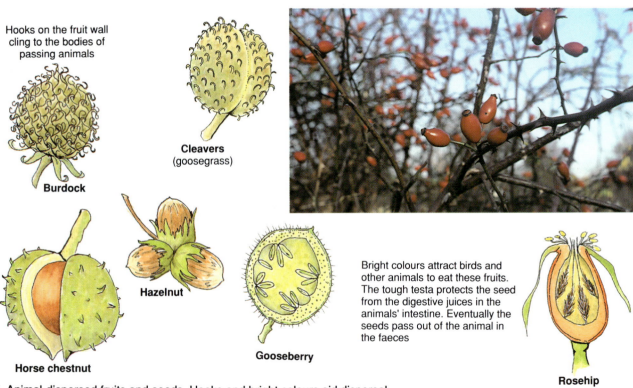

Hooks on the fruit wall
cling to the bodies of
passing animals

Burdock

Cleavers
(goosegrass)

Hazelnut

Horse chestnut

Gooseberry

Bright colours attract birds and other animals to eat these fruits. The tough testa protects the seed from the digestive juices in the animals' intestine. Eventually the seeds pass out of the animal in the faeces

Rosehip

(b) Animal-dispersed fruits and seeds. Hooks and bright colours aid dispersal. Hazelnuts and horse chestnuts are often stored by squirrels and then forgotten. The photographs show the brightly coloured fruits of the honeysuckle and the wild rose.

The fruit wall of some plants dries and splits open. As it does so the seeds are thrown out. Examples of fruits that disperse seeds in this way are shown in Figure 15.3E. A few fruits are dispersed by water currents. Examples are the waterlily and the coconuts (see Figure 15.3F).

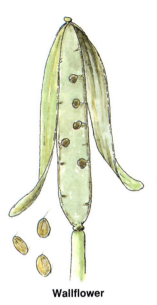

Wallflower

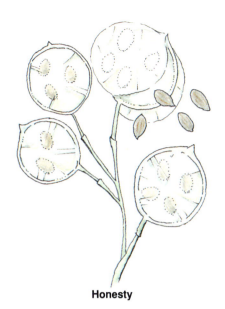

Honesty

Figure 15.3E ● Self-dispersal. The fruit dries and splits and the seeds fall out. The photograph shows the fruit of broom, split open with some of the seeds still in place.

Figure 15.3F ● Coconuts float and are dispersed by water currents. The outer fibres of the coconuts pictured have been removed. The coconut is thought to have come originally from South America and the water currents in the Pacific Ocean could have dispersed the coconut from South America to the South Sea Islands.

CHECKPOINT

❶ Which part of the flower usually develops into the fruit?

❷ What is the fruit wall called?

❸ Describe which parts of the flower have developed into the fruits illustrated in Figures 15.3B and 15.3C.

❹ Why is dispersal of seeds important?

15.4 Asexual reproduction in plants

This section deals with asexual reproduction in flowering plants. As you read it think about:
• vegetative parts of flowering plants,
• organs of asexual reproduction and organs which store food,
• cuttings and grafting.

Many different plants can reproduce asexually. The root, leaf or more often the stem may grow into new plants. These parts are called **vegetative parts** and asexual reproduction in flowering plants is sometimes called **vegetative reproduction**. Since new plants come from one parent plant, they are genetically the same.

The strawberry plant shown in Figure 15.4A reproduces asexually by stems called runners which grow horizontally on the surface of the soil.

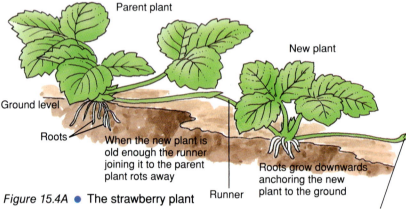

Parent plant

New plant

Ground level

Roots

When the new plant is old enough the runner joining it to the parent plant rots away

Roots grow downwards anchoring the new plant to the ground

Figure 15.4A ● The strawberry plant

Runner

Terminal bud – produces shoots which grow upwards

Many plants reproduce vegetatively underground. Figure 15.4B shows the thick, fleshy iris rhizome which is an example of a stem that grows horizontally underground. It branches at intervals from buds growing next to small leaves called scale leaves (leaves without chlorophyll). Because of the branching growth and the eventual dying of the old part of the rhizome, iris plants seem to move their position from year to year in the soil.

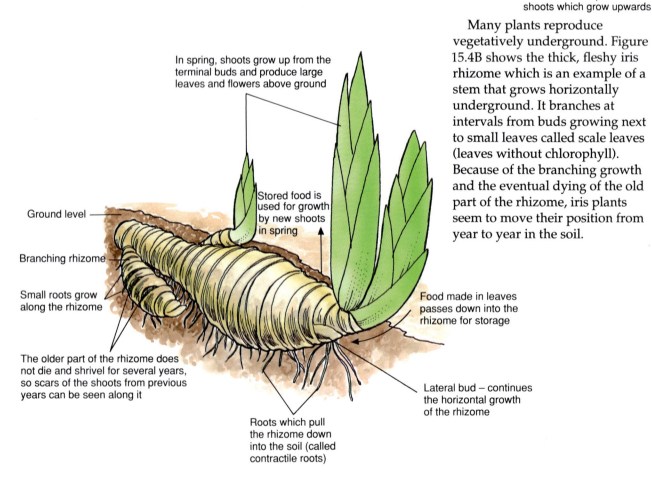

In spring, shoots grow up from the terminal buds and produce large leaves and flowers above ground

Stored food is used for growth by new shoots in spring

Ground level

Branching rhizome

Small roots grow along the rhizome

Food made in leaves passes down into the rhizome for storage

The older part of the rhizome does not die and shrivel for several years, so scars of the shoots from previous years can be seen along it

Roots which pull the rhizome down into the soil (called contractile roots)

Lateral bud – continues the horizontal growth of the rhizome

Figure 15.4B ● The thick, fleshy iris rhizome grows and branches horizontally underground. The arrows show the movement of stored food. *In which directions is the rhizome growing underground?*

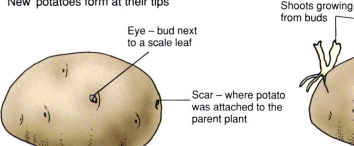

Rhizomes – grow from buds near to the soil surface

Old potato

Roots

New potato – swollen stem (tuber) on the end of the rhizome

Figure 15.4C shows the potato plant. The potato is a swelling of stored food at the end of a rhizome. It is called a stem tuber. Leaves on shoots which grow above ground make starch which passes down the stem into the rhizomes. The old tuber rots away at the end of the growing season.

(a) Rhizomes grow from the buds nearest to the soil's surface. 'New' potatoes form at their tips

Eye – bud next to a scale leaf

Scar – where potato was attached to the parent plant

Shoots growing from buds

Roots

(b) A resting potato in winter showing 'eyes' which are buds next to scale leaves
Figure 15.4C ● The potato

(c) A sprouting potato in spring showing the growth of new shoots and roots from the 'eyes'

Dry bases of last year's leaves

Flower bud

This year's young leaves

Bud

Remains of last year's corm

Roots

(a) Resting corm, showing the parts which will grow next season

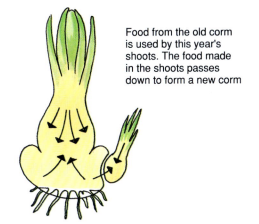

Food from the old corm is used by this year's shoots. The food made in the shoots passes down to form a new corm

(b) Growing leaves and flowers use up food stored in the old corm and a new corm begins to form on top of it (arrows show movement of stored food)

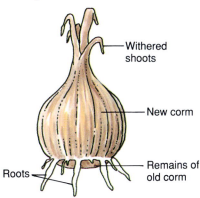

Withered shoots

New corm

Roots

Remains of old corm

(c) New corm which stores food for the next year – the new corn separates from the remains of the old corn at the end of the growing season
Figure 15.4D ● The crocus corm

A corm is a short, swollen, underground stem and a bulb is a large underground bud. At the end of a growing season the leaves make food which is stored in the corm or bulb underground until it is used the next year for the growth of new leaves and flowers. Daughter corms or bulbs develop from buds on the side of the parent organ and when they are large enough they break off and become independent plants. (See Figure 15.4D.)

Bulbs, corms, tubers and most rhizomes are not only organs of asexual reproduction but also organs which store food. They fill up with starch during the summer when the plants which have grown from them are in leaf and making food by photosynthesis. In the autumn the plants die down but the organs underground, full of stored food, survive the winter and produce new plants the following year. Because these organs asexually reproduce new plants year after year they are called **perennials**.

New plants grow at the points where old leaves are stripped off

Cutting from geranium plant stripped to two leaves

Pot of peat and sand

Figure 15.4E ● A cutting from a geranium plant

Artificial vegetative reproduction

Gardeners and farmers need to produce fresh stocks of plants with desirable characteristics like disease-resistance, colour of fruit or shape of flower. They exploit vegetative reproduction to achieve these aims.

● *Cuttings*

Figure 15.4E shows a geranium cutting. Short pieces of stem are cut just below the point where a leaf joins the stem. Most of the leaves are stripped from the stem. New shoots grow at the points where the old leaves are stripped off.

● *Grafting*

Grafting is often used for reproducing roses and fruit trees. A twig is cut from the tree to be reproduced and replaced in a slit in the stem of an already well-rooted tree so that the cut surfaces are brought together. The tissues of the twig and the tree join together and the graft grows on the rooted tree.

Budding is another type of grafting. A bud is used instead of a twig. Several buds can be grown on one root stock, each growing into an individual plant.

The plants produced by these methods are the same genetically as the parents from which they are taken. This means that the desirable qualities of the parent are preserved in the offspring, guaranteeing plant quality from one generation to the next.

CHECKPOINT

❶ What are the vegetative parts of flowering plants?

❷ Briefly compare the similarities and differences between a runner and a rhizome.

❸ What is a tuber?

❹ Look at Figure 15.4D. Describe the movement of food in a corm from the beginning of the growing season to the end.

15.5 Human reproduction

This section begins by asking you to explore the development of a sexual relationship. This puts in context the facts on human reproduction.

Getting to know you

Getting to know you charts the relationships between two imaginary people. Look at the different stages of their developing relationship.

Who is this person?
• Handsome • Pretty
What makes you notice someone of the opposite sex?

Attraction
• Caring • Easy to talk to
• Similar interests and opinions
Why do you want to see more of the other person?

Getting to know you
• You want to touch and kiss
• Are you nervous?
Why do you want to get closer to the other person?

Loving you
• Sharing your life with someone else
• Having children is a possibility
Why do you want to commit yourself to the other person?

• What points do you think are important at each stage of the developing relationship?
• Do you agree with the order of the different stages of *Getting to know you*?
  How would you describe each stage?
• What do you think the role of love is in a relationship?
• How do you think the responsibility of children should be shared between partners?

Sexual feelings for someone of the opposite sex are called heterosexuality. Homosexuality is having sexual feeling for someone of your own sex. Homosexual feelings are not uncommon in adolescents who, as a result, often feel guilty and different from other people. Talking about homosexual feelings to parents, friends or a counsellor can help to keep problems in proportion. What is important about relationships is that they should be based on trust, love and understanding of one another.

The human reproductive system

The visible parts of the reproductive system are called the **genitalia**. A man's genitalia consist of the **penis** and the **testes** which are contained in a bag-like scrotum which hangs down between the legs. This position protects the testes from injury. It also keeps them about 3 °C lower than body temperature. This is important because sperms only develop properly inside the testes in these slightly cooler conditions. The reproductive system of a man is illustrated in Figure 15.5A.

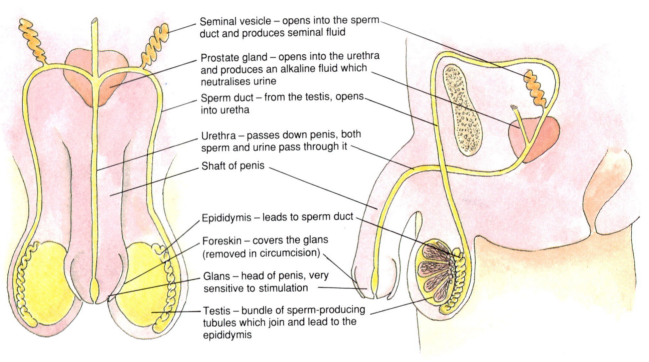

Seminal vesicle – opens into the sperm duct and produces seminal fluid

Prostate gland – opens into the urethra and produces an alkaline fluid which neutralises urine

Sperm duct – from the testis, opens into uretha

Urethra – passes down penis, both sperm and urine pass through it

Shaft of penis

Epididymis – leads to sperm duct

Foreskin – covers the glans (removed in circumcision)

Glans – head of penis, very sensitive to stimulation

Testis – bundle of sperm-producing tubules which join and lead to the epididymis

(a) Front view

(b) Side view

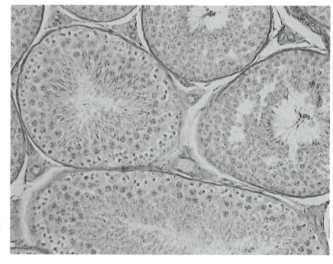

(c) Cross section through a sperm-producing tubule showing sperms clustered inside. Stretched out end-to-end the tubules inside each testis are more than 500 m long. (x200)

Figure 15.5A ● The male reproductive system

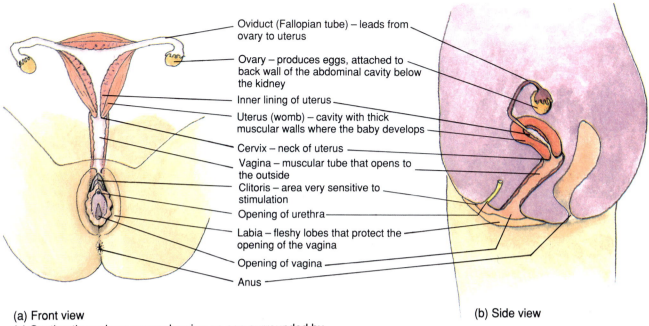

Oviduct (Fallopian tube) – leads from ovary to uterus

Ovary – produces eggs, attached to back wall of the abdominal cavity below the kidney

Inner lining of uterus

Uterus (womb) – cavity with thick muscular walls where the baby develops

Cervix – neck of uterus

Vagina – muscular tube that opens to the outside

Clitoris – area very sensitive to stimulation

Opening of urethra

Labia – fleshy lobes that protect the opening of the vagina

Opening of vagina

Anus

(a) Front view

(b) Side view

(c) Section through an ovary showing an egg surrounded by the cells which form the follicle (x200). Each follicle contains one egg. At birth each ovary contains approximately 300 000 follicles. Of these, approximately 300 complete their development in a woman's lifetime.

Figure 15.5B ● The female reproductive system

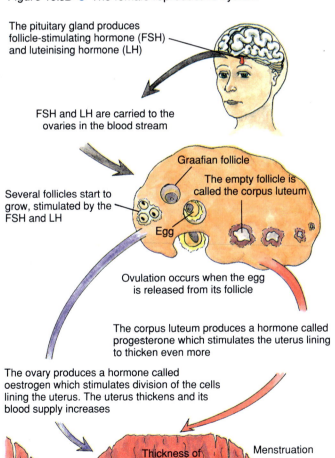

The pituitary gland produces follicle-stimulating hormone (FSH) and luteinising hormone (LH)

FSH and LH are carried to the ovaries in the blood stream

Several follicles start to grow, stimulated by the FSH and LH

Graafian follicle

The empty follicle is called the corpus luteum

Egg

Ovulation occurs when the egg is released from its follicle

The corpus luteum produces a hormone called progesterone which stimulates the uterus lining to thicken even more

The ovary produces a hormone called oestrogen which stimulates division of the cells lining the uterus. The uterus thickens and its blood supply increases

Thickness of uterus lining

Menstruation

0   14   28

Days

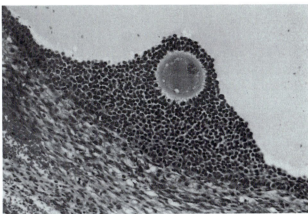

The reproductive system of a woman is illustrated in Figure 15.5B. The genitalia cover and protect the opening to the rest of the reproductive system inside her body.

Menstrual cycle

The human female usually produces one mature egg each month from the onset of puberty (age 11–14 years) to the beginning of the menopause (age about 45 years). This monthly cycle is called the **menstrual cycle** (from the Latin *mensis* meaning month). Egg production becomes more and more irregular during the menopause and stops altogether usually by about the age of 50. Figure 15.5C shows how the different events of the menstrual cycle fit together.

Figure 15.5C ● The menstrual cycle. A sharp increase in the level of luteinising hormone causes ovulation. The intervals of time for each stage may vary depending on the individual. For example, ovulation may occur earlier or later in the cycle than shown.

277

LOOK AT LINKS
for **hormones**
See Topic 16.5.

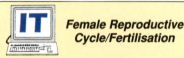

**Female Reproductive
Cycle/Fertilisation**

The disk shows the stages in the
menstrual cycle. You can also
investigate the process of
fertilisation and implantation.

The changes occuring during the menstrual cycle prepare the uterus to receive an egg if it is fertilised. If the egg is not fertilised, the production of the **hormones** oestrogen and progesterone tails off and the thick lining of the uterus begins to break down as Figure 15.5C shows. The release of blood and tissue through the vagina is called menstruation and is what is meant by 'having a period'. It lasts for several days. A new menstrual cycle then begins.

Blood released during a period can be absorbed by a sanitary towel, which a woman wears as a lining to her underwear, or as an alternative she can put a tampon made of cotton wool in her vagina (see Figure 15.5D).

Figure 15.5D ● Hygiene during a period. Sanitary towels come in different thicknesses. The woman can choose which type suits her best depending on how much blood and tissue is released during her period. The tampon is removed from the vagina by its thread.

The menstrual cycle can affect a woman's emotions. How a woman feels depends on the individual. Some have few problems but others feel irritable and below their best just before and during menstruation.

How do sperms meet eggs?

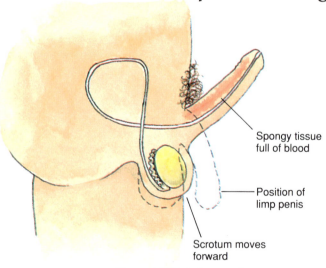

Spongy tissue
full of blood

Position of
limp penis

Scrotum moves
forward

Figure 15.5E ● An erection

An erect penis is a sign that a man is sexually excited. The penis stiffens and lengthens as blood fills the spongy tissue of the shaft (see Figure 15.5E). Signs of sexual excitement in women are less obvious. The labia fill with blood and swell a little. All of these changes in a man and woman help to prepare them for sexual intercourse.

The swollen labia help to guide the erect penis into the vagina. The muscles of the vagina wall relax, helping entry. Fluid produced by the vaginal wall lubricates the

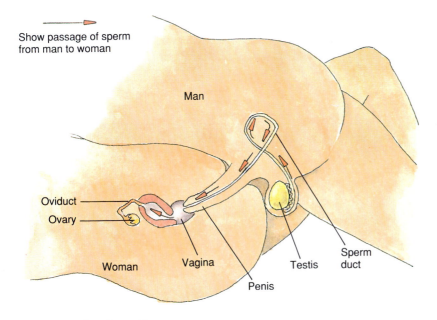

Show passage of sperm from man to woman

Figure 15.5F ● Sexual intercourse

movements of the penis during sexual intercourse. These movements stimulate the muscles in the scrotum and around the epididymis and sperm ducts to contract, pushing sperm from the testes along the sperm ducts to the urethra. During this journey the sperm mix with fluids from the seminal vesicles and prostate gland. These fluids and the sperm form semen. Continuing stimulation results in ejaculation which is a series of contractions that propel semen through the urethra into the vagina (see Figure 15.5F).

Semen is white and sticky. It contains sugars which are an energy source for the sperms as they swim up through the vagina and uterus to the oviducts.

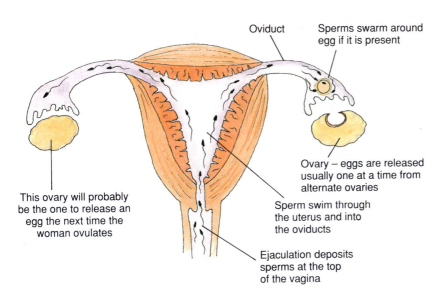

Figure 15.5G ● How sperms meet an egg. Sperms reach the top of the oviduct approximately 40 minutes after ejaculation

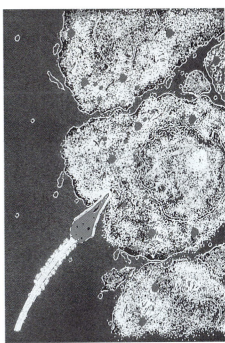

Figure 15.5H ● One sperm penetrates the membrane surrounding an egg

During ejaculation the man experiences a pleasant feeling called an orgasm. The woman may experience an orgasm as well. The muscles of the vagina gently relax and contract around the penis. The woman's orgasm is usually caused by gentle pressure stimulating the clitoris.

We have seen that ovulation releases an egg from the ovary into the opening of the oviduct (see Figure 15.5C). If a sperm is to meet an egg it must make its way from the vagina, through the uterus to the oviduct. Figure 15.5G shows the distance a sperm must travel. The journey is not an easy one. Of the hundreds of millions of sperm deposited in the vagina only a million or so make it through the cervix into the uterus. Of these, only a few thousand arrive at the end of the opening of the oviduct. Here they swarm around the egg if one is present (see Figure 15.5H).

279

Fertilisation and development of the embryo

Although thousands of sperms may reach an egg only one enters it. The tail of the successful sperm is left outside as the head travels through the cytoplasm of the egg to the nucleus. Fertilisation occurs when the sperm nucleus fuses with the egg nucleus to form a zygote (see Figure 15.5I). This is the moment of conception and the woman is now pregnant.

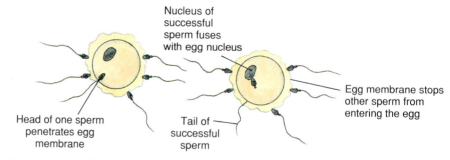

Figure 15.5I ● Fertilisation of the egg

After fertilisation, the developing embryo takes over the function of the corpeus luteum and produces progesterone. As a result:
• the lining of the uterus remains intact
• production of FSH and LH is inhibited
• the ovaries stop producing eggs and follicles.
When the baby is born progesterone levels fall, FSH and LH are released once more and the menstrual cycle restarts. Looking again at Figure 15.5C will help you remember the events of the menstrual cycle.

Figure 15.5J shows what happens next. The zygote travels down the oviduct, dividing by mitosis as it goes, forming a ball of cells. The journey may take up to seven days. By the time the ball of cells reaches the uterus it has formed an **embryo**. Remember that at this stage of the menstrual cycle (see Figure 15.5C) the wall of the uterus has a thick lining. The embryo sinks into it – a process called implantation.

Finger-like extensions called villi project from the embryo into the lining of the uterus. The surfaces firmly bind together forming a region called the placenta. In the next few weeks the embryo develops into a foetus which is attached to the placenta by the umbilical cord. An artery and a vein run through the umbilical cord and connect the foetus' blood system to the placenta. Figure 15.5K shows that the foetus' blood system is not directly connected to the blood system of the mother. The exchange of oxygen, food and wastes between mother and foetus depends on diffusion across the thin wall of the placenta.

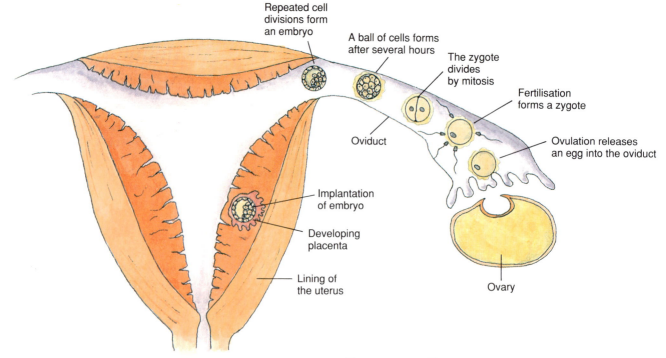

Figure 15.5J ● The stages from ovulation to implantation

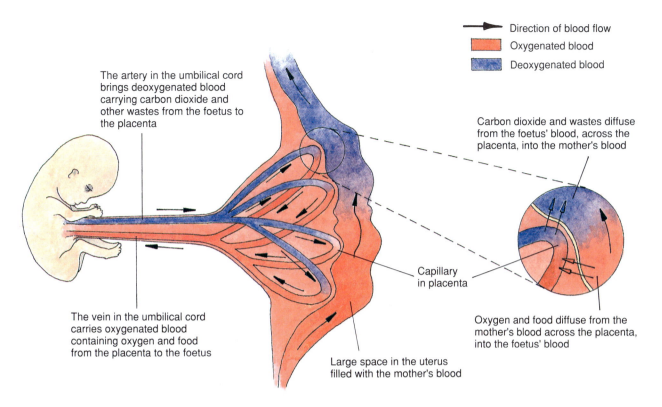

The artery in the umbilical cord brings deoxygenated blood carrying carbon dioxide and other wastes from the foetus to the placenta

Direction of blood flow
Oxygenated blood
Deoxygenated blood

Carbon dioxide and wastes diffuse from the foetus' blood, across the placenta, into the mother's blood

Capillary in placenta

The vein in the umbilical cord carries oxygenated blood containing oxygen and food from the placenta to the foetus

Oxygen and food diffuse from the mother's blood across the placenta, into the foetus' blood

Large space in the uterus filled with the mother's blood

Figure 15.5K ● The blood systems of mother and foetus exchange material by diffusion across the placenta

The time taken for the foetus to develop from conception into a baby is called the **gestation period**. It usually lasts nine months in humans. Figure 15.5L shows the growth and development of the foetus from the early stages of pregnancy to just before the time the baby is due to be born.

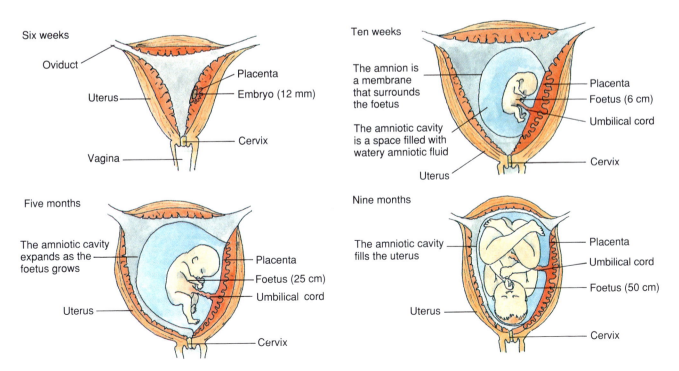

Six weeks
Oviduct
Uterus
Placenta
Embryo (12 mm)
Cervix
Vagina

Ten weeks
The amnion is a membrane that surrounds the foetus
The amniotic cavity is a space filled with watery amniotic fluid
Placenta
Foetus (6 cm)
Umbilical cord
Cervix
Uterus

Five months
The amniotic cavity expands as the foetus grows
Placenta
Foetus (25 cm)
Umbilical cord
Uterus
Cervix

Nine months
The amniotic cavity fills the uterus
Placenta
Umbilical cord
Foetus (50 cm)
Uterus
Cervix

Figure 15.5L ● Growth and development of the foetus in the uterus. The amniotic fluid in the amniotic cavity cushions the baby from bumps as the mother moves

Figure 15.5M ● Chartered physiotherapist teaching a group of antenatal mothers

Birth

The mothers pictured in Figure 15.5M are heavily pregnant. It will not be long before they give birth to their babies. They are visiting an antenatal clinic where a doctor will check that all is well with each mother and her baby. The mothers also receive advice on how best to prepare for the baby's birth.

Figure 15.5N shows childbirth in progress. Safely delivered, the baby starts to breathe, sometimes helped by a tap on the back which causes a surprised intake of breath. Now that the baby can breathe for itself, the placenta and umbilical cord are no longer needed. The placenta comes away from the uterus wall and passes out through the vagina as the afterbirth. The umbilical cord is clamped near to where it joins the baby and is cut. This does not hurt the baby because there are no nerves in the cord. The stump that remains becomes the baby's navel.

Babies born before the ninth month of pregnancy are described as premature. They have a good chance of survival providing they are not too small and weak. Premature babies are often kept in incubators. An incubator is a cabinet with a controlled environment that keeps the baby warm and provides extra oxygen to help with breathing. The baby stays in the incubator until he or she is strong enough to survive independently.

In humans, pregnancy usually results in the birth of only one baby. However, sometimes two babies are born one after the other. They are called twins and Figure 15.5O shows how this arises. The twins develop together in the uterus, each with its own placenta and umbilical cord.

Occasionally ovulation releases three or more eggs into the uterus at the same time, especially if the woman has been given a fertility drug to help her to become pregnant. These multiple pregnancies can be difficult because of the space taken up in the uterus by the growing foetuses. Very often the mother gives birth early at around the seventh month and some of the babies may die.

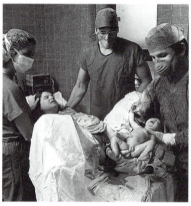

Figure 15.5N ● Childbirth

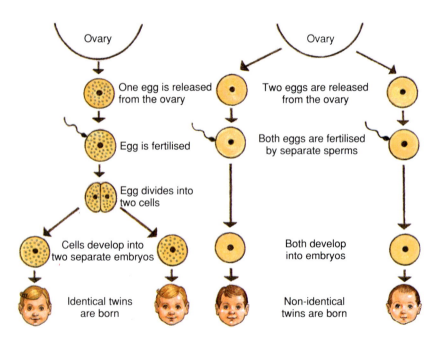

Figure 15.5O ●
(a) Identical twins are alike because they each have the same genes

(b) Non-identical or fraternal twins are different from one another because they do not have the same genes

Gestation periods are:
• Mouse – 18 days
• Cat – 2 months
• Horse – 11 months
• Elephant – 20 months
Usually the larger the animal, the longer the gestation period.

Animal mothers very often eat the afterbirth because the smell of blood might attract hungry predators. The afterbirth contains a lot of iron compounds and other nutritious substances.

Looking after baby

A newborn baby will naturally suck at the nipple of the mother's breast. Figure 15.5P shows that glands (called mammary glands) inside the breast secrete milk. This happens soon after birth and is called lactation. Mother's milk is perfect food for the baby. It contains all the necessary nutrients as well as the mother's antibodies which help to protect the baby from diseases during the first few months of life. Sometimes a mother does not produce enough milk, so the baby has to be bottle-fed (see Figure 15.5Q).

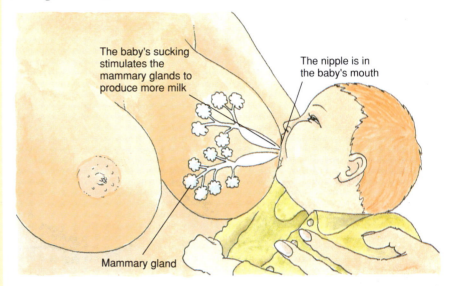

The baby's sucking stimulates the mammary glands to produce more milk

The nipple is in the baby's mouth

Mammary gland

Figure 15.5P ● A baby feeding from the mother's breast

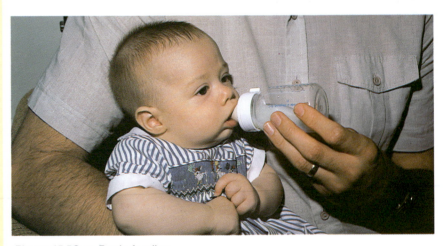

Figure 15.5Q ● Bottle feeding

A newborn baby cannot take in solid food because he or she has no teeth to chew it with. Also the digestive system is unable to deal with solid food. After about six months the first teeth appear and solid food can now be added to the baby's diet. At this stage the baby's milk intake decreases. **Weaning** is the word used to describe the change from a diet of milk to one of solid food.

Looking after a baby is time-consuming and exhausting. Apart from feeding, the baby must be kept clean and warm. It is also very important that the emotional needs of the baby are cared for. Keeping the baby interested, stimulated and happy is just as important as looking after the baby's physical needs.

LOOK AT LINKS
for **antibodies** and **immunity**
See Topic 14.2.

LOOK AT LINKS
for **teeth**
See Topic 11.5.

Preventing pregnancy

Whether to have children or not is a choice open to everyone involved in sexual relationships. If a couple want to have sexual intercourse but do not want to have children they must use some form of contraception to prevent pregnancy. Using methods of contraception enables people to choose when they want children and how many children they want. This choice is called **family planning** or **birth control**.

To prevent pregnancy, the method of contraception must either:
* stop sperm from reaching the egg,
* stop eggs from being produced,
* or stop the fertilised egg from developing in the uterus.

Six methods of contraception are described below and illustrated in Figure 15.5R.

* **Intrauterine devices (IUDs)**. IUDs are fitted inside the uterus by a doctor. The IUD touches the inner wall of the uterus and prevents implantation of the embryo. The IUD can be removed by a doctor by pulling on the strings attached to it which pass through the cervix. IUDs are usually only used by women who have already had a child. (See Figure 15.5R(a)).
* **Diaphragms**. The diaphragm or cap is a dome-shaped device which fits over the cervix. They come in different sizes and a woman must be taught to insert the diaphragm by her doctor or at a family planning clinic. A diaphragm can be inserted immediately before intercourse but must remain in place for sometime after intercourse. Diaphragms offer more protection when used with a spermicide. (See Figure 15.5R(b)).
* **Sheaths (condoms)**. The condom is a thin sheath which is rolled on to the erect penis before intercourse. The penis must be removed from the woman's vagina immediately after ejaculation to avoid any spillage of sperm. (See Figure 15.5R(c)).
* **Spermicides**. Spermicidal creams kill sperms. A woman uses an applicator to put spermicide inside her vagina just before intercourse. Spermicides are not very effective on their own and are more often used as 'back-up' for other methods. (See Figure 15.5R(d)).

Figure 15.5R ● Methods of contraception

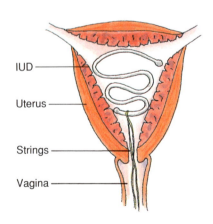

(a) Intrauterine device

LOOK AT LINKS
for **hormones**
See Topic 16.5.

Contraception Update

The female condom is a thin sheath which lines the vagina. It is closed at one end and open at the other. A ring at the closed end helps the woman to insert the condom before sexual intercourse. A ring at the open end remains outside the body, pushed flat against the labia. The penis is guided into the sheath and moves inside the lining during intercourse. When the penis is removed after ejaculation; the ring at the open end is twisted to make sure no sperm is spilt, and the sheath gently pulled from the vagina.

Injected contraceptives contain the hormone progesterone. Once injected into the arm, the hormone is slowly released into the body over the next two or three months. Like the mini-pill (a make of pill that contains only progesterone), it stops the ovaries from producing eggs. Injected contraceptives are useful for women who find it difficult to take the pill or experience problems with other methods of contraception.

- **The pill**. The contraceptive pill contains one or both of the hormones oestrogen and progesterone. The concentration of the hormones stops the ovaries from producing eggs. The woman takes a pill every day for 21 days of her menstrual cycle. When she stops taking the pill menstruation occurs. The woman begins taking the pill again on day one of her next menstrual cycle. If the woman forgets to take a pill on one day then the protection is not complete and another form of contraception must be used until the woman's next menstrual cycle begins.

    Another type of pill is the morning-after pill. It delivers a large dose of hormones which prevents implantation of the embryo. The morning-after pill must be taken within three days of intercourse to be effective.

- **Sterilisation**. Both men and women can be sterilised. Sterilisation involves a minor operation. In a man, the sperm ducts are tied off and cut by the surgeon. The man can still ejaculate as the ducts are cut below the seminal vesicles which produce seminal fluid, but his semen will not contain any sperms. In a woman, the oviducts are tied off and cut. This prevents the sperm from reaching the egg. Although some sterilisation operations can be reversed, this is not usually the case so a man or woman who is sterilised has to be very sure that he or she does not want any more children. (See Figure 15.5R(e)).

All the methods of contraception described above, and shown in Figure 15.5R depend on surgery, chemicals or a mechanical device to be effective. The rhythm method does not use any of these aids but depends on the woman (and possibly her partner) understanding how her menstrual cycle works. Look at Figure 15.5C once more and calculate at what time in the menstrual cycle intercourse is most likely to lead to pregnancy. *Why is pregnancy not possible at other times – the so-called safe period?*

Figure 15.5S (overleaf) shows the changes in a woman's body temperature during her menstrual cycle. Notice the slight increase in body temperature when she ovulates. *How do you think the woman can use the information to help her prevent pregnancy?*

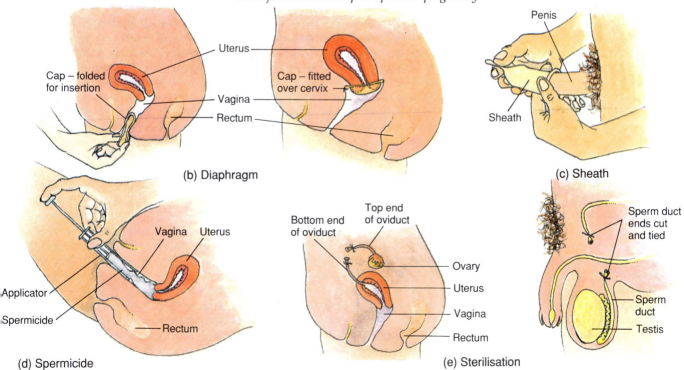

(b) Diaphragm

(c) Sheath

(d) Spermicide

(e) Sterilisation

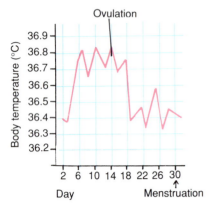

Figure 15.5S ● The temperature rise at ovulation is about 0.5 °C

Unfortunately the menstrual cycle is not always predictable. The cycle can vary a good deal, especially in teenagers, which makes it difficult to predict whether or not it is safe to have intercourse. However, the rhythm method is a natural form of contraception and it is used mostly by people whose religion does not allow other methods.

Table 15.3 compares the reliability of different methods of contraception. However you should not think that it is a matter of just looking down the list, choosing one that suits best and having intercourse when you please and with whom you please. A close physical relationship is only part of the relationship you have with a member of the opposite sex. It means that you must be old enough to take responsibility not only for your own actions and feelings but also for your partner's actions and feelings as well. Look back at *Getting to know you* on page 275 and think again about your answers to the questions now that you have read most of this section. *Have your opinions changed?*

Table 15.3 ● The reliability of contraceptive methods

Method	Percentage of pregnancies	How reliable is the method?
No method	54	Very unreliable
Rhythm method	17	Unreliable without expert help
Diaphragm with spermicide	12	Quite reliable when fitted well
Sheath	8	Quite reliable if used properly
IUDs	2	Reliable
The pill	0	Very reliable

Sexually transmitted diseases

Sexually transmitted diseases (sometimes called venereal diseases or VD) are a group of diseases which can pass from person to person during sexual activity. Syphilis and gonorrhea are both sexually transmitted.

Syphilis is caused by the bacterium *Treponema pallidum*. The bacterium can live for only a short time outside the body and is very quickly killed by heat, lack of water and antiseptics. The bacterium that causes gonorrhea is quickly killed in similar fashion. This is why it is very difficult to pick up syphilis and gonorrhea other than by the intimate contact between two people having sex. Table 15.4 summarises the symptoms of the two diseases in men and women.

Table 15.4 ● Syphilis and gonorrhea in men and women

Syphilis		Gonorrhea	
Men	Women	Men	Women
Sores appear on the genitals weeks or sometimes months after sexual intercourse		Becomes painful to pass urine; yellow discharge from penis	Many women show no symptoms but it may become painful to pass urine and there may be a yellow discharge from the vagina
Symptoms disappear		Symptoms disappear	
If syphilis is not treated, years later it can cause blindness, heart trouble, insanity and eventually leads to death	Same effects as in men In addition, babies can be badly affected in the uterus	In the long term, sperm ducts become blocked leading to sterility. May also lead to bladder problems	Oviducts become blocked resulting in sterility. Babies affected in the uterus may be born blind

Other diseases that may be passed from person to person by sexual activity include:

- Herpes – which is caused by a virus similar to the kinds that cause cold sores and chicken pox. Blisters appear, usually on the glans of the penis and inside the vagina. Unfortunately, once infected a person remains infected for life and the blisters often recur.
- AIDS – which is caused by the human immunodeficiency virus (HIV).

LOOK AT LINKS
for **AIDS**
See Topic 14.2.

These diseases are not always caught through having sex with an infected person. AIDS, for example, is spread when a person's blood infected with HIV mixes with someone else's blood. This is how many of the people suffering from haemophilia have become infected with HIV. They picked up the virus from the blood clotting agent factor VIII which had been donated by HIV-infected people. Now blood, and blood products like factor VIII are screened for HIV before being given to patients.

Avoiding sexually transmitted diseases means avoiding sexual intercourse with a person who is infected. *How can you tell if someone has a sexually transmitted disease?* The short answer is, you cannot, but the chances of becoming infected are considerably reduced if you only have sex within the context of a stable relationship and do not have a lot of sexual partners.

People who are worried that they have been infected with a sexually transmitted disease can go to a special clinic (most large hospitals have one) where they are examined and if necessary treated. Antibiotic drugs like penicillin and streptomycin are used to treat syphilis and gonorrhea. They will cure the disease providing treatment is started early enough. Nobody need know that treatment has been given; the hospital keeps the visit to the special clinic confidential.

Viral diseases like herpes cannot be cured with antibiotics. AIDS is a special problem for which there is no known cure at present. There are drugs that slow down the progress of HIV and scientists world-wide are trying to find new drugs and vaccines to fight the disease.

CHECKPOINT

❶ Briefly describe the route taken by ejaculated sperm from where they are produced to the oviduct of the woman.

❷ The uterus can be the most powerful muscle in the body. Why does it need to be so powerful?

❸ (a) A diaphragm is fitted over the cervix: how does it work as a contraceptive?
   (b) How do condoms (or sheaths) work as contraceptives?

❹ Complete the following paragraph using the words below. Each word may be used once, more than once or not at all.

fertilisation vagina testes weeks semen bladder urethra
sperm duct seven sperm seminal cervix penis uterus sexual
ovaries prostate oviduct

An egg is released by one of the two _____ about every four _____ . It passes into the _____ . It may take up to _____ days to reach the _____ . If _____ is to take place the egg must be met by _____ before, or just after, it reaches the _____ . Sperm are produced in the tubules of the _____ in vast quantities. Ejaculation forces the sperm from the epididymis, into the _____ . The _____ vesicle, and _____ gland add their secretions to the sperm, forming _____ . This leaves the body through the urethra running through the _____ .

❺ Name three substances which show a net movement into foetal blood across the placenta, and three substances that show a net movement out of the foetal blood across the placenta.

❻ Why is birth a considerable shock to the baby?

❼ Complete the following paragraph using the words provided below. Each word may be used once, more than once or not at all.

cervix uterus placenta oxygen oviducts oxygenated wastes
amniotic vagina muscles implants

Once fertilisation has occurred, normally in one of the _____ the embryo grows, moves into the uterus and _____ into its wall. The _____ develops which provides a surface for the exchange of materials with the mother's blood. _____ and food cross into the foetal blood, whereas carbon dioxide and other _____ enter the mother's circulation. The developing foetus is surrounded and protected by the _____ fluid. At birth the _____ dilates, and powerful contractions of the _____ of the uterus push the baby out through the _____ .

15.6 Growth and development

The cotyledons of plants like the broad bean stay below ground during germination. During germination of plants like the French bean the cotyledons are carried above ground by the developing shoot.

How plants grow

Trees and other living things grow as the number of cells making up the body increases. Growth can be measured as an increase in an organism's mass, length or number of cells and the data can be represented as a growth curve. Figure 15.6A shows the growth curve of a broad bean seed as the embryo inside it develops into a new plant.

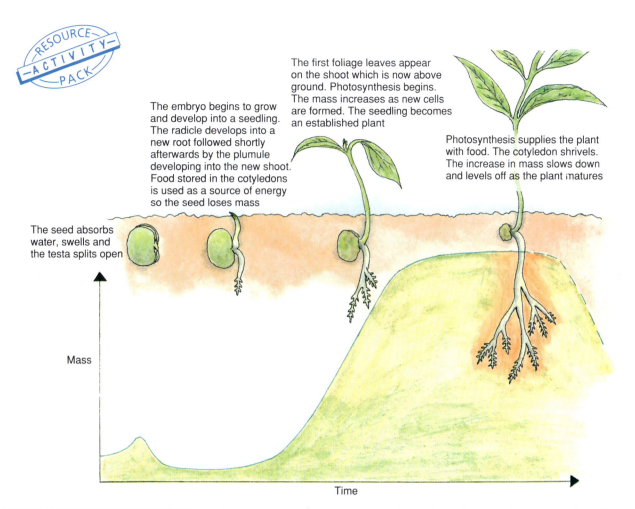

The embryo begins to grow and develop into a seedling. The radicle develops into a new root followed shortly afterwards by the plumule developing into the new shoot. Food stored in the cotyledons is used as a source of energy so the seed loses mass

The first foliage leaves appear on the shoot which is now above ground. Photosynthesis begins. The mass increases as new cells are formed. The seedling becomes an established plant

Photosynthesis supplies the plant with food. The cotyledon shrivels. The increase in mass slows down and levels off as the plant matures

The seed absorbs water, swells and the testa splits open

Mass

Time

Figure 15.6A ● Growth curve of a broad bean from seed to mature plant. The stages of growth from the embryo to the time when the seedling no longer depends on stored food is called germination

LOOK AT LINKS
for **xylem** and **phloem**
See Topic 14.1.

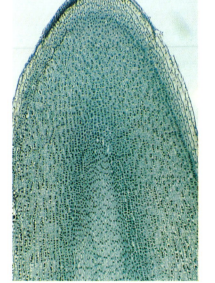

Figure 15.6B ● The growing root tip

The growth curves of nearly all annual plants look like Figure 15.6A. The dotted line represents flowering time followed by the formation of fruits and seeds which are then dispersed. The plant then quickly loses mass and eventually dies.

Figure 15.6B shows a root tip. Behind the tip of the developing root of a germinating seed there is a region where cells divide very quickly. Behind the region of dividing cells is another region where cells become longer, growing to ten times or more their original length.

As the cells get longer they become different from one another: we say that they become **differentiated**. Some form the sieve cells of the phloem, others form xylem cells. Differentiation in the shoot is more complicated than in the root since the developing shoot produces leaves and flowers.

Differentiation therefore, lays down the tissues of the plant body. The word **primary** is used to describe these tissues because they are the first tissues to develop. Plants which have only primary tissues are called **herbaceous** plants.

● *Where does growth occur in plants?*
In animals most cells can divide by mitosis. This means that nearly all parts of the body can grow. In plants only some cells divide by mitosis. These cells form the growing points of a plant, mostly at the tips of the shoots and roots.

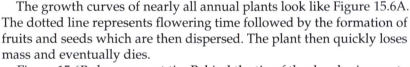

289

IT
Seed Germination
(program)

The program shows experiments to find out the best conditions for seeds to germinate. The factors that matter are light level, temperature, oxygen level and the presence of water. Find out how well the seeds develop under conditions which you select.

LOOK AT LINKS
Buttercups are herbaceous plants; beech trees are perennials. Wood makes the difference.
See Topic 1.5.

Figure 15.6C ● Annual rings in the cut end of the trunk of a beech tree

Trees grow year after year because they contain cells which divide to produce more and more xylem. Such growth is called secondary growth because the new xylem develops after differentiation during germination has laid down the primary tissues. This is how wood forms. It is impregnated with lignin which gives it extra strength.

In countries like the UK with a temperate climate, wood grows seasonally. The cycle is repeated each year to form cylinders of wood, which in cross-section appear as annual rings (see Figure 15.6C).

What controls growth and development?

The zygote of a multicellular organism looks very simple under the microscope but it is the starting point for the development of a new individual. The instructions for development are contained in its genes. These are duplicated many times as the embryo develops. Even though all the cells of the multicellular individual have the same set of genes, different types of cell develop (differentiate) to do different biological jobs. *How does the same set of genes produce the many different kinds of cell which make up the individual organism?* It seems that part of the time only some genes are active in each cell type.

Development is an exact sequence of events which lays down the different features of the embryo in the right place at the right time. This means that the genes which produce these features must be switched on and off in the right order. **Hormones** seem to play an important part in switching genes on and off. The hormone thyroxine, for example, controls the growth and development of amphibian tadpoles into adults. The change is called **metamorphosis**. When the production of thyroxine in a tadpole is stopped, the tadpole does not metamorphose into an adult but grows into a giant tadpole.

Hormones also control insect development. The hormone ecdysone makes the young insect grow and moult its outer body covering, the exoskeleton. It does this in the correct sequence by switching on the genes which control these processes at the right time in development.

LOOK AT LINKS
for **hormones**
See Topic 16.5.

How insects grow

The growth curve for insects is not like the smooth growth curve for plants shown in Figure 15.6A or that for humans shown in Figure 15.6G. It is stepped (see Figure 15.6D).

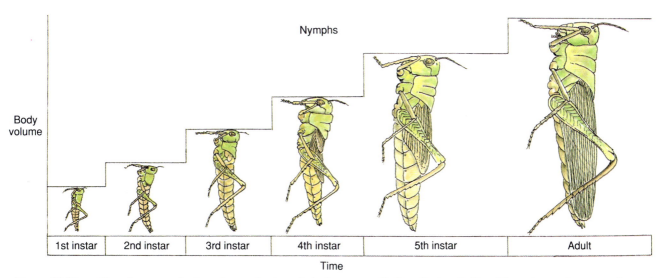

Figure 15.6D ● Growth curve of a locust – each stage in between moults is called an instar. A locust nymph passes through five instars before it becomes an adult. It increases in size at every moult.

Insects grow like this because the hard exoskeleton that surrounds the body cannot stretch. An increase in length occurs only when the old exoskeleton is removed (called moulting or ecdysis) and replaced by a new one. The body tissue expands while the exoskeleton is still soft and stretchy.

Insects change a lot as they moult and grow into adults. The changes are another example of metamorphosis. In insects like locusts, metamorphosis is gradual. Young locusts (called hoppers) look rather like miniature adults except that the wings and sex organs are not developed. With each moult (five in all) the hoppers get bigger as Figure 15.6D shows. At the last moult they become adults with fully developed wings and sex organs. This gradual change is called **incomplete metamorphosis** and the young of insects which show incomplete metamorphosis are called nymphs (see Figure 15.6E).

Figure 15.6E ● Nymphs and adults (a) Locust   (b) Earwig

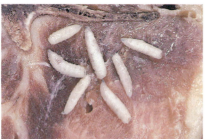

Figure 15.6F ● Larvae, pupae and adults. The larvae of different insects do not look alike. Butterfly and moth larvae have a distinct head and short, stumpy legs. They are called caterpillars. Blowfly larvae are much simpler; they are called maggots.

Metamorphosis is more dramatic in insects like butterflies, moths and flies. The young are called larvae and do not look at all like the adult. They moult and grow but then turn into pupae. The final changes into the adult take place inside the pupa. When the changes are complete the adult emerges from the pupa, dries off and flies away. This dramatic change from young to adult is called **complete metamorphosis** (see Figure 15.6F).

How humans grow

A baby grows to become a child; children develop into adolescents who become adults at around the age of 20 years. These are the stages signposting the route of human growth and development. Figure 15.6G shows how different parts of the body increase in size during the first twenty years of a person's life. Notice that different parts of the body grow at different rates because cell division occurs more quickly in some parts than in others. A child's head is bigger in proportion to the rest of the body compared with an adult's. Growth of arms and legs speeds up during adolescence resulting in the proportions of head and body we see in adults.

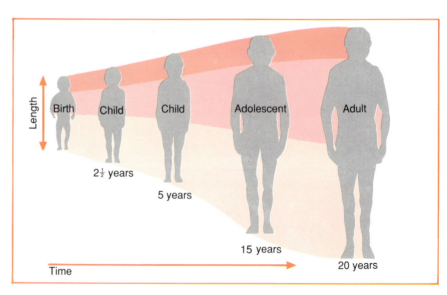

Figure 15.6G ● Human growth

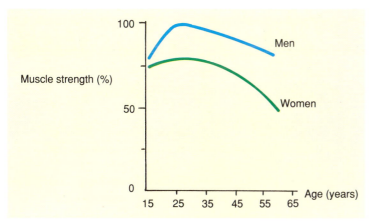

Figure 15.6H ● Muscle strength in men and women of different ages

Ageing begins in the middle twenties. By ageing we mean not only advancing years but also loss of fitness. Different aspects of fitness peak at different times, and individuals vary, especially between the sexes. For example Figure 15.6H shows the peak and decline of average muscular strength in men and women as they grow older.

Exercise helps to slow the ageing process so that early death from heart disease, for example, is less likely. Slowing down ageing does not mean living longer (although this may happen) but rather that good health is enjoyed for a much greater part of life.

CHECKPOINT

❶ Which region of the root tip contributes most to the extension of the root through the soil? Give reasons for your answer.

❷ Where in the root tip are you likely to find:
(a) the smallest cells,
(b) cells containing the highest percentage of water?

❸ Complete the following paragraph using the words provided below. Each word may be used once, more than once or not at all.

nucleus expand mitosis grow newly differentiate transport large water dividing small osmosis elongation

Roots _____ at their tips. The cells of the region immediately behind the root tip divide by _____ . Repeated divisions mean that there are _____ numbers of _____ cells. Further back these _____ formed cells begin to _____ due to the uptake of _____ by _____. This expansion produces rapid _____ of the root through the soil. We say that they _____ .

❹ Figure 15.6A shows the growth curve of a broad bean during germination. Study it and answer the following questions.
(a) Define the word 'germination'.
(b) Why does the mass of seed increase as germination begins?
(c) After the initial increase, why does the germinating plant then lose mass?
(d) After losing mass why does the germinating plant then increase in mass?

❺ The diagram below shows changes in the proportions of the human body from birth to adulthood. Analyse the changes in proportions of the head and the rest of the body in relation to total body length at each stage of development. Write a brief report of your analysis suggesting reasons for the changes.

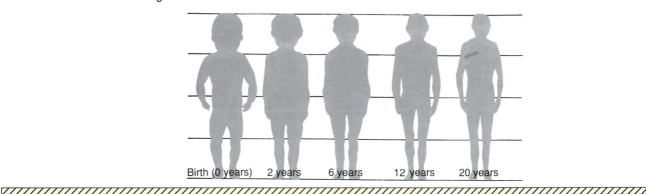

Birth (0 years)   2 years   6 years   12 years   20 years

TOPIC 16 ● RESPONSES AND CO-ORDINATION

FIRST THOUGHTS

16.1 ● Plant responses: growth movements

> Plants really do move – not in the same way as animals which use muscles, but by the way they grow in response to stimuli. Read on and find out about this.

'Responses' and 'stimuli' are everyday words but what do we mean when we talk about living things responding to stimuli? Think of it like this. The environment is changing all the time. Some changes are long-term, others short-term. Because these changes cause plants and animals to take action, the changes are called **stimuli**. The actions which plants and animals take are called **responses**. Being able to respond to stimuli means that living things can alter their activities according to what is going on around them (see Figure 16.1A).

Figure 16.1A ● Responses to stimuli
(a) Some species of the mimosa plant close up when touched. *What is the stimulus: what is the response?*

(b) People respond to loud noises by trying to shut them out. *Why can loud noise be dangerous?*

(c) Stimuli come from the environment inside the body as well as outside. *What is the stimulus: what is the response?*

Plant growth movements

Do you keep plants in the house? If you do you may have noticed that they bend towards the window. Plants do this because light is a stimulus. Plants respond by growing towards the light (see Figure 16.1B). The benefit to the plant of this response is clear: the leaves receive as much light as possible for photosynthesis. Stems twist and turn and flowers and leaves move in daily rhythms to follow the light.

Plant movements are growth movements in response to stimuli. There are two kinds of growth movement:

- **Nastic movements** are responses to stimuli which come from all directions. For example, temperature change is the stimulus for tulip and crocus flowers to open and close. They open when the temperature rises and close when it falls.
- **Tropic movements (tropisms)** are responses to stimuli which come from one direction. Tropisms are positive if the plant grows towards the stimulus; negative if it grows away (see Figure 16.1C).

Figure 16.1B ● Responding to light – these plants have bent towards the window

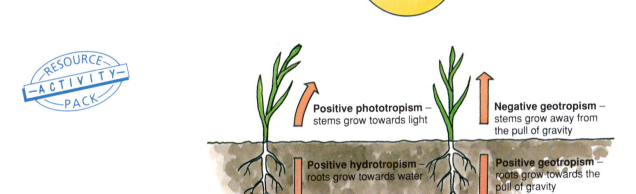

Figure 16.1C ● Different tropisms

Positive phototropism – stems grow towards light

Negative geotropism – stems grow away from the pull of gravity

Positive hydrotropism – roots grow towards water

Positive geotropism – roots grow towards the pull of gravity

● *Control of growth movements*

Many factors affect the growth of plants. Important among them are compounds called **plant growth substances**. Auxin was the first plant growth substance to be discovered. It is produced in shoot tips. Under the influence of auxin the cellulose walls of plant cells become more elastic and the cells elongate rapidly. Figure 16.1D shows how auxin enables the plant to grow towards the light.

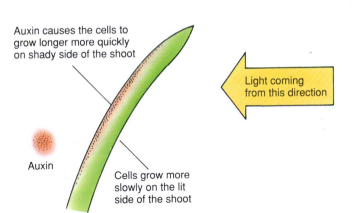

Auxin causes the cells to grow longer more quickly on shady side of the shoot

Light coming from this direction

Auxin

Cells grow more slowly on the lit side of the shoot

Figure 16.1D ● Auxin distribution. There is more auxin in the side of the shoot tip in the shade. The shoot bends towards the light because auxin makes the cells grow faster.

Experiment 1

Key

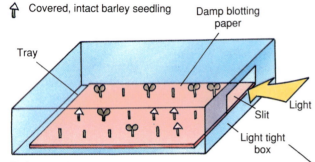

🌱 Uncovered, intact barley seedling

▯ Uncovered barley seedling with its shoot tip (5 mm) removed

⬆ Covered, intact barley seedling

Tray

Damp blotting paper

Light

Slit

Light tight box

Barley seedlings about 25 mm high are grown in a light tight box with a slit at one end. Foil caps on some seedlings exclude light from the shoot tip

Two days later

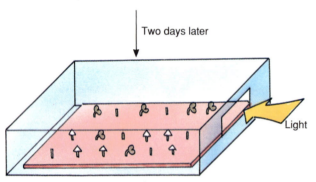

Light

The seedlings with their shoot tips removed or covered by foil caps do not grow towards the light. The uncovered, intact seedlings grow towards the light

Figure 16.1E ● (a) Does the whole shoot or just the shoot tip respond to light?

Experiment 2

Key

🌱 Intact barley seedling

🌱 Barley seedling with tip cut off Tip placed on a slip of metal foil and replaced on the rest of the shoot. The metal foil prevents diffusion of chemicals from the shoot tip to the rest of the shoot

🌱 Barley seedling with tip removed. Tip placed on agar block and replaced on the rest of the shoot.
The agar block allows diffusion of chemicals from the shoot tip to the rest of the shoot.

🌱 Barley seedling with tip cut off. An agar block is placed on the remaining part of the shoot

🌱 Barley seedling with tip cut off. An agar block is soaked in a mash made of the shoot tip and then placed on top of the remaining part of the shoot

🌱 Barley seedling with tip cut off. An agar block is soaked in a solution of auxin and then placed on top of the remaining part of the shoot

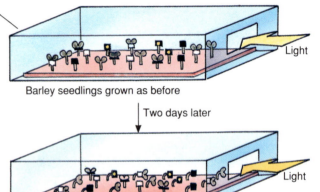

Light

Barley seedlings grown as before

Two days later

Light

The seedlings with their tips cut off and replaced with plain agar blocks and the seedlings with their tips separated from the rest of the shoot by metal foil do not grow towards the light. The other seedlings grow towards the light.

(b) Is it a chemical produced in the shoot tip or just the presence of the shoot tip itself which causes the response to light?

Two experiments designed to investigate the responses of plants to light are illustrated in Figure 16.1E. Examine the diagrams carefully. *What conclusions do you draw from the results of the experiments? Is it the shoot tip that responds to light? Is there a chemical in the shoot tip which controls growth?*

After the discovery of auxin the search was on for other plant growth substances. Today we know that a range of substances control plant growth.

The **herbicide** 2, 4-D is a synthetic auxin. It kills plants by making them grow too fast. However, narrow-leafed plants, including the food crops barley, wheat and oats are not affected by 2, 4-D at concentrations which destroy broad-leaved plants like docks, daisies and dandelions. Farmers spray wheat with 2, 4-D. The treatment kills the broad-leaved weeds which would otherwise compete with the wheat crop for growing space, nutrients and water. The yield of wheat increases and the extra money the farmer receives more than offsets the cost of buying and using the chemical.

IT'S A FACT

Growers control ripening by keeping fruit in sheds in an atmosphere that contains ethene. As little as one part of ethene per million parts of air is enough to speed up ripening.

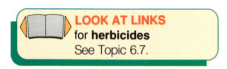

LOOK AT LINKS
for **herbicides**
See Topic 6.7.

❶ Comment critically on the evidence that shoot tips respond to light.

❷ Complete the paragraph below using the words provided. Each word may be used once, more than once or not at all.

tip phototropism auxin geotropism elongate nastic towards faster

The _____ of the shoot produces a growth substance called _____ which causes cells behind the tip to _____ . When the shoot is lit from one side, _____ accumulates on the shaded side. The cells on the shaded side grow _____ and the shoot bends _____ the light. The response of the plant is called positive _____ .

16.2 Senses and the nervous system

FIRST THOUGHTS

Stimuli are converted by receptors into signals to which the body can respond. The signals are called nerve impulses. Neurones (nerve cells) conduct nerve impulses to muscles which respond by contracting. Muscles are called effectors. Nerves are formed from bundles of neurones and are the link between stimulus and response. The sequence reads: stimulus → receptor → nerves → effector → response. Remember the sequence as you read this section.

Figure 16.2A ● Dog and man respond vigorously to stimuli

IT The Nervous System (program)

The Nervous System 1 and 2 contain diagrams of the central nervous system, the structure of the brain, nerves and neurones.

Look at Figure 16.2A – it illustrates important points about the way animals respond to stimuli.

Eyes and ears contain specialised **sensory receptor** cells which convert stimuli into signals the body can respond to. The signals are minute electrical disturbances called **nerve impulses**. They are messages for the man's leg muscles to start working hard. The muscles are called **effectors** because they respond to nerve impulses. Specialised cells called **neurones** (nerve cells) conduct nerve impulses to their destination. Each second, thousands of nerve impulses arrive at the muscle cells making them contract vigorously.

The sensory cells of the man's eyes and ears in Figure 16.2A are linked by neurones to the cells of his leg muscles. Sensory cells and muscle cells are at the beginning and end of the process that allows the man to respond to the fierce dog. The process runs

Stimulus ➜ Receptor ➜ Nerves ➜ Effector ➜ Response

LOOK AT LINKS

The structure and function of the eye is examined in Topic 16.4.

LOOK AT LINKS

The structure and function of the ear is examined in Topic 16.3.

Who's behind the science

In 1829 Louis Braille, who was blind from the age of three, invented a system of writing for the blind. Letters are represented by different combinations of raised dots on paper. The dots are then read by touch.

IT'S A FACT

The fastest nerve impulses in humans travel at 8 m/s.

SCIENCE AT WORK

Computer scientists developing artificial intelligence are studying the pathways of the human brain. A computer that thinks for itself is not likely to be developed for many years yet!

The senses

Our senses keep us in touch with the world around us. Each one consists of sensory cells adapted to detect a particular type of stimulus. The sensory cells of the eye detect light. The sensory cells of the ear detect sound. The sensory cells of the nose and tongue detect different chemicals. These different types of sensory cell are parts of complex organs located in the head. These organs are called organs of special sense. The rest of the body has sensory cells too. Different sensory cells detect pressure, pain, cold and heat.

Neurones

Neurones link sensory cells with effectors. In all animals they are similar in structure and in the way they work. Figure 16.2B shows a human motor neurone. A nerve impulse can travel along an axon in milliseconds. It takes more than one neurone to link the cell of a sensory receptor with an effector cell. A number of neurones link up to form a route to take nerve impulses to their destination.

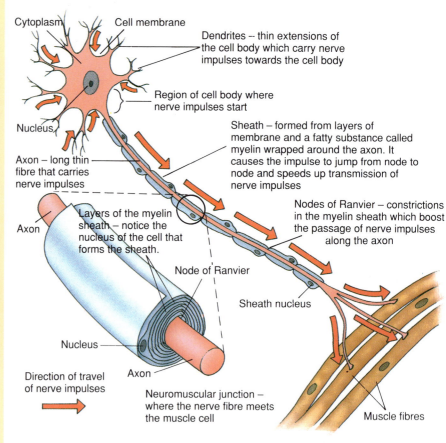

Figure 16.2B ● Human motor neurone causing a muscle to contract. The axons passing to the leg muscles can be up to a meter in length

Minute gaps called **synapses** separate neurones from one another. Each synapse separates the ends of the axon of one neurone from the dendrites of the next (see Figure 16.2C). When nerve impulses arrive at the end of the axon they stimulate the production of a special chemical called neurotransmitter which diffuses across the synapse to the dendrites of the neighbouring neurone. The neurotransmitter stimulates the dendrites to fire off a new nerve impulse.

The dendrites of different neurones may form synapses with many incoming axons. This allows for an enormous number of linkages. Nerve impulses, therefore, may be switched from one pathway to another within the billions of neurones that make up the nervous system.

● *Nerves and nervous systems*

Neurones are grouped together into bundles called nerves (see Figure 16.2D) which pass to all parts of the body forming a nervous system. Most animals have a front end and a rear end. The front end forms a head which leads the rest of the body into new environments. The ability to detect changes in front of the body, therefore, is especially important. In *Planaria*, for example, sensory cells grouped together into simple organs are concentrated in the head for this purpose (see Figure 1.5J).

Figure 16.2E shows that in different animals some nerves form cords of tissue which run the length of the body. The plan of the nervous system is similar for all the animals shown although they are classified in different phyla. At the head end the nerve cord expands into a brain or a brain-like swelling which makes numerous connections with the sense organs in the head. The sense organs feed information as nerve impulses to the brain. The brain interprets the information and sends nerve impulses through the nerve cords and nerves to effectors. This is how information about its environment travels through an animal so that its effectors can respond in a useful way. The process is called **co-ordination**.

The nerve cord(s) and brain form the **central nervous system**. The nerves that join the central nervous system form the **peripheral nervous system**.

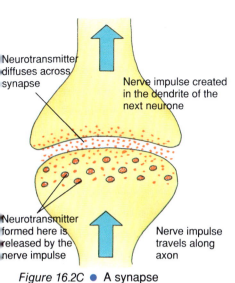

Neurotransmitter diffuses across synapse

Nerve impulse created in the dendrite of the next neurone

Neurotransmitter formed here is released by the nerve impulse

Nerve impulse travels along axon

Figure 16.2C ● A synapse

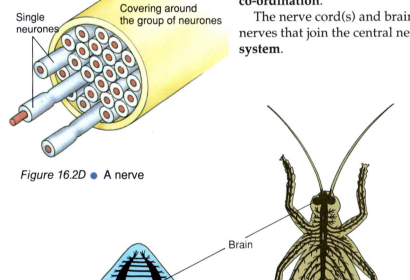

Single neurones

Covering around the group of neurones

Figure 16.2D ● A nerve

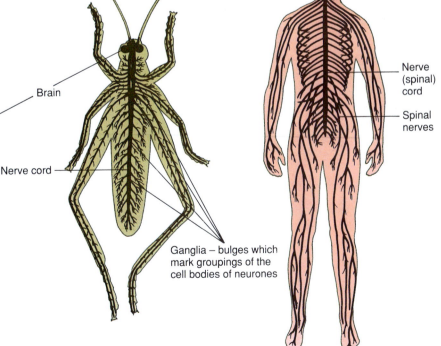

Brain

Nerve cord

Nerve cord

Ganglia – bulges which mark groupings of the cell bodies of neurones

Brain

Cranial nerves

Nerve (spinal) cord

Spinal nerves

Planaria   **Locust**   **Human**

Figure 16.2E ● Plans of the nervous systems of *Planaria*, the locust and the human

Nerve
(program)

The basic processes of the nervous system are shown in the program. The second part contains simulations of sciatic nerve experiments.

● *Reflex responses*

If you touch a hot stove you automatically move your hand away. We call the response a **reflex response** and the nerves involved form a **reflex arc** (see Figure 16.2F).

Different types of neurone form a reflex arc. A synapse separates each type of neurone from the next neurone in the arc.

- **Sensory neurones** transmit nerve impulses from sensory receptors to the central nervous system. (Figure 16.2F.) When you touch a hot object a pain-sensitive receptor cell in your finger detects the stimulus – heat – which triggers off nerve impulses. These are transmitted to the nerve cord by sensory neurones.
- **Relay neurones** receive nerve impulses from the sensory neurones and pass them to the motor neurones.
- **Motor neurones** receive nerve impulses from the central nervous system and transmit them to the effector. (Figure 16.2F.) In this case your arm muscles then contract, lifting your finger out of harm's way.

Reflex actions often occur before the brain has had time to process the information. However, when the brain catches up with the events it then takes over and brings about the next set of reactions. These reactions could be a shout of pain or a decision to switch off the stove.

The bundles of neurones running up and down the nerve cord are called ascending and descending fibres. They form a zone of tissue called **white matter**. The white colour comes from the pale myelin sheaths that cover the axons. In the core of the nerve cord lies an H-shaped mass of **grey matter** that consists mainly of the cell bodies and axons of relay neurones.

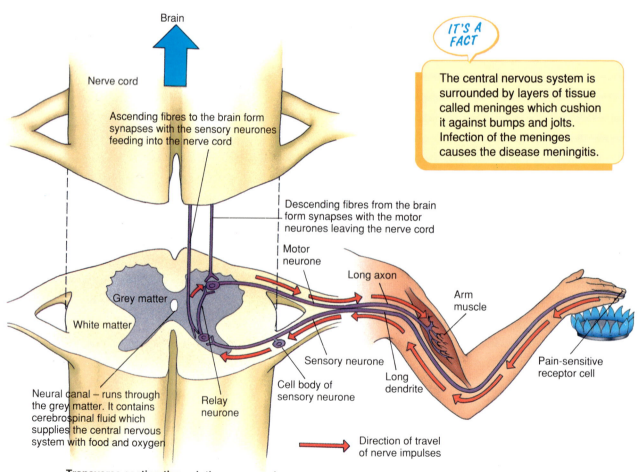

IT'S A FACT

The central nervous system is surrounded by layers of tissue called meninges which cushion it against bumps and jolts. Infection of the meninges causes the disease meningitis.

Transverse section through the nerve cord

Figure 16.2F ● A reflex arc

● Conditioned reflexes

The influence of the brain on reflex responses was first investigated in a scientific way by the Russian physiologist Ivan Pavlov (1849–1936).

Pavlov noticed that when food was placed in a dog's mouth the flow of saliva increased. He also noticed that the flow of saliva increased as soon as the animal smelt his hand, even before the food was placed in its mouth. The salivary reflex was made stronger by following Pavlov's personal smell with the taste of food. After a period of presenting the dog with both the personal smell and the taste of food, the personal smell alone was enough to make the dog produce as much saliva as if it were given food.

Pavlov used the word 'conditioned' to describe the dog's response because it could be switched on by a non-food stimulus which the dog associated with the meal. Later work showed that in conditioned dogs new nerve pathways had been made. These connected the salivary reflex with other nerve circuits of the nerve cord.

Pavlov also conditioned dogs to salivate in response to other stimuli – the ringing of a bell for example. Conditioning fades unless it is periodically reinforced. Conditioning by personal smell, therefore, must be reinforced from time to time with a meal if the dog's salivary reflex is to remain conditioned. Try out the experiment on your dog or cat if you have one.

Figure 16.2G ● Ivan Pavlov

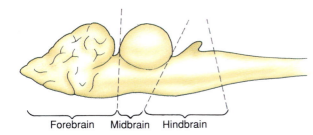

Forebrain   Midbrain   Hindbrain

Figure 16.2H ● Section lengthways through a frog's brain

● How the brain works

In vertebrates the brain consists of three regions: forebrain, midbrain and hind-brain. This three-part structure can be seen clearly in adult amphibia (see Figure 16.2H) but is less obvious in adult mammals because of the expansion of the forebrain. In humans the forebrain forms the cerebrum which is so large that it almost covers the rest of the brain (see Figure 16.2I).

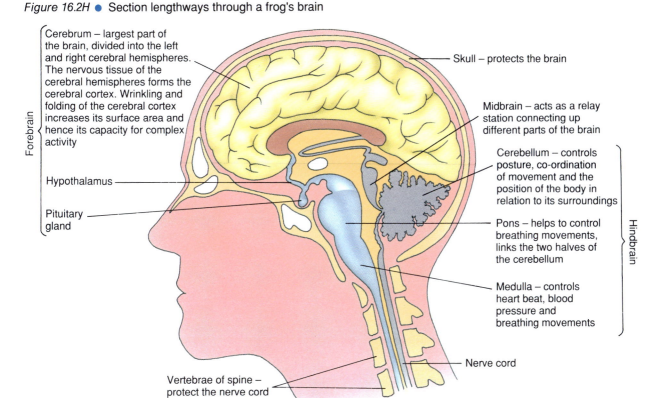

Forebrain

Cerebrum – largest part of the brain, divided into the left and right cerebral hemispheres. The nervous tissue of the cerebral hemispheres forms the cerebral cortex. Wrinkling and folding of the cerebral cortex increases its surface area and hence its capacity for complex activity

Hypothalamus

Pituitary gland

Skull – protects the brain

Midbrain – acts as a relay station connecting up different parts of the brain

Cerebellum – controls posture, co-ordination of movement and the position of the body in relation to its surroundings

Pons – helps to control breathing movements, links the two halves of the cerebellum

Medulla – controls heart beat, blood pressure and breathing movements

Hindbrain

Nerve cord

Vertebrae of spine – protect the nerve cord

Figure 16.2I ● Section through the human head showing the different parts of the brain

Figure 16.2J ● Multipolar neurones in the brain cortex (x70)

Figure 16.2J shows some of the neurones in the brain. They are called multipolar neurones because each one has numerous dendrites which can form synapses with incoming axons. Scientists estimate that up to six million cell bodies make up 1 cm³ of brain matter and that each neurone is connected to as many as 80 000 others.

The human brain weighs approximately 1.3 kg. It is the body's thinking and control centre. Reactions under the brain's control are called voluntary reactions. Memory and learning are also under the brain's control (see Figure 16.2I).

The different regions of the cerebral cortex each have different functions (see Figure 16.2K).

● **Intelligence**

What is intelligence? A difficult question. It includes the ability to decide how to tackle a problem and the ability to change your approach if it does not work.

Intelligence is not determined by a fixed centre in the association cortex. It depends on the way nerve fibres connect together in the different parts of the cortex and the way they connect the cortex with the rest of the brain. These nerve fibres form association pathways.

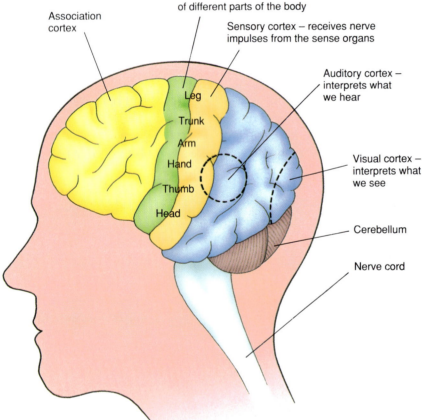

Figure 16.2K ● Functions of the cerebral cortex

Association cortex

Motor cortex – controls movement of different parts of the body

Sensory cortex – receives nerve impulses from the sense organs

Auditory cortex – interprets what we hear

Leg
Trunk
Arm
Hand
Thumb
Head

Visual cortex – interprets what we see

Cerebellum

Nerve cord

> **═══ SCIENCE AT WORK ═══**
>
> The brain is insensitive to pain. This means that during surgical operations which expose the brain, the patient needs only a local anaesthetic and can therefore report the effects of stimulating different areas of the cortex with, for example, very tiny electric shocks.

Ethanol (alcohol in beers, wines and spirits) depresses the activity of the nervous system. Small amounts affect the association cortex of the brain, which controls judgement. Larger quantities affect the motor cortex, which controls movement. Even more impairs memory. More and more alcohol affects further areas of the brain until vital brain centres that keep us alive are affected. Death may follow (see p. 185). Glue sniffing (**solvent abuse**) is a widespread problem. Glues, paints, nail varnish and cleaning fluids contain volatile solvents like esters and ethanol. Breathing them in gives a warm sense of well-being but also produces dangerous disorientation. For example, people may think that they can jump out of high windows without falling. Solvents slow down bodily functions, affecting, for example, the nerve centres which control breathing and heart rate. Long-term solvent abuse can lead to damaged liver and kidneys.

CHECKPOINT

❶ (a) What is a response?

(b) The diagram below shows a reflex arc.
   (i) Explain briefly what is happening at the points labelled A–G on the diagram.
   (ii) Name the parts numbered 1–3.

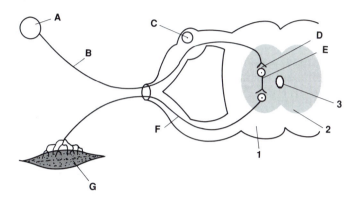

❷ Complete the following paragraph using the words provided below. Each word may be used once, more than once or not at all.

largest   nerve cord   axons   skull   vertebrae   cerebral hemispheres

The brain and _____ make up the central nervous system. They control many bodily activities and are well protected. The brain is enclosed in the _____ and the spinal cord runs through a channel in the _____ . The cerebrum is divided into two _____ . It is the _____ part of the human brain.

❸ The diagram opposite shows a motor nerve cell. Name the parts labelled A–F on the diagram and explain their function.

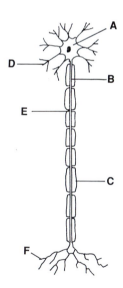

❹ We respond to stimuli all the time. When a door bell rings (stimulus), we answer the door (response): when we feel hot (stimulus), we take jumpers and coats off (response).
(a) List at least seven stimuli to which you have responded today and state your responses.
(b) Name the sensory receptors responsible for detecting the stimuli.
(c) Briefly state the role (i) sensory receptors, (ii) nerves, (iii) muscles or glands, play in the chain of events from stimulus to response.

❺ Match each of the biological terms in Column A with their functions in Column B.

Column A	Column B
Cerebellum	Transmits nerve impulses from the central nervous system to a muscle
Medulla	Transmits nerve impulses from a sense receptor to the central nervous system
Relay neurone	Controls learning and memory
Cerebrum	Controls pulse, breathing movements and other involuntary actions
Sensory neurone	Controls balance
Motor neurone	Links a sensory neurone with a motor neurone or relay neurone

16.3 The ear

Your ears are vital organs that enable you to receive information from other people. In this section you will find out how the ear works and how to test your own hearing response.

 IT'S A FACT

The fleshy lobe of the outer ear is called the **pinna**. It funnels sound waves down the ear tube to the ear drum. Cats, dogs and other mammals can adjust the pinna and cock it towards sources of sound. In most humans it is fixed. The walls of the ear tube produce wax which keeps the ear drum soft and supple.

Have you ever listened to the sound of your own voice played back on a tape recorder? Try it and you will hear yourself as others hear you. To you, your voice will sound different on the recording. When you hear yourself speak, sound waves from your voice travel through your head as well as through the air to reach your ears. Other people only receive the sound of you speaking through the air.

The human ear is a remarkable organ that can detect an enormous range of sound waves. The loudest sounds it can withstand carry over one million million times more energy than the quietest sounds it can hear. It can detect frequencies from about 20 Hz (a bee buzzing) to about 18 000 Hz (a very high-pitched whistle).

The ear sends signals to the brain in response to sound waves arriving at the **eardrum**. Sound waves arriving at the ear make the eardrum vibrate. These vibrations are passed through the middle ear by three tiny bones, the **hammer**, **anvil** and **stirrup**, to reach the **oval window** of the inner ear. The vibrations of the oval window are transmitted through the fluid of the **cochlea**, making the **basilar membrane** vibrate. Tiny hair cells, which are sound-sensitive receptors, are lined up on the basilar membrane. The vibrating membrane activates the hair cells, which fire off nerve impulses to the brain along the **auditory nerve**.

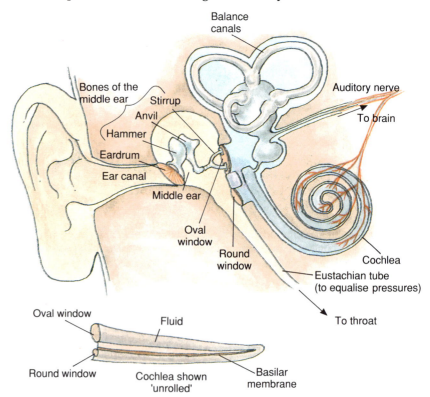

Figure 16.3A ● The human ear

What happens to the ear if very loud sound falls on it? If this happens too often, the ear becomes less and less sensitive and deafness can occur. One reason for this is that the bones of the middle ear vibrate too much and get worn down. They then become less effective at passing the vibrations from the ear drum to the oval window. Operators of noisy machines must wear ear pads or they suffer permanent loss of hearing. Protect your ears at noisy discos by putting cotton wool in your ears.

 LOOK AT LINKS
find out more about the response of the human nervous system in Topic 16.2.

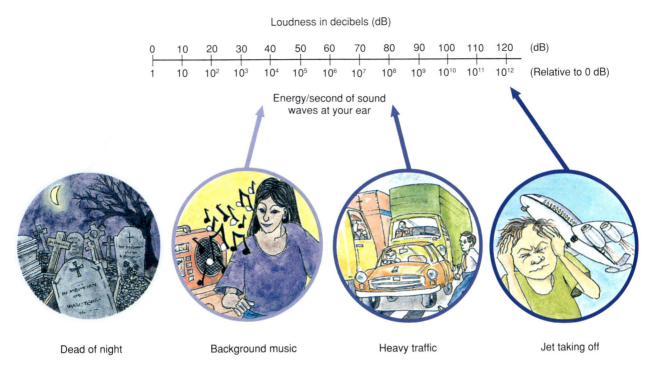

Loudness in decibels (dB)

0	10	20	30	40	50	60	70	80	90	100	110	120	(dB)
1	10	10^2	10^3	10^4	10^5	10^6	10^7	10^8	10^9	10^{10}	10^{11}	10^{12}	(Relative to 0 dB)

Energy/second of sound waves at your ear

Dead of night   Background music   Heavy traffic   Jet taking off

Figure 16.3B ● Decibel levels of everyday sounds

Loudness is measured in **decibels** (dB). The faintest sound that the ear can hear is defined as zero decibels (0 dB). Imagine steadily increasing the loudness of a radio from zero until it becomes too loud to bear. For every ten decibel (10 dB) increase in loudness, the energy of the sound waves is increased by a factor of 10. The sound would become too loud to bear at about 120 dB. Since this is 12 steps at 10 dB for each step, sound waves at this loudness carry $10 \times 10 \times 10 \times 10 \times 10 \times 10 \times 10 \times 10 \times 10 \times 10 \times 10 \times 10 = 10^{12}$ times as much energy as the faintest sound waves. Figure 16.3B shows the decibel levels of some everyday sounds.

The response of the ear to different levels of loudness varies with frequency, as shown in Figure 16.3C. The ear is most sensitive and can detect the softest sounds at about 3000 Hz. It is completely insensitive and cannot detect any sound over 18 000 Hz.

SUMMARY

The ear converts signals carried by sound waves into nerve impulses that it sends to the brain. The ear cannot detect frequencies above 18 000 Hz. Loudness levels above 120 decibels can cause deafness if the ear is not protected.

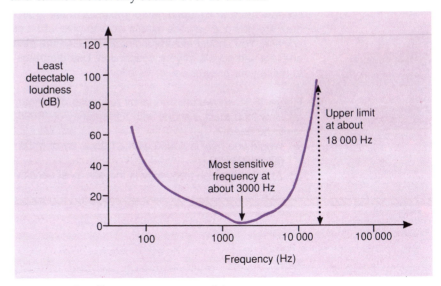

Figure 16.3C ● Frequency response of the ear

CHECKPOINT

1 (a) One of the first hearing aids was the 'ear trumpet', which was a large hollow horn held to the ear. Why do you think this device improves hearing ability?

(b) Modern hearing aids are so small that they can be worn behind the ear. Such a device contains an electronic amplifier, a tiny microphone and an earpiece speaker. The amplifier makes electrical signals bigger without changing the frequency of the signal. What is the purpose of the microphone and what is the earpiece speaker for?

2 What is the function of each of the following parts of the ear:
(a) the eardrum,
(b) the pinna,
(c) the bones of the middle ear,
(d) the oval window,
(e) the hair cells?

3 Play back your own voice using a tape recorder. How does it differ from what you hear when you speak? Compare the voices of your friends on a tape recorder. Do their voices seem different from when they speak directly to you?

4 Here is a passage from Claire's diary describing part of an evening out with her friends.

'It was very noisy and hot in the disco. We could only hear each other when the music stopped. I got a lift home with Michelle and her dad. There was a thunderstorm on the way home. When I got home, the TV was on very loud so I went to my bedroom for some peace and quiet.'
(a) When was the loudness level greatest and when was it least?
(b) Estimate the loudness when it was greatest.
(c) Which do you think was most damaging to the ears: the thunderclap or the disco noise?

5 If you are in an aeroplane coming in to land, or in a fast train entering a tunnel, you may feel 'popping' sensations in the ears. Look at Figure 16.3A and describe how the Eustachian tubes help overcome this uncomfortable feeling.

6 Notice in Figure 16.3A that the balance canals are arranged three-dimensionally, each at right-angles to the others. In each canal there is a swelling at one end called an **ampulla**, which contains a structure called the **cupula**. Liquid inside the balance canals pushes the cupula one way or the other depending on the position of the head. The cupula pulls on sensory hairs which send nerve impulses to the brain. Find out and describe how the balance canals work to give you a sense of balance and position. You should include a structure called the **utriculus** as well as the ampulla and cupula in your description. Use diagrams wherever they help to make your answer easier to follow.

7 Figure 16.3C shows the frequency response of the ear.
(a) How loud must a sound with a frequency of 18 000 Hz be in order for it to be heard?
(b) Would you hear a sound with a frequency of (i) 50 Hz (ii) 1000 Hz (iii) 25 000 Hz?
(c) At around which frequency is the ear most sensitive?

16.4 The eye

This section concentrates on the eye as an optical instrument and on its biological details. Read on to find out about sight defects and how they are corrected.

Have you ever been told to 'use your eyes' when you complain that you cannot find something or other? Think about what we use our eyes for. They tell us about colour, shape, position and movement. When you look at an object, each eye forms an image of it and sends signals to your brain. Your brain 'reads' these signals and you 'see' the object. Figure 16.4A shows four unusual views of everyday objects. Can you recognise them?

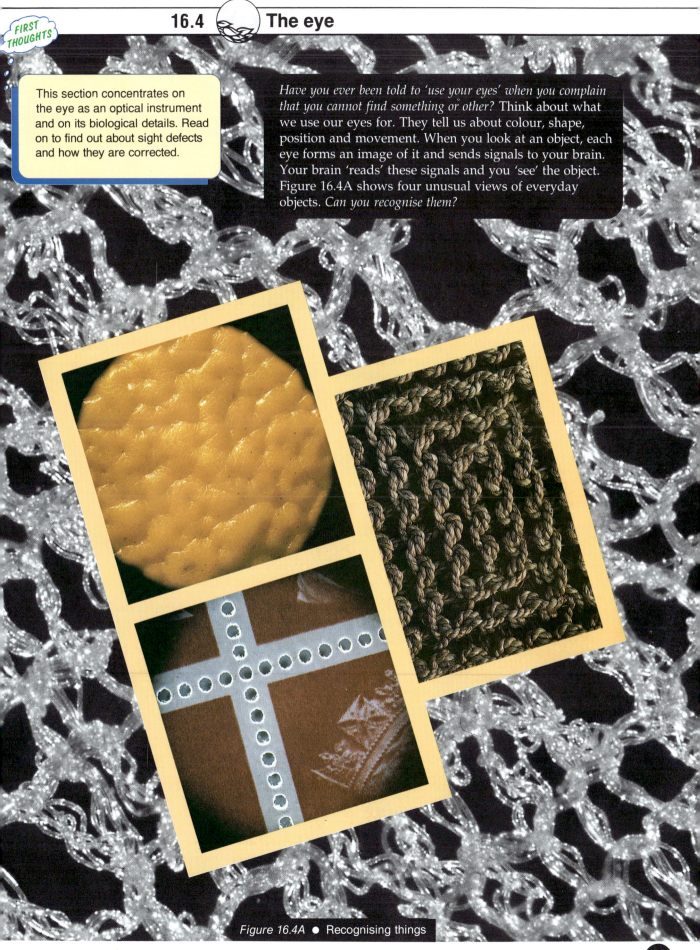

Figure 16.4A ● Recognising things

The front of the eye is covered by a thin transparent membrane called the conjunctiva. Dust particles that collect on the conjunctiva are washed away by a watery fluid from the tear glands, which are under the eyelids. This fluid contains lyosyme – an enzyme that destroys bacteria. Blinking helps to spread the fluid across the conjunctiva. When the fluid reaches the lower part of the eye, it drains into a tube and goes down into the nose.

Figure 16.4B explains how the parts of the eye work. Light enters the eye through a tough transparent layer called the **cornea**. This protects the eye and it helps to focus the light onto the **retina**, the layer of light-sensitive cells at the back of the inside of the eye. The amount of light entering the eye is controlled by the iris, which adjusts the size of the pupil – the circular opening in its centre. The eye lens focuses the light to give a sharp image on the retina. Although the image on the retina is inverted, the brain interprets it so you see it the right way up.

1. Conjuctiva membrane
2. Cornea – helps to focus light on to retina
3. Aqueous humour – transparent watery liquid that supports the front of the eye
4. Iris– coloured ring of muscle that controls the amount of light entering the eye
5. Pupil – the central hole formed by the iris. Light enters the eye through the pupil
6. Eye lens – focuses light on to the retina
7. Ciliary muscles – attached to the lens by suspensory ligaments. The muscles change the thickness of the eye lens
8. Vitreous humour – transparent jelly-like substance that supports the back of the eye
9. Retina – the light-sensitive layer around the inside of the eye
10. Fovea – region of the retina where the retinal cells are densest
11. Blind spot – region where the retina is not sensitive to light (no light-sensitive cells present)
12. Choroid – black layer of blood vessels that carry food and oxygen to the eye and remove waste products
13. Eye muscles – move the eye in its socket
14. Optic nerve – carries nerve impulses from the retina to the brain

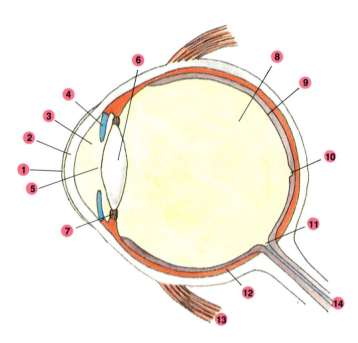

Figure 16.4B ● The human eye

Focusing

How does the eye focus on objects at different distances? If you look up from this book and gaze out of the window, your eye lens becomes thinner to keep your vision in focus. This is called **accommodation**. Your eye muscles alter the thickness of your eye lens. The muscles fibres run round the eye lens, so when they contract they shorten and squeeze the eye lens, making it thicker.

Human eyes are damaged by ultraviolet light, but insects' eyes can see in ultraviolet light. The honey guides on the petals of flowers stand out when photographed on film sensitive to ultraviolet light. Bees searching for nectar can see the honey guides very clearly.

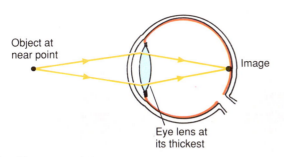

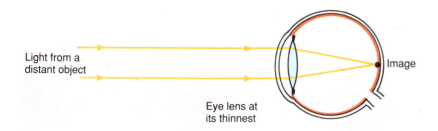

Figure 16.4C ● The near point

Figure 16.4D ● The far point

What is your range of vision? A normal eye can see clearly any object from far away to 25 cm from the eye.

- The **near point** of the eye is the closest point to the eye at which an object can be seen clearly. The eye lens is then at its thickest (Figure 16.4C).
- The **far point** of the eye is the furthest point from the eye at which an object can be seen clearly. The eye lens is then at its thinnest (Figure 16.4D).

LOOK AT LINKS
for more information about the region of the human brain responsible for interpreting visual signals
See Topic 16.2.

IT'S A FACT

There are more rods than cones. Reliable estimates are about 130 million rods and 7 million cones in a pair of human eyes.

LOOK AT LINKS
for **colour addition**
See *KS: Physics*, Topic 14.1.

Seeing shape and colour

How do we recognise the shape of an object? An image of the object is formed on the retina, which consists of lots of light-sensitive cells. When light falls on a cell, the cell sends an electrical impulse as a signal to the brain. The brain recognises the pattern of the signals from the cells covered by the image, and so recognises the object's shape.

How do we tell the colour of an object? There are two types of cells on the retina – **rods** and **cones** (Figure 16.4E). Rods occur mostly near the edges of the retina. They are not sensitive to colour and only respond to the brightness of light. Ask a friend to test you to see if you can tell the colour of something at the edge of your field of vision.

Cones are packed densely together at the middle of the retina. This area is called the **fovea**. Each cone is sensitive to red or blue or green light. For example, when red light falls on the retina, it 'activates' the red-sensitive cones, so you see red. Other colours activate more than one type of cone. For example, yellow light activates the red and the green cones, so they send messages to the brain. When the brain receives signals from adjacent red and green cones, it knows yellow light is on that part of the retina.

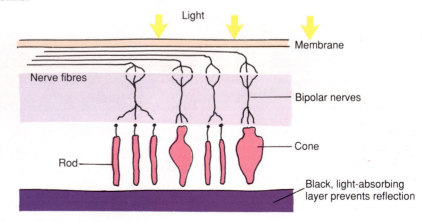

Figure 16.4E ● Rods and cones

IT'S A FACT

Some scientists think that reading in poor light harms your eyes. Poor light causes the retinal cells to expand a little, effectively making the retina a bit closer to the eye lens, so the lens needs to be stronger to see near objects clearly. If the eye lens can not become strong enough, the objects become blurred. Make sure you don't read in poor light!

IT **Blindspot** (program)

The *Blindspot* program shows screen models of the basic structure of a human eye. Using this program you can investigate where the blind spot in your right eye is and why it is called the blind spot.

Sight defects

Sight defects occur when the eye lens cannot form a sharp image on the retina. Spectacles contain lenses that compensate for sight defects.

Short sight is caused by over-strong eye muscles. A short-sighted eye cannot see far away objects clearly because the eye muscle cannot relax enough to make the eye lens thin enough. A suitable concave lens in front of the eye counteracts the effect of the over-strong eye lens (Figure 16.4F).

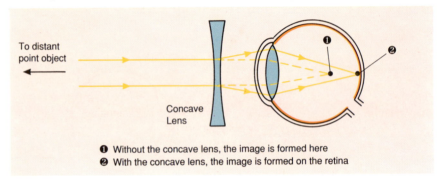

To distant point object

Concave Lens

❶ Without the concave lens, the image is formed here
❷ With the concave lens, the image is formed on the retina

Figure 16.4F ● Short sight

Long sight is caused by weak eye muscles, and often develops as the muscles weaken with age. The muscles are unable to contract enough around the lens to make it thick enough to focus near objects. A suitable convex lens in front of the eye helps the eye lens to form a clear image on the retina (Figure 16.4G).

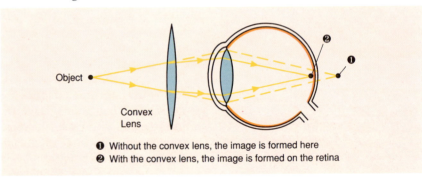

Object ●

Convex Lens

❶ Without the convex lens, the image is formed here
❷ With the convex lens, the image is formed on the retina

Figure 16.4G ● Long sight

IT'S A FACT

Contact lenses are a popular alternative to spectacles. The photograph shows a man fitting a contact lens to his eye.

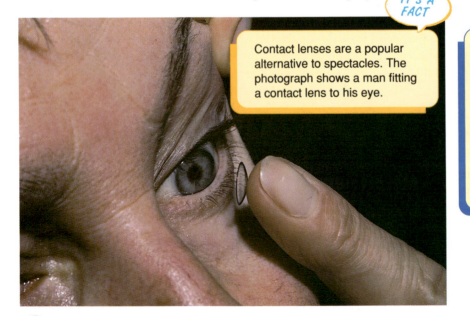

SUMMARY

The eye lens forms a clear image on the retina. The iris controls the amount of light entering the eye. The thickness of the eye lens changes to accommodate (focus) objects at different distances. Spectacle lenses compensate for defects in the eye muscles controlling the eye lens, to give the wearer normal vision.

❶ (a) Why do your eye muscles relax when you look at a distant object and contract when you look at something close to you?
(b) Why is it easier to study an object in detail if you look straight at it?
(c) Why does the eye pupil dilate (i.e. widen) in dim light?

❷ Explain how (a) short sight and (b) long sight are caused and how they are corrected.

❸ Try these 'eye tests', which are explained below.
(a) The blind spot test.
(b) The sausage test.
(c) The birdcage test.
(d) The dark room test. Sit in a dark room for twenty minutes or more and you will discover your eyes can see in the dark. The rods become much more sensitive than normal in dark conditions and they make the most of whatever light there is.

(a)

The blind spot test
❶ Position the black spot in front of your left eye. Cover your right eye
❷ Move the book closer and keep staring at the spot
❸ The X disappears when its image falls on the blind spot of your left eye

(b)

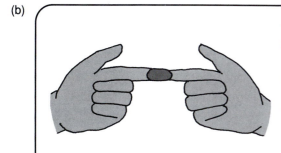

The 'sausage' test
❶ Hold your hands in front of your face with the tips of your index fingers touching
❷ Stare past your hands at a distant object and move them towards you
❸ You should see a 'sausage' between the tips of your index fingers caused by overlapping images from each eye

(c)

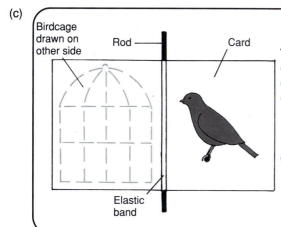

Birdcage drawn on other side
Rod
Card
Elastic band

The birdcage test
❶ Make the test card as shown
❷ Spin the card about the rod
❸ The bird appears in the cage because the images last for a fraction of a second and hence overlap. This is called 'persistence of vision'
❹ Try spinning the card at different speeds. What difference does it make?

16.5  Hormones

> Hormones are chemicals produced by animals to regulate their bodies' activities.

LOOK AT LINKS
Hormones control metamorphosis in amphibia and insects. For more information see Topic 15.6.

Each meal or snack that you eat provides your body with a surge of glucose. Glucose is an important source of energy but high levels seriously disrupt the body's cells and cause them to function very inefficiently.

Your body keeps the level of glucose constant by means of **hormones**. These substances regulate the concentration of glucose in the blood and cope with the surge of glucose at meal-times. The different ways in which the body keeps its internal environment from changing is called **homeostasis**. The control of glucose levels is one example of homeostasis.

Hormones are chemicals produced by animals and plants to regulate the organisms' activities. In animals they are produced in the tissues of ductless glands called **endocrine glands** and released into the blood system. Hormones circulate in the blood and cause specific effects on the body. (See Figure 16.5A.)

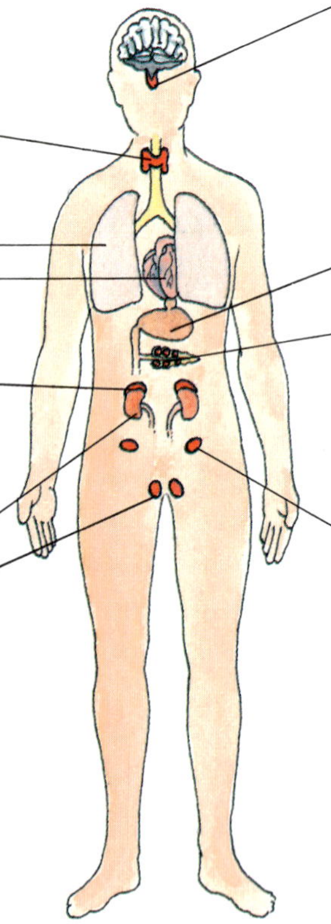

Pituitary gland (connected to the hypothalamus at the base of the brain) – produces nine different hormones which affect:
• water reabsorption from kidney tubules
• growth
• sperm and egg production
• release of hormones by other endocrine glands

Thyroid – produces **thyroxin** which affects the rate of metabolism

Lungs

Heart

Stomach

Islets of Langerhans – groups of cells in the pancreas which produce **insulin** and **glucagon**. These hormones help to regulate glucose levels in the blood

Adrenal gland – produces **adrenalin** which prepares the body for sudden action (fight, fright or flight hormone)

Kidney

Ovary (female) – produces **oestrogen** and **progesterone** which regulate the menstrual cycle and help to develop and maintain secondary sexual characteristics

Testis (male) – produces **testosterone** which helps to develop and maintain secondary sexual characteristics

> Hormone-like substances called pheromones carry information between individuals. Different species of insect, for example, produce their own types of pheromone. Individuals release pheromones into their environment where they are smelt or tasted by fellow individuals close-up. For example, worker bees release a pheromone called geraniol which attracts other workers to them; ants release an 'alarm' pheromone which warns fellow ants when their nest is invaded by ants from another colony.

Figure 16.5A ● Hormones are made in tissues called endocrine glands – they are released into the blood which transports them around the body

The tissue on which a particular hormone (or group of hormones) acts is called a **target tissue**. A hormone affects its target tissue more slowly than a nerve impulse affects a muscle. This is because nerve impulses move rapidly along neurones. Muscles therefore can respond very quickly to changing circumstances. The action of most hormones is longer-term.

Insulin
(program)

Use the program to find out how human blood-sugar level is controlled. You can take control of the system by changing sets of hormone levels.

In teenagers the body matures under the control of hormones over several years. In girls, oestrogen controls broadening of the hips and breast development. In boys, testosterone controls beard growth, broadening of the shoulders and the deepening of the voice. These developments (called secondary sexual characteristics) mark the start of puberty and continue through adolescence. Puberty begins in girls at age 11–13 and in boys at age 13–14. Because of the changing balance of hormones in the body, adolescents may experience swings of mood and also skin troubles such as spots and acne. Usually these problems have cleared up by the early twenties.

Hormone regulation of blood glucose

To keep the level of glucose in the blood constant hormones balance the glucose-producing and glucose-using processes of the body.

- The hormone thyroxine increases the rate at which glucose is oxidised in **cellular respiration**. This **decreases** the level of glucose in the blood.
- The hormone insulin also **decreases** the level of glucose in the blood. It does this by promoting the conversion of glucose into glycogen.
- The hormone glucagon **increases** the level of glucose in the blood by promoting the conversion of glycogen into glucose.

If the pancreas does not produce enough insulin a condition called **diabetes mellitus** occurs. The glucose level in the blood becomes dangerously high and can cause blindness or kidney failure. Concentrations of glucose become so high that the kidneys cannot reabsorb all the glucose and glucose is excreted in the urine. A simple test for glucose in the urine of a patient can tell a doctor if a patient is **diabetic.**

Diabetics suffer from thirst and tiredness. If the diabetes is not too severe, a carefully chosen low-sugar diet can control the condition. If the diabetes is severe, diabetics are taught to inject themselves regularly with insulin to lower their blood glucose levels (see Figure 16.5B). Getting the dose of insulin right is not always easy. If too much insulin is injected, the glucose level in the blood falls too low and diabetics can suffer from unpleasant side effects. Diabetics soon learn to recognise the symptoms and eat a little sugar to boost blood glucose to the right level.

Figure 16.5B ● A diabetic child injecting herself with insulin

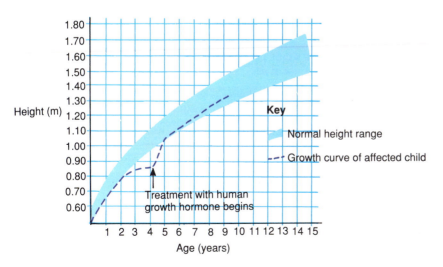

Hormones, growth and muscle building

If a child's pituitary gland does not produce enough growth hormone then the child will not grow to a normal height. Providing the condition is diagnosed at an early age, the affected child can be given a growth hormone to make up for the deficiency. The child puts on a growth spurt and soon 'catches up' with other children of the same age (see Figure 16.5C).

Figure 16.5C ● Human growth curve. How long after treatment begins does it take for the affected child to 'catch up' with other children?

The hormone thyroxine contains the element iodine. If a person's diet does not contain enough iodine the thyroid gland does not produce enough thyroxine. As a result the person's metabolic rate is lowered and the person feels sluggish and tired. Under-production of thyroxine also causes the thyroid gland to enlarge – a condition called **goitre**.

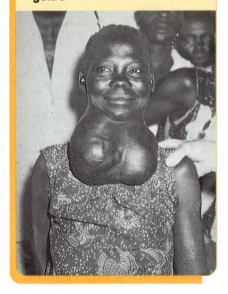

Figure 16.5D ● The muscles you need for strength events

At some time or other you have probably seen athletes straining to lift heavy weights or put the shot long distances. Athletes train very hard to build up the muscles needed to compete in these 'strength' events (see Figure 16.5D). A few, however, cheat and inject themselves with hormones called **anabolic steroids** which develop their muscles even further, giving them an unfair advantage over other competitors.

Anabolic steroids mimic the effect of the hormone testosterone which in men controls the development of secondary sexual characteristics. Testosterone also increases the rate of protein synthesis (in women as well as men). This is why athletes who take anabolic steroids develop bigger and stronger muscles. The side-effects of anabolic steroids are very unpleasant. Liver damage is possible and women can develop male secondary sexual characteristics. Men can become sexually impotent.

Anabolic steroids are on the list of drugs banned by the International Olympic Committee. Any athlete caught using them can be prohibited from taking part in future sporting events.

CHECKPOINT

❶ What are hormones and how are they transported round the body?

❷ Complete the following paragraph using the words provided below. Each word may be used once, more than once or not at all.

oestrogen   target   long-term   testosterone   nervous   short-term
transports   progesterone   adrenalin   thyroxine   blood   sexual

The endocrine tissues secrete hormones directly into the _____ which _____ them all over the body. The tissues affected by hormones are called _____ tissues. The uterus, for example, responds to the hormones _____ and _____ . The action of most hormones is _____ . Female secondary _____ characteristics, for example, develop under the influence of _____ , male secondary _____ characteristics develop under the influence of _____ . The adrenal gland secretes _____ which prepares the body for sudden action. The rapidity of its effect is more like that of the _____ system than that of other hormones.

SUPPORT AND MOVEMENT

17.1 Support in plants and animals

FIRST THOUGHTS

All living things are supported in some way. Think what would happen to your body without the support of the skeleton. It would collapse into a shapeless heap.

RESOURCE -ACTIVITY- PACK

📖 **LOOK AT LINKS**
for **xylem** and **phloem**
See Topic 14.1.

📖 **LOOK AT LINKS**
for **turgor pressure**
See Topic 9.2.

Support in plants

Figure 17.1A ● Bundles of xylem and phloem in the stem of an oak

Plants are supported by the strands of **xylem** and **phloem** that run in vascular bundles from the roots, through the stem to the leaves and flowers (see Figure 17.1A). Rings of lignin strengthen the walls of the xylem cells. Non-woody (herbaceous) plants are supported by the firmness of their cells. The pressure of the cell contents against the cell wall makes a plant cell **turgid**. Individual cells press against each other and hold the plant upright. Wood is formed from lignified xylem and gives even more support. This is why trees can grow to be tall and heavy.

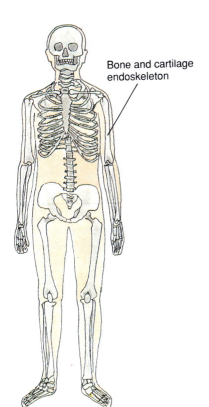

Bone and cartilage endoskeleton

Most animals are supported by a skeleton. There are three main types of skeleton:

- **Endoskeletons** are found in vertebrates. The skeleton, which lies inside the body surrounded by the soft tissue, is made of hard bone and cartilage. (See Figure 17.1B(a)).
- **Exoskeletons** are found in insects and their relatives. The skeleton is made of hard chitin segments which surround the body like armour plate. (See Figure 17.1B(b)).
- **Hydrostatic skeletons** are found in the larger worms. They consist of body spaces filled with fluid under pressure. Hydrostatic skeletons are firm but flexible. (See Figure 17.1B(c)).

Chitin exoskeleton

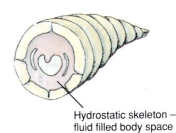

Hydrostatic skeleton – fluid filled body space

Figure 17.1B ● (a) The endoskeleton of a human

(b) The exoskeleton of an insect

(c) The hydrostatic skeleton of a worm

17.2 How muscles work

Hydrostatic skeletons are firm but flexible. The parts of exoskeletons and endoskeletons, however, are rigid. Their flexibility comes from **joints** which are made wherever two parts of a skeleton meet. The joints are pivots for limbs which form a system of levers. Muscles move limbs by pulling on them like a set of pulleys.

Up to 2000 parallel strands called **myofibrils** pack the cytoplasm of a muscle cell. Myofibrils consist of units called **sarcomeres**. Each sarcomere is composed of filaments of two types of protein parallel to one another. The thicker filament is made of the protein **myosin**; the thinner filament is made of the protein **actin**. Filaments from neighbouring sarcomeres link together in a region called the **Z line**. When nerve impulses stimulate the muscle cells, the thin actin filament slides over the thicker myosin filament and cross bridges between the filaments rapidly form, break and reform. Because filaments are anchored at the Z line, the sarcomere shortens, the myofibrils shorten and the muscle cell as a whole contracts. The interaction between myosin and actin releases energy from **ATP** (see p. 212) for muscle contraction.

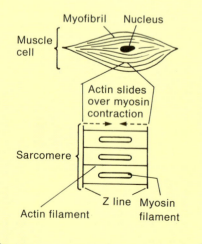

Figure 17.2A compares the arrangement of muscles in the locust leg and the human arm. The muscles are in pairs which stretch across the joint. One muscle of a pair has the opposite effect to that of its partner. We call them **antagonistic** pairs. For example, contraction of the biceps muscle lifts the lower arm (flexing); contraction of the triceps muscle straightens (extends) the arm (see Figure 17.2B). When the biceps contracts the triceps relaxes and vice versa. The antagonistic pairs of muscles in the locust leg work in a similar way.

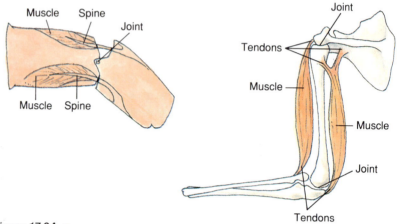

Figure 17.2A ●
(a) Locust leg – notice that the exoskeleton surrounds the muscles which are attached to its inside surface by inward projecting spines

(b) Human arm – notice that the endoskeleton is surrounded by the muscles which are attached to the outside surface by tough tendons

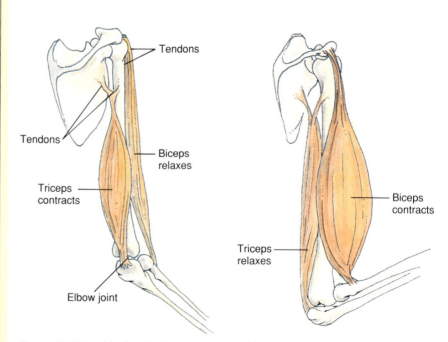

Figure 17.2B ● Moving the lower arm – the biceps and the triceps are an antagonistic pair of muscles

(a) Smooth muscle (x20)

(b) Skeletal muscle (x20)

(c) Cardiac (heart) muscle (x20)

Figure 17.2C ● The three types of muscle viewed under a microscope

Types of muscle

Animals are able to move from place to place because of the action of muscles which pull on the skeleton. Moving from place to place is called locomotion. Muscular contractions are also responsible for other forms of movement. As heart muscle contracts and relaxes rhythmically it propels blood through the blood vessels. **Peristalsis** is brought about by contraction and relaxation of the muscles in the wall of the intestine. There are three types of muscle in the human body – smooth muscle, skeletal muscle and cardiac muscle (see Figure 17.2C).

● Smooth muscle

Smooth muscle contracts and relaxes slowly and steadily and does not become tired. These properties are ideal for the continuous movement of substances through the organs of the body. In mammals, smooth muscle is found in the walls of the intestine, blood vessels and air passages. Smooth muscle receives nerve impulses from the **autonomic nervous system** (which acts without conscious control from the brain). Contractions of smooth muscle therefore occur automatically. This is why smooth muscle is called **involuntary** muscle.

● Skeletal muscle

Skeletal muscle is often called striated (striped) muscle. It consists of fibres which are crossed with alternate light and dark bands. Skeletal muscle becomes tired after prolonged periods of activity but is otherwise ideal for moving parts of the skeleton. Its contractions are quick, strong and usually voluntary. Conscious messages from the brain control the strength and speed of contractions. The fibres of skeletal muscle receive branches from the axons of **motor neurones**. The muscle fibres contract when nerve impulses reach them.

● Cardiac muscle

Cardiac (heart) muscle is striated, like skeletal muscle. Its fibres are branched and connect with one another. This structure lets nerve impulses spread throughout the whole tissue, co-ordinating its contractions. Cardiac muscle never becomes tired. Its action is **involuntary**.

How fish swim

Fish swim by pushing their bodies and fins against the water. Their stream-lined shape helps to reduce resistance to the movement of their bodies through the water (see Figure 17.2D overleaf).

Blocks of muscle attached to each side of the fish's vertebral column move the body from side to side and drive the fish forward (see Figure 17.2E overleaf). The blocks of muscle on either side of the vertebral column form antagonistic pairs. The vertebral column itself is flexible and acts like a lever.

A gas-filled sac called the swim bladder helps to control buoyancy. When the swim bladder is full of gas the density of the fish decreases and the fish rises: when gas is removed, the density of the fish increases and the fish sinks. Cartilaginous fish such as sharks do not have swim bladders. If they stop swimming they slowly sink to the bottom.

LOOK AT LINKS
for **buoyancy**
See *KS: Physics*, Topic 16.6.

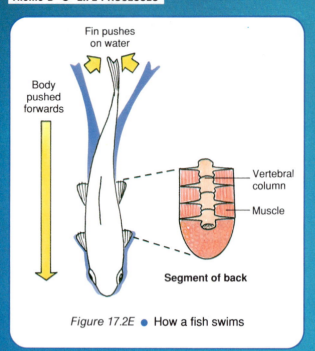

Fin pushes on water

Body pushed forwards

Vertebral column

Muscle

Segment of back

Figure 17.2E ● How a fish swims

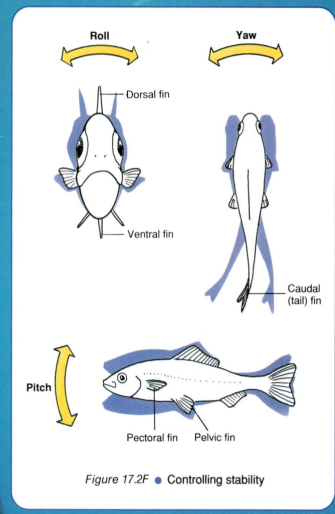

Roll

Yaw

Dorsal fin

Ventral fin

Caudal (tail) fin

Pitch

Pectoral fin   Pelvic fin

Figure 17.2F ● Controlling stability

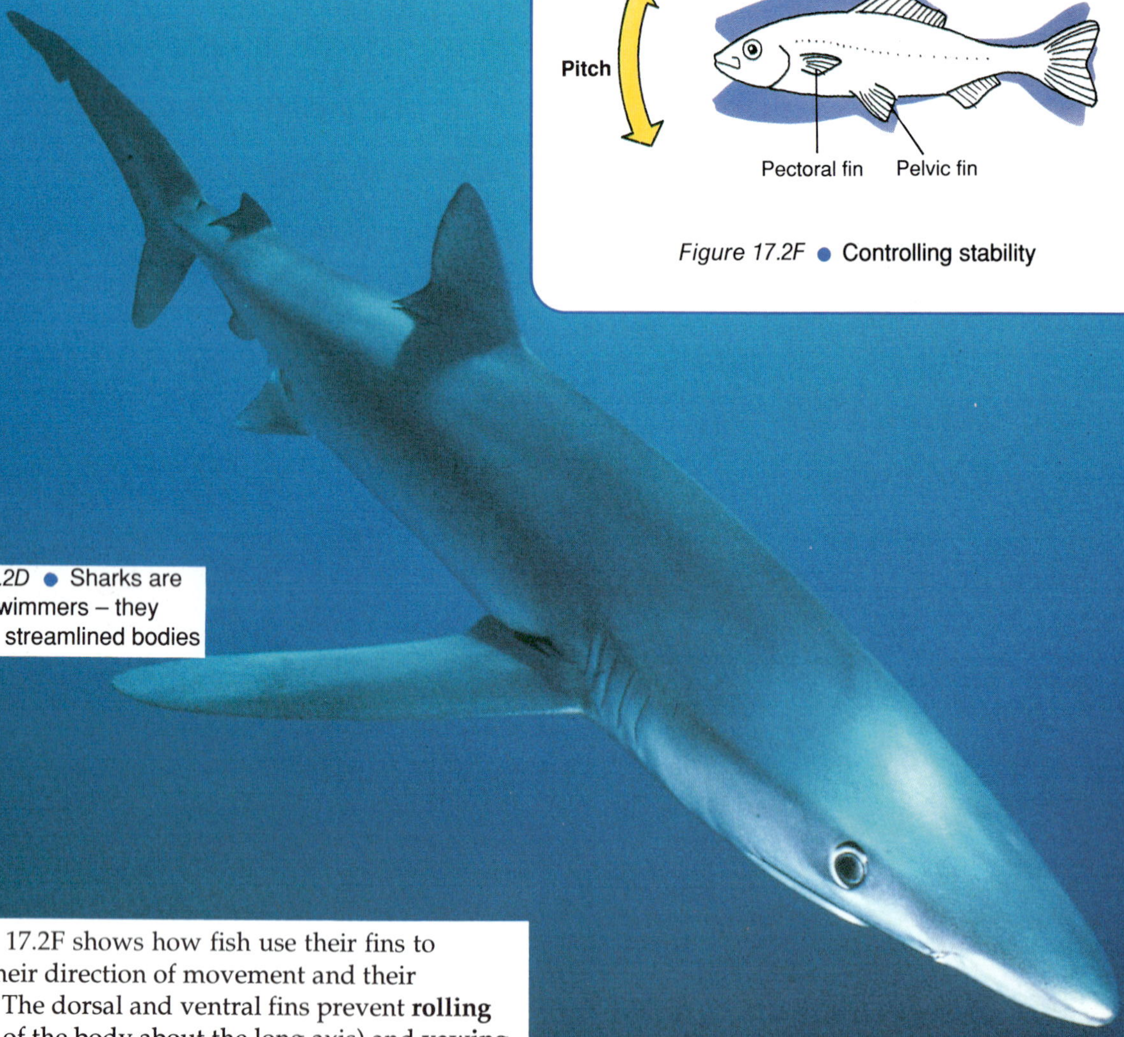

Figure 17.2D ● Sharks are efficient swimmers – they have very streamlined bodies

Figure 17.2F shows how fish use their fins to control their direction of movement and their stability. The dorsal and ventral fins prevent **rolling** (rotation of the body about the long axis) and **yawing** (side to side movement of the front part of the body). The pectoral and pelvic fins prevent **pitching** (the tendency to nose dive).

IT'S A FACT

Skeletons and locomotion: Flexibility v rigidity
The fluid-filled **hydrostatic** skeleton of the earthworm is surrounded by an outer layer of circular muscle and an inner layer of longitudinal muscle (see Figure 11.6H and Figure 17.1B(c)). Contractions of the muscles send pressure waves through the fluid of the hydrostatic skeleton, causing changes in the shape of the body. These shape changes and the chaetae (bristles – see Figure 1.5K) projecting from the body help the earthworm to burrow through soil.

Limbs and a rigid skeleton make walking possible. An insect at rest is supported by all six legs attached to the rigid **exoskeleton** (see Figure 1.5N). When walking, the insect supports itself on three legs at a time: the first and third legs of one side, and the second leg of the other side. These form a tripod while the other three legs are carried forward.

LOOK AT LINKS
for **warm-bloodedness**
See Topic 21.5.

How birds fly

Birds are adapted to fly:
- **hollow bones** reduce weight
- **flight muscles** move the wings up and down
- **feathers** smoothly shape the body.

Feathers insulate the body. At around 41 °C the body temperature of a bird is higher than that of most other warm-blooded animals. The high temperature means that the flight muscles work more efficiently. Feathers make flying possible: they keep in heat, they keep out water and their colour is used either to attract mates or for protective camouflage. Feathers are arranged in layers over the body. The outer layers give a smooth shape to the body and are used for flight. The inner layers, called **down**, give extra insulation (Figure 17.2G).

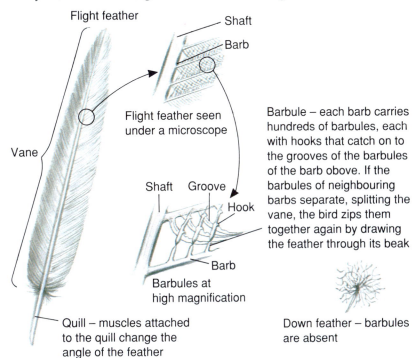

Figure 17.2G ● Flight feather and down

Wing feathers are of different sizes. The longest are the stiff flight feathers. Figure 17.2H shows that those on the outer edge are called **primaries**; those on the inner edge are called **secondaries**. Rows of smaller feathers called **wing coverts** smoothly overlap the flight feathers on top and underneath the wing. The **alula** is formed from a group of small feathers on the outer front edge.

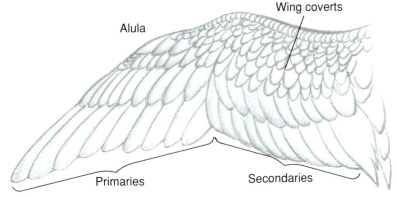

Figure 17.2H ● Wing feathers

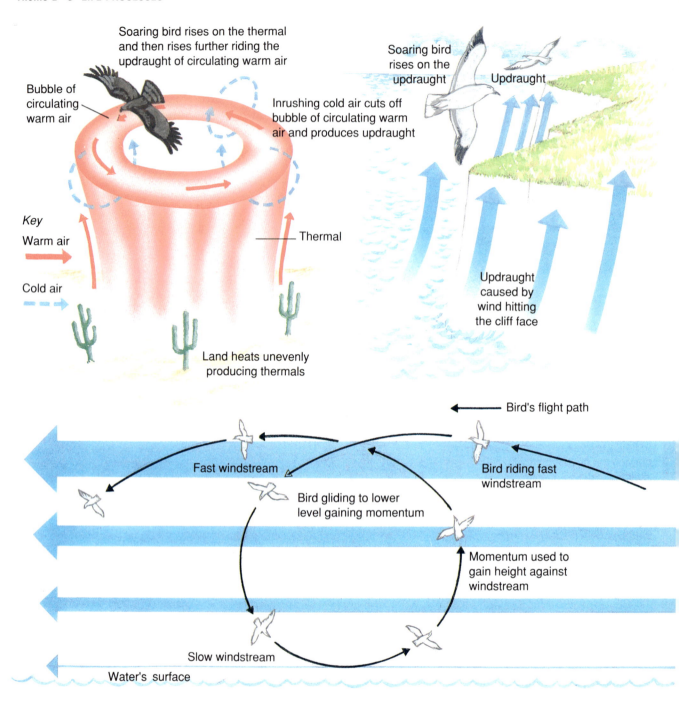

Soaring bird rises on the thermal and then rises further riding the updraught of circulating warm air

Bubble of circulating warm air

Inrushing cold air cuts off bubble of circulating warm air and produces updraught

Soaring bird rises on the updraught

Updraught

Key
Warm air

Cold air

Thermal

Updraught caused by wind hitting the cliff face

Land heats unevenly producing thermals

Bird's flight path

Fast windstream

Bird riding fast windstream

Bird gliding to lower level gaining momentum

Momentum used to gain height against windstream

Slow windstream

Water's surface

(a) Land heats up unevenly producing rising columns of warm air called **thermals**. Vultures and other soaring birds circle on the updraught and rise with it

(b) Seagulls gain height by riding the **updraughts** when wind hits the cliffs

(c) Wind near the surface of the sea is slowed by friction with the waves. Wind speed increases with height reaching its maximum 15 m above the water's surface. The albatross uses the **variation in windspeed** to pick up momentum on the high, fast windstream, gliding down to a lower level then using its momentum to gain height against the wind. The bird travels thousands of kilometres across the southern oceans without flapping its wings

Figure 17.2l ● Gliding and soaring

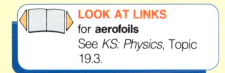

LOOK AT LINKS
for **aerofoils**
See *KS: Physics*, Topic 19.3.

Figure 17.2J shows that the wing is in the shape of an **aerofoil**. The upper surface is more strongly curved than the lower surface. Air moving over the top has further to go than the air moving underneath and therefore travels faster. This increases air pressure beneath the wings and creates a reduced pressure above them. The result is an upward force and the wings lift.

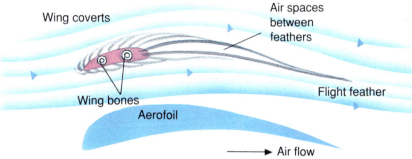

Figure 17.2J ● Section of wing showing the shape of an aerofoil. Notice that the flow of air is smooth and free of turbulence

Table 17.1 sets out the forces acting against a bird in flight and how the design of the body and wings is adapted to overcome them.

Table 17.1 ● Counteracting forces

Force	Adaptation
Drag caused by the resistance of air	Tips of feathers point backward in the same direction as the flow of air. The **leading edge** of the wing is blunt and rounded: wing feathers taper toward the **trailing edge**
Gravity causing a downward pull	Lift on both wings opposes the force of gravity and the bird stays aloft
Turbulence (uneven flow of air) caused by the bird's movements	The alula forms a slot in front of the wing, reducing turbulence

Have you ever watched a seagull gliding and soaring? Gliding is the simplest way of staying aloft. It makes the smallest demands on muscle power and energy reserves. Figure 17.2I gives you the idea.

Figure 17.2K shows that to climb higher a bird tilts the leading edge of each wing upwards. Notice, however, that the more the wings are tilted, the greater the turbulence as the smooth airflow over the wings breaks up into swirling eddies. The wings lose lift and the bird is in danger of stalling. At the critical moment the bird spreads the alula on the leading edge of each wing. Each alula acts as a slot through which air rushes, keeping the airflow over each wing smooth and turbulence-free (Figure 17.2L).

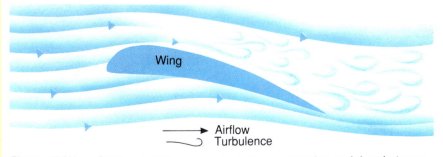

Figure 17.2K ● Gaining height – the angle between the wing and the airstream is increased. Notice the turbulence developing over the wing's surface

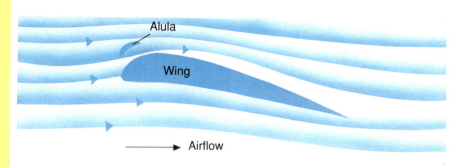

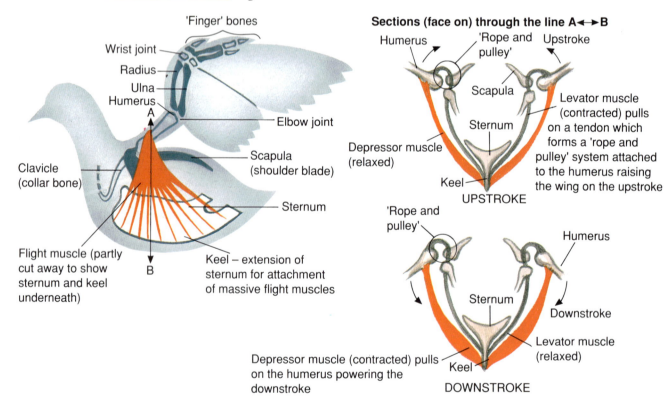

Figure 17.2L ● Controlling the stall – the alula opens, restoring a smooth airflow over the wing.

LOOK AT LINKS
for the names of bones
See Topic 17.3 and the
Resource Bank activity.

Flapping flight requires more effort than gliding and soaring. Figure 17.2M shows how muscles and bones work together to move the wings. Multi-flash photography allows us to see the air movements as a bird flies (Figure 17.2N). The down-stroke is the power stroke. Air is pushed downwards supporting the weight of the bird. The primary feathers lie flat against one another forming an unbroken surface which provides maximum resistance to the air underneath. As the wing moves downwards and forwards, the ends of the primaries curve back forming a propeller shape, pulling the bird forward.

The upstroke produces little air movement. The primary feathers twist open and air passes through the gaps between them. Minimum air resistance means that the upstroke needs less effort. It is a recovery stroke in preparation for the next powerful downstroke. Even so the upstroke still helps to drive the bird forward. The primaries curl back, each becoming a small propeller, and the wings tilt to maintain lift (see Figure 17.2K).

Figure 17.2M ● Moving the wings – massive flight muscles (depressors and levators) are attached to the wing bones and deep keel-like extension of the sternum. These muscles may account for up to a third of the bird's body weight and they form the white breast meat we eat

Fast-flying birds usually have long narrow wings. The tail feathers stabilise the body in flight and help in braking and landing. Racing pigeons have been timed at speeds of 80 km/h, pheasants at 100 km/h and swifts at 160 km/h. The peregrine is one of the fastest birds of all. It can chase prey in level flight at 80 km/h and dive on its victim at a speed of 260 km/h.

FIRST THOUGHTS

Propulsion pushes an animal in the direction it wants to go. **Support** comes from the environment in which the animal lives (water gives more support than air because it is denser). **Stability** is controlled by fins, wings and tail feathers or legs. Animals are adapted to overcome the problems of propulsion, support and stability.

LOOK AT LINKS
for **joints** and **antagonistic pairs of muscles**
See p. 316 and Topic 17.3.

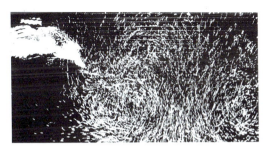

Figure 17.2N ● Soap bubbles filled with the gas helium are as dense as air (ordinary bubbles are denser than air and slowly sink). Movements of the bubbles show how the air is moving as the bird flies through the bubble cloud. Notice the circular vortex rings of moving air produced by the down-stroke generating lift and thrust

How horses walk and run

Think how animals move from place to place. To swim, fish push against water; to fly birds push against air. *How do land animals move when walking and running?*

The hind legs of a horse push against the ground and the body moves forward. Analysing movement, therefore, is best begun with the hind legs because that is where the force is applied. A complete cycle of movement is called a **stride**. All four legs complete their motion and move the horse.

Figure 17.2O shows the arrangement of muscles and bones in the horse's hind leg. Muscles stretch across joints and pull on the bones. One group of muscles bends (flexes) a joint; an opposing group straightens (extends) it. *Which groups of muscles are antagonistic pairs?*

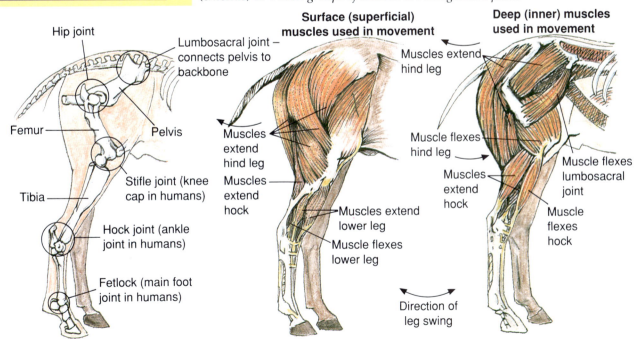

Figure 17.2O ● (a) Joints and bones of the hind leg of the horse

(b) The muscles used to move the hind leg of the horse. Two diagrams are needed to show the main muscles for movement because the muscles are layered over each other

Look at Figure 17.2P. Propulsion begins at the moment when the leg is vertically beneath the hind quarters and continues until the foot leaves the ground. Notice the ligaments which bind the bones together. Also notice the tendons which attach the muscles to the bones. At the end of each swing the leg briefly stops. At this moment the tendons store energy like a coiled spring, only to release it in the elastic recoil helping the leg to start moving the other way.

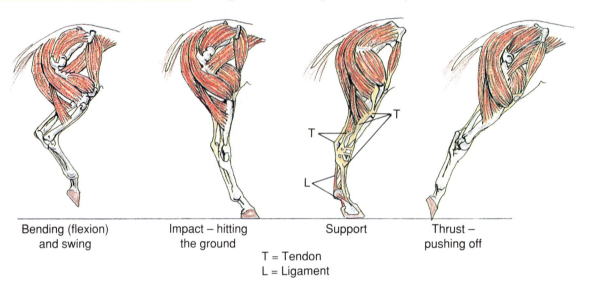

Bending (flexion) and swing

Impact – hitting the ground

Support

Thrust – pushing off

T = Tendon
L = Ligament

Figure 17.2P ● Stages of the stride. Ligaments and tendons are shown at the support stage. They keep the bones in place even when the muscles are relaxed

LOOK AT LINKS
for **ligaments** and **tendons**
See Topic 17.3.

Next time you see a horse walking, listen to the cycle of its hoof beats. You should hear a repeating pattern of four distinct beats as each foot strikes the ground in succession. Figure 17.2Q shows the sequence.

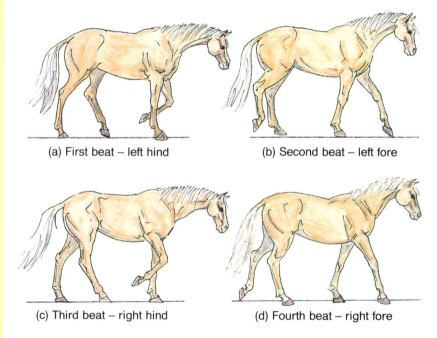

(a) First beat – left hind

(b) Second beat – left fore

(c) Third beat – right hind

(d) Fourth beat – right fore

Figure 17.2Q ● The walk is a cycle of four beats. Two or three legs are always on the ground giving stability. Although the foreleg appears to move first from the halt, the walk sequence starts with a hind leg because the power comes from the hindquarters

Galloping is also a four-beat cycle. The strides are long and the hind legs reach forward under the body. Figure 17.2R shows the sequence. Notice that when the horse pushes off with its fore-leg it is suspended in the air for a moment before the hind foot strikes for the first beat. The head and neck move forward and back balancing the horse and carrying it on to the next stride. The speed of the gallop varies: a horse galloping cross-country can maintain a speed of 30 km/h for long distances; a race horse can clock 70 km/h sprinting at top speed.

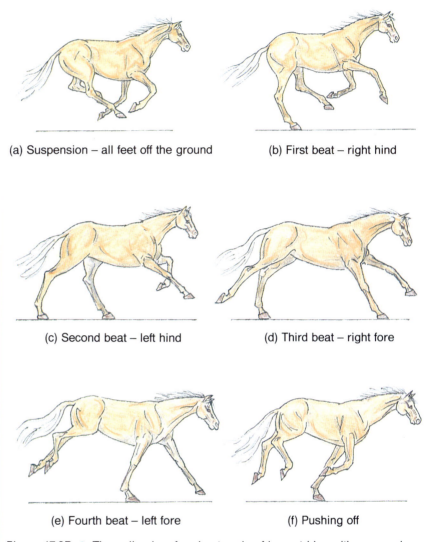

(a) Suspension – all feet off the ground

(b) First beat – right hind

(c) Second beat – left hind

(d) Third beat – right fore

(e) Fourth beat – left fore

(f) Pushing off

Figure 17.2R ● The gallop is a four-beat cycle of long strides with suspension. The abdominal muscles help the hind legs to reach forward under the body. Breathing is in rhythm with the strides

SUMMARY

Muscular contractions are responsible for movement. Smooth muscle moves substances through the organs of the body; cardiac muscle propels blood through the blood system; skeletal muscle moves the skeleton and is responsible for locomotion (moving from place to place). Propulsion, support and stability are the main components of locomotion and are analysed in fish, birds and horses.

17.3 The skeleton

The vertebrate skeleton is made mostly of bone (see Figure 17.3A). It also contains cartilage which is softer than bone because it contains less **calcium** and phosphate. The amount of cartilage in the skeleton depends on the species and the age of an animal. The skeleton of sharks is made entirely of cartilage. In adult mammals, however, cartilage is found in only a few places. Cartilage covers the ends of limb bones where it helps to reduce friction in the joints as bones move upon one another.

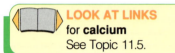
LOOK AT LINKS
for **calcium**
See Topic 11.5.

Pixel Perfect
(DTP)

Use the *Pixel Perfect* system to create your own wall display of human anatomy. With the class working in groups you could cover topics such as the eye and ear, teeth, the skeleton, digestion, respiration, blood circulation, excretion, muscles and joints and the nervous system.

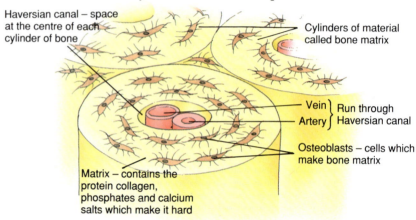

Haversian canal – space at the centre of each cylinder of bone

Cylinders of material called bone matrix

Vein ⎱ Run through
Artery ⎰ Haversian canal

Osteoblasts – cells which make bone matrix

Matrix – contains the protein collagen, phosphates and calcium salts which make it hard

Figure 17.3A ● The structure of living bone (x200)

The human skeleton

The human skeleton is illustrated in Figure 17.3B. The bones of the skeleton are connected to one another at joints. Strap-like ligaments hold joints together. Tendons attach muscles to the skeleton. As muscles contract and relax across joints they move the different parts of the skeleton.

The human skeleton consists of two parts: the **axial skeleton** and the **appendicular skeleton**.

If the bones of a joint separate, the joint is said to be **dislocated**.

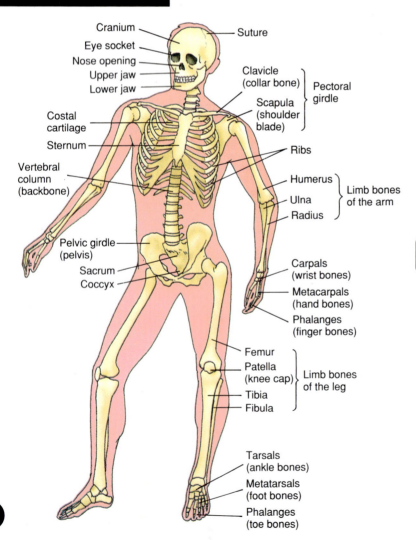

Cranium
Eye socket
Nose opening
Upper jaw
Lower jaw
Suture
Clavicle (collar bone)
Scapula (shoulder blade)
Pectoral girdle
Costal cartilage
Sternum
Vertebral column (backbone)
Ribs
Humerus
Ulna
Radius
Limb bones of the arm
Pelvic girdle (pelvis)
Sacrum
Coccyx
Carpals (wrist bones)
Metacarpals (hand bones)
Phalanges (finger bones)
Femur
Patella (knee cap)
Tibia
Fibula
Limb bones of the leg
Tarsals (ankle bones)
Metatarsals (foot bones)
Phalanges (toe bones)

Figure 17.3B ● The human skeleton

● *The axial skeleton*

The axial skeleton consists of the skull, vertebral column and the ribs. These bones protect vital organs and tissues and form a strong support for the body. The skull and the vertebral column form a protective sheath of bone around the brain and spinal cord.

- **The skull** consists of plates of bone fused together to form the cranium which encloses and protects the brain. A hinged joint allows powerful muscles to move the lower jaw against the fixed upper jaw.
- **The vertebral column** (backbone) consists of a series of bones called **vertebrae** arranged like a curved rod (see Figure 17.3C). The vertebral column supports the skull and the limb girdles (pectoral and pelvic girdles). A cavity called the **neural canal** runs through the centre of each vertebra forming a continuous space in the vertebral column through which the spinal cord runs. At the top of the vertebral column the spinal cord passes through a hole in the skull and expands into the brain.
- **The ribs** form a curved bony cage around the heart and lungs (see Figure 17.3B). At the front they are attached by the costal cartilages to a bone called the sternum. At the back they form joints with the thoracic vertebrae. Each rib has ridges to which the intercostal muscles are attached. The costal cartilages and the joints make the ribs flexible. The intercostal muscles, the diaphragm and the muscles of the abdomen move them. Breathing depends on these movements.

IT'S A FACT

A slipped disc happens when the intervertebral disc bursts or is forced out from its position between the centrums. The person suffers considerable pain because of the pressure on the spinal cord.

LOOK AT LINKS
for more about breathing movements
See Topic 12.2.

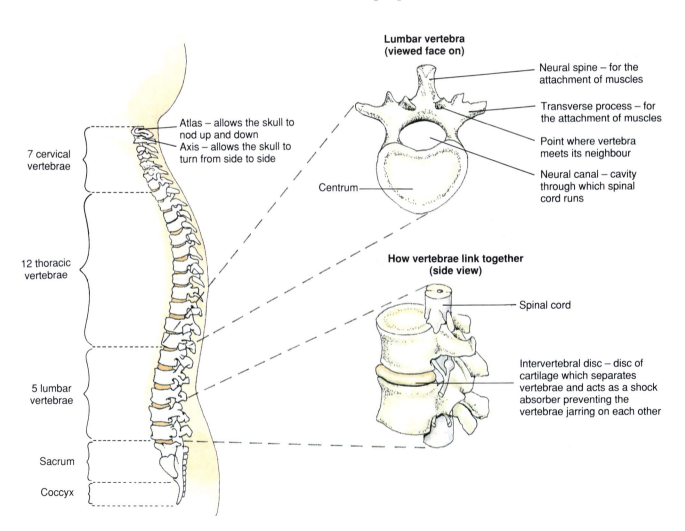

Figure 17.3C ● The vertebral column

IT'S A FACT

The fossil remains of vertebrates are usually parts of skeletons. Bone is very hard and therefore preserves well.

X-ray photographs help doctors locate broken (fractured) bones. A plaster cast or supporting bandage holds the broken bone (broken end to broken end) in position. Bone cells divide and produce new bone tissue which hardens and heals the break. If the fracture is severe, the broken ends of the bones are lined up and pinned to help them grow together. Children's bones are softer, more flexible and less likely to break. However, an accident can splinter the bone causing a **green stick** fracture.

LOOK AT LINKS
for **artificial hip joints**
See *KS: Physics*, Topic 15.4.

● The appendicular skeleton

The appendicular skeleton consists of the limb girdles and limb bones. We use the hind limbs (legs and feet) for walking and running. Our upright position means that the forelimbs (arms and hands) are free for other activities.

- **The pelvic girdle** links the legs with the vertebral column (see Figure 17.3B). The pelvic girdle, which is made of two bones fused together, gives a solid framework that helps to bear the weight of the body. It is joined rigidly to the base of the vertebral column. The rigid arrangement of bones allows forces on the leg to be transmitted to the rest of the body.
- **The pectoral girdle** links the arms with the vertebral column. It is made up of the scapulas (shoulder blades) and clavicles (collar bones). This arrangement is not as effective as the pelvic girdle in transmitting force from the limbs to the body but it gives the shoulders and arms great freedom of movement.
- **The limb bones** in humans consist of long bones which form joints at the elbow in the arm and the knee in the leg. The upper long bone of each limb is attached to a limb girdle. The lower long bones are attached to the hand or the foot by a set of bones which form the wrist or the ankle. The joints with the pectoral and pelvic girdles and the joints at wrist, ankle, elbow and knee enable the limbs to move freely.

Joints

A joint is a meeting of two bones. Some joints are fixed. The bones of the skull, for example, meet at fixed joints called **sutures** (see Figure 17.3B). Other types of joint are not fixed:

- **Ball and socket joints** are formed where the upper long bones of the arms and legs meet their respective girdles. Figure 17.3D shows how the rounded end of the femur fits into a cup-shaped socket in the pelvic girdle.
- Figure 17.3E shows the **hinge joint** of the elbow. Hinge joints are also found in the fingers and knee.

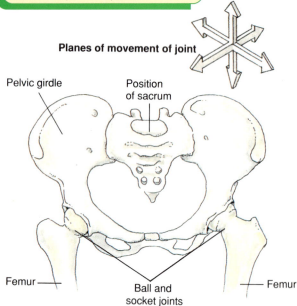

Figure 17.3D ● Ball and socket joint (hip joint)

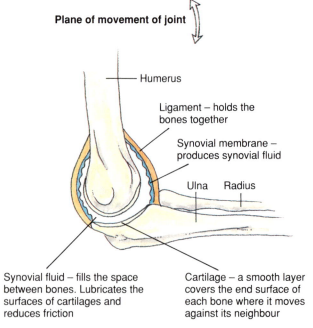

Figure 17.3E ● Hinge joint (elbow joint)

CHECKPOINT

❶ Study the diagram of the arm below and answer the following questions.

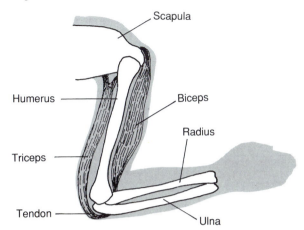

(a) Which muscle contracts when the arm is raised?
(b) Which muscle contracts when the arm is straightened?
(c) Why are the biceps and triceps called an antagonistic pair of muscles?
(d) How are the arm muscles attached to the arm bones?
(e) What type of joint is formed between the humerus and the scapula?

❷ Match each of the bones in Column A with its description in Column B.

Column A	Column B
Cranium	Consists of vertebrae fused together
Rib cage	Upper arm bone
Sternum	Protects the brain
Sacrum	Moves as you breathe
Clavicle	Ribs are attached to it
Femur	Called the collar bone
Humerus	Upper leg bone

❸ (a) Distinguish between the hydrostatic skeleton, exoskeleton and endoskeleton.
(b) Say what features the hydrostatic skeleton, exoskeleton and endoskeleton have in common.

❹ Give explanations for the following statements.
(a) Pregnant women should be encouraged to drink milk.
(b) The sutures of the skull are fixed joints.
(c) The human femur is stronger than the humerus.
(d) Breathing depends on the movement of the ribs.
(e) Women usually have broader hips than men.

❺ Distinguish between smooth muscle, cardiac muscle and skeletal muscle. Explain how each type of muscle is suited to the job it performs.

❻ Complete the following paragraph using the words listed below. Each word may be used once, more than once or not at all.

levers calcium support collagen joint rigidity muscles movement contract food bones protect

The _____ of the skeleton _____ and _____ the soft tissues of the body. The limb bones are _____ which move because _____ attached to them _____ . Where bones meet a _____ is formed. Bone consists of the protein _____ and _____ salts. Blood vessels in the bone supply _____ and oxygen.

7 Name the bones of a human arm. Do they resemble the bones of a bird's wing?

8 List the different functions of feathers. Discuss **two** of the functions in more detail.

9 Different forces act against a bird in flight. Identify the forces and show with the help of diagrams how the design of the body and wings is adapted to overcome them.

10 Look at Figure 17.2M. Summarise how muscles and bones work together to move the wings.

11 Compare the arrangement of muscles and bones in the horse's hind leg with the arrangement of muscles and bones in the human. (Identify the similarities and comment on the differences.)

12 Describe the stages in a horse's stride.

13 What is the difference in the cycle of hoof beats between a horse walking and a horse galloping?

 DISEASE

18.1 **What is disease?**

FIRST THOUGHTS

Think about these figures:
- in developing countries more than 4 million children die each year from the effects of diarrhoea
- parasitic worms infect around 2400 million people world-wide
- cancer accounts for 15% of deaths in developed countries but only 4% in developing countries
- influenza ('flu) and hepatitis B viruses infect enormous numbers of people world-wide. Millions die from these diseases each year.

This section explores the background to these startling facts.

LOOK AT LINKS
Living sensibly
for **healthy diet**
See Topic 11.3.
for **smoking**
See Topic 12.2.
for **exercise and heart disease**
See Topic 14.5.

Myxomatosis is a viral disease of rabbits, which affects the skin. In the 1950s, the virus was deliberately introduced into Australia, where the European rabbit had previously been released and had become a serious pest. Few rabbits survived. By accident the virus came to the UK and the wild rabbit population crashed. Now the virus seems to cause less serious symptoms and rabbits are more resistant to it. As a result the rabbit population in the UK is increasing.

LOOK AT LINKS
The protozoan *Trypanosomas brucei* is a zoonosis. It causes sleeping sickness See Topic 3.7.

For most of us the word 'disease' conjures up the idea of feeling unwell. Figure 3.7A on page 61 shows a range of organisms that cause different diseases. We blame **bacteria** for most of our ailments but **viruses** are the most important disease-causing agents. Notice that **protists** and **fungi** are culprits as well, and that different animal **parasites** also cause disease as a result of their activities inside our bodies. Disease organisms are called **pathogens**. Diseases are said to be **infectious** if the organisms can be passed from one person to another.

Not all diseases are infectious. Many **non-infectious** diseases develop because the body is not working properly (see Table 18.1). The way we treat our bodies affects the onset of non-infectious diseases. Many disorders can be avoided or at least delayed providing we live sensibly. You can find more about disease from the index and good advice on looking after yourself by checking the first **LOOK AT LINKS** feature.

Table 18.1 ● Non-infectious diseases

Type of disease	Description	Examples
Cancer	Cell division produces new cells to replace old cells as they wear out and die. If cell division runs out of control then a cancerous growth (called a tumour) develops	Lung cancer, leukaemia, cancer of the cervix
Degenerative ('wear and tear')	Organs and tissues 'wear' and work less well with age	Sight and hearing may deteriorate, heart attacks, arthritis
Deficiency	Poor diet may deprive the body of vitamins and other essential substances	Scurvy, rickets, kwashiorkor
Allergies	Reaction to a substance (antigen) which is normally harmless but which the body regards as harmful	Hayfever
Psychological/ mental illness	A wide range of disorders from feeling more than normally 'fed up' to more severe conditions due to brain damage or the brain not working properly	Depression, paranoia, schizophrenia
Metabolic	Defect (often inherited) that prevents an organ or tissue from working properly	Diabetes, dwarfism, haemophilia, sickle cell anaemia

Zoonosis

We share many pathogens and the diseases they cause with other animals. A pathogen which infects people and other animals is called a **zoonosis**. Animals may form a reservoir of zoonoses from which infection may spread to more and more people. Effective treatment depends on controlling the pathogen in animals as well as the human victims.

Strict quarantine laws have so far helped to keep Britain free of rabies. Animals brought into Britain from other countries are kept away from people and other animals until it is certain that they do not carry the disease. Cats and dogs, for example, are put into quarantine for six months from the time of arrival. An animal showing signs of rabies is destroyed to prevent the disease from spreading. Anyone smuggling animals into the country is prosecuted for breaking the quarantine laws. *Why should we obey the quarantine laws?*

IT'S A FACT

Treatments to control plant disease cost about six billion dollars per year world-wide. Scientists estimate that without treatment farmers would lose about one-third of crops to disease.

IT'S A FACT

In 1845–7 *Phytophthera infestans* devastated the potato crop in Ireland. More than a million people died of starvation. A further million escaped famine by emigrating, mainly to the USA. Within a few years the population of Ireland was halved.

LOOK AT LINKS

for **spores** and **hyphae**
See Topic 1.5;
for **stoma**
See Topic 14.1;
for **fungicides**
See Topic 6.9;
for **monoculture**
See Topic 6.4;
for **potato tubers**
See Topic 15.4.

Rabies virus is a zoonosis. It infects warm-blooded animals and is found most often in wolves, jackals and other members of the dog family. Contact with domestic pets (biting, licking) is the most usual route by which rabies virus passes to people. However, in South America blood sucking vampire bats are a common source of infection.

Rabies virus attacks the central nervous system (see p. 299) causing 'mad' behaviour. At this time the virus is present in the saliva. A rabid dog often runs and bites at random, passing on the infection to people and other animals.

The virus destroys nerve cells but takes time to establish itself in the central nervous system of the victim. If cuts and bite marks are washed with disinfectants and treatment with rabies serum (serum is plasma with fibrinogen removed – see p. 235) and vaccination is started immediately after infection, then there is a good chance of preventing the disease. However, once symptoms develop (from 2–3 weeks up to 5 months after infection) then death is nearly always inevitable within 10 days.

Fungal diseases

Plant disease is caused by as wide a range of pathogens as animal disease. However, the impact of particular types of pathogen is different. Bacteria for example are a major cause of animal disease but pose few problems for plants. Fungi on the other hand are a major hazard. More than 1000 species cause plant disease with serious effects on crops and food production.

Phytophthera infestans causes **late blight** of potatoes. In the 1840s, the fungus wiped out the European harvest. Control measures were inadequate and potato plants had little resistance to the disease.

Phytophthera passes from plant to plant owing to:
- planting infected potatoes
- infected left-overs from the previous crop in the ground
- spores carried by wind or rain to other plants in the crop.

The spores germinate and fungal hyphae grow between the potato leaf cells. Side branches penetrate and feed on the cells, destroying them. The parts of the plant above ground eventually die. As a result little starch is produced through photosynthesis for storage by the potato tubers underground. The tubers fail to develop.

Before the potato dies, some hyphae grow through the stomata of the leaves to the outside. They carry spores which break off and infect other plants.

Good hygiene helps break the chain of infection. Burning infected plants and killing the plant above ground before harvest reduces the risk of infecting the potato tubers below ground. Other control measures include using disease-free seed potatoes and spraying plants with fungicides to prevent spores that settle on the leaves from germinating.

Rust disease of wheat is caused by *Puccinia graminis*. This fungus lives in the stem and leaves. The weakened plant produces less grain and poor quality straw.

Wheat rust is persistent, helped by farming practices such as **monoculture** (see Figure 6.1A). Pathogens able to infect the crop spread rapidly through the hectares of identical plants causing huge losses of grain. Breeding rust-resistant strains of wheat is the best form of defence, but rust fungus quickly mutates. Scientists are in a race to discover new resistant varieties of wheat before new strains of fungus overwhelm the crop. *Do you think this is another example of the Red Queen Effect (see p. 366)?* Rust disease also affects other cereal crops.

LOOK AT LINKS

for **world hunger**
See Topic 6.11.

Disease world-wide

In the developing countries of Africa and Asia infectious diseases are responsible for 40% of all deaths. For example, each year more than 14 million children under the age of five years die from diarrhoea, measles and other diseases associated with inadequate medical services, poverty, poor housing and malnutrition.

In sharp contrast, relatively few people living in the developed countries of Japan, Europe and North America, where there are good medical services, die from infectious diseases. Instead many illnesses are related to our being better-off. We can afford to smoke, drink too much alcohol and eat too much of the wrong sort of food. Unhealthy life-styles are partly responsible for the onset of heart disease and cancer, which are the leading causes of death in many developed countries.

Life expectancy and the infant mortality rate reflect the difference in causes of death between countries. In developed countries the low level of infectious diseases means that many people reach old age and most children survive to become adults. In developing countries infectious diseases make life shorter and the number of children who die is much greater. Despite the gap, progress is being made. The World Health Organisation has established minimum objectives for the year 2000 of a life expectancy at birth of 60 years and an infant mortality rate of less than 50 deaths per 1000 births. Education, good medical services, the relief of poverty and improvements in living conditions are all essential for success.

SUMMARY

Pathogens are organisms that cause infectious diseases. Non-infectious diseases arise when the body is not working properly. Zoonoses are pathogens which people share with other animals. Plant pathogens damage crops, reducing yields. Infectious diseases are responsible for the majority of deaths in developing countries.

CHECKPOINT

❶ Distinguish between infectious diseases and non-infectious diseases. Give examples of each category.

❷ What is a zoonosis? Discuss how the biology of a **named** zoonosis helps its transfer between people and other animals.

❸ Discuss how the biology of the fungus *Phytophthera infestans* makes it a serious pest of the potato.

❹ Distinguish between the different causes of death in the developing countries of Africa and Asia and the developed countries of Japan, Europe and North America.

18.2 Influenza: a case study

FIRST THOUGHTS

Every few years influenza ('flu) breaks out in the winter affecting thousands of people. Every 10 to 20 years 'flu sweeps the world affecting millions. This section explains the outbreaks and why preventing them is difficult.

A sneeze produces a jet of moisture droplets which shoot out of the nose and mouth. Microorganisms infecting the mucous membranes lining the nose and the rest of the **upper respiratory tract** are carried in the droplets and spread to other people nearby. This is how diseases like colds and 'flu pass from person to person, especially in crowded places like schools and hospitals. Remember to turn away from people when you sneeze or, if in time, sneeze into a handkerchief. It is not only good manners but also helps to cut down the spread of air-borne diseases.

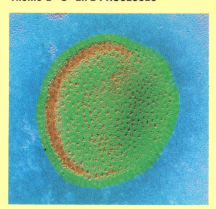

Figure 18.2A ● 'Flu virus particle

LOOK AT LINKS
Interferons are responsible for some 'flu symptoms
See Topic 18.2.

LOOK AT LINKS
for **antigens** and **antibodies**
See Topic 14.2.

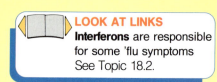
LOOK AT LINKS
for **vaccines**
See Topic 20.5.

How many times have you caught a 'cold'? Sneezing, a runny nose, a cough, a sore throat and a raised temperature make you feel miserable. Many different viruses cause these symptoms. Some resistance to colds may develop with frequent exposure to the viruses but their variety means that lifelong immunity is impossible.

'Flu is caused by a virus (Figure 18.2A). Notice the spikes and knobs projecting from the surface of the virus. They are proteins which bind the virus to, and help it penetrate, the cells of the mucous membrane. The infected cells are destroyed, and after 48 hours the symptoms of 'flu appear. Muscles ache and overproduction of mucus causes a runny nose with sneezing and coughing. A sore throat and fever develop, and the victim feels generally unwell. Symptoms usually subside within 2–7 days unless the infection spreads to the lungs. Then viral pneumonia may set in making the patient seriously ill. The lungs fill with fluid and bacteria may set up other infections.

Treatment

Normally treating the patient amounts to little more than bed rest and good care. Very few drugs are effective against viral diseases (including 'flu virus), although **amantadine** is available for patients who are at risk of serious illness. It seems to prevent the 'flu virus reproducing in the cells of the mucous membrane. A person who is ill with 'flu is easily attacked by bacteria. Antibiotics may be used to treat the bacterial infection.

Vaccines

The surface proteins of 'flu virus (see Figure 18.2A) are **antigens** against which an infected person produces **antibodies**. However, people may catch 'flu more than once during their lifetime. *Why are the antibodies ineffective against 'flu virus when it next invades the body?* After all you rarely catch diseases like chicken pox and measles more than once (see p. 237).

Unfortunately, 'flu virus antigens often change shape. Minor changes are called **antigenic drift**. They produce new **strains** of virus which are probably responsible for the frequent occurrence of 'flu **epidemics**. Major changes are called **antigenic shift** and result in new **types** of virus. They are less frequent and seem to be linked to the 10–20 year cycle of world-wide **pandemics** (Figure 18.2B). Antigenic drift and antigenic shift means that antibodies produced against a particular type of 'flu virus do not protect the person from a new infection. They also mean that producing an effective vaccine is difficult.

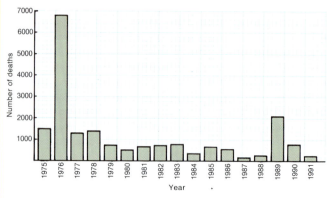

Figure 18.2B ● Deaths from 'flu in England and Wales 1975–1991. A disease is said to be an **epidemic** if it affects a significant proportion of a country's population for a limited period. A series of epidemics affecting countries world-wide is called a **pandemic**. In 1918 a 'flu pandemic killed 20 million people: more than those killed in the Great War which had just ended. *In which years do you think England and Wales were probably part of a 'flu pandemic?* (Source: Office of Population Censuses and Surveys)

SUMMARY

'Flu is caused by a virus. Epidemics and pandemics of the disease are related to frequent changes in 'flu antigens. To be effective, new vaccines must be developed quickly in response to changes in the virus.

Prevention

The **World Health Organisation** monitors the **antigenic changes** in 'flu virus and recommends the strains of virus which should be included in vaccines. The recommended strains are grown in fertilised chicken eggs, extracted and inactivated with methanal.

The vaccine usually consists of several strains and is used to protect young children, the old and other people who may be at risk of serious illness if they catch 'flu. Protection from a particular strain of 'flu virus is effective providing the vaccine for the strain is given each year and that its antigens do not change. When antigenic changes occur new techniques allow vaccine to be redeveloped quickly to keep pace.

CHECKPOINT

❶ Describe the symptoms of 'flu and discuss how they are caused.

❷ After many years of research why is an effective vaccine against the common cold still not available?

❸ Why does the World Health Organisation monitor antigenic changes in the 'flu virus?

FIRST THOUGHTS

18.3 Fighting disease

0
0
0

With so many pathogens capable of causing disease why do we stay healthy most of the time? Reading this section will add to what you have read on this subject elsewhere in this book. Use the Look at Links feature to help you find the extra information.

LOOK AT LINKS

Treating disease is covered in the following topics:

Saturated/unsaturated fats and heart disease	10.3, 11.3
Immunisation	14.2, 20.5
Antibiotics	15.5, 11.4
	4.5, 20.5
Hormones	16.5, 20.5
Anti-malarials	4.4
Oral rehydration therapy	4.5
Influenza	18.2
Transgenic sheep as source of Factor 8	20.3
Monoclonal antibodies and cancer	20.5
Interferon	20.5

Figure 18.3A shows the body's natural defences against disease. Notice the **physical** and **chemical** barriers to infection. They help to keep most of us healthy for most of our lives. Looking after ourselves and maintaining good hygiene help the defences to do their job.

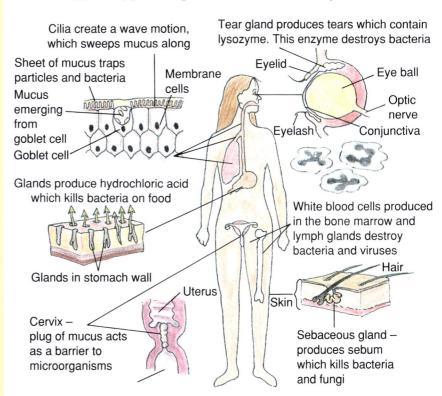

Cilia create a wave motion, which sweeps mucus along

Tear gland produces tears which contain lysozyme. This enzyme destroys bacteria

Sheet of mucus traps particles and bacteria

Membrane cells

Eyelid

Eye ball

Mucus emerging from goblet cell

Optic nerve

Goblet cell

Eyelash

Conjunctiva

Glands produce hydrochloric acid which kills bacteria on food

White blood cells produced in the bone marrow and lymph glands destroy bacteria and viruses

Glands in stomach wall

Hair

Uterus

Skin

Cervix – plug of mucus acts as a barrier to microorganisms

Sebaceous gland – produces sebum which kills bacteria and fungi

Figure 18.3A ● The body's natural defences

335

In 1865, the Edinburgh doctor Joseph Lister (1827–1912), working as professor of surgery at King's College Hospital, London, heard of Pasteur's experiments (see p. 359). Pasteur had shown that airborne microbes turn meat broth 'bad'. Lister wondered if infection with airborne microorganisms was responsible for **sepsis** developing after surgery, turning wounds 'bad'. At that time many patients died from septic wounds so operations were carried out only on the most serious cases.

Lister thought that carbolic acid might kill bacteria infecting a wound. He invented a spray which directed a fine mist of carbolic acid over the area of the patient where the surgeon was working. The reduction in infection and sepsis was dramatic. Lister's invention marked the beginning of **antiseptic surgery**.

Today the aim is to prevent microorganisms from infecting the wound. This is the basis of **aseptic surgery**:

- sterilised gowns and gloves are worn by surgeons and their assistants
- air entering the operating theatre is filtered
- equipment, furniture and surfaces are easy to clean.

Disinfectants and antiseptics

Disinfectants are chemicals that kill microorganisms; **antiseptics** stop them from multiplying. We use disinfectants and antiseptics to attack microorganisms outside the body.

Disinfectants help to keep surfaces clean and free from microorganisms. For example, bleach poured down the lavatory kills microorganisms lurking around the bowl. Bleach is a strong disinfectant which contains calcium chlorate(I). It is effective but corrosive, so use it carefully.

Antiseptics are usually weaker than disinfectants. They can be used to swab a wound or before an injection to clean the area of skin where the hypodermic needle is to be inserted.

The skin is one of the main barriers against infection. Washing regularly reduces the chances of infection. Some soaps and skin cleansers have antiseptic properties. They are particularly effective against the bacteria that cause body odour. They also help prevent fungal infections such as ringworm and athlete's foot (see Figure 3.7A, p. 61). Sweaty feet are an ideal environment for fungi. So, wash and dry carefully between your toes. Also, borrowing other people's footwear is not a good idea. The fungi are spread from person to person by physical contact.

Chemotherapy

The word **chemotherapy** describes treatments which use drugs to attack pathogens. There is a lot more about drugs and treatment of disease elsewhere in the book. Use the LOOK AT LINKS feature on p. 337 to look up the information.

In the fight against disease, antibiotic drugs have been the success story of the twentieth century. Before their discovery, bacterial diseases were major killers, especially of children. Now the majority of infectious diseases can be cured – thanks to antibiotics.

There are two sorts of antibiotic:

- **bactericides** like penicillin (see p. 373), which kill bacteria
- **bacteristats** like tetracycline (see p. 373), which prevent bacteria from multiplying.

Figure 18.3B shows how different antibiotics affect bacteria. Notice that those which damage the structure of the cell are mostly bactericidal. Those which disrupt metabolism (see p. 156) are mostly bacteristatic.

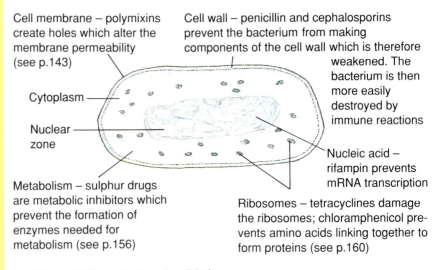

Cell membrane – polymixins create holes which alter the membrane permeability (see p.143)

Cytoplasm

Nuclear zone

Cell wall – penicillin and cephalosporins prevent the bacterium from making components of the cell wall which is therefore weakened. The bacterium is then more easily destroyed by immune reactions

Nucleic acid – rifampin prevents mRNA transcription

Metabolism – sulphur drugs are metabolic inhibitors which prevent the formation of enzymes needed for metabolism (see p.156)

Ribosomes – tetracyclines damage the ribosomes; chloramphenicol prevents amino acids linking together to form proteins (see p.160)

Figure 18.3B ● Action of antibiotics

LOOK AT LINKS

for more about **drug resistance in bacteria**
See Topic 21.4;
for the **structure of bacteria**
See Topic 9.1;
for the **Red Queen Effect** and **resistance**
See Topic 20.3.

Resistance

Since the 1940s, widespread use of antibiotics has led to the development of **resistance** (see Figure 18.3C). Bacteria multiply very quickly (in some species a new generation is produced every 20 minutes) and resistance spreads quickly. To be effective, dosage has to be increased step-by-step until the drug becomes so inefficient or poisonous to the patient that an alternative has to be found.

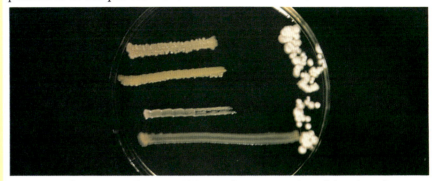

Figure 18.3C ● *Penicillium* mould is growing at one edge of the plate; four different bacteria have been streaked on the plate; three are sensitive to the penicillin diffusing from the mould and show a space; the fourth bacterium is not sensitive and the streak therefore grows up to and in contact with the mould.

Even so, new drugs are only part of the answer to the problem. Resistance to each new drug soon develops. Precautions can be taken in an attempt to slow the development of resistance:

- *avoid* using antibiotics by practising good hygiene to prevent the spread of infection
- antibiotics should be used *sparingly* and not prescribed unnecessarily
- a patient should always *finish* the prescribed course of antibiotics
- *reduce* the antibiotics given to farm animals to improve production. Meat contains the antibiotics. Eating meat, therefore, means that people carry traces of antibiotic.

Resistance develops in different ways. For example some bacteria are resistant to penicillin G (see p. 373 – one of the first of the family of penicillins, and still a useful antibiotic) because they produce **penicillinase**. This enzyme breaks down the penicillin, making the drug ineffective. A particular strain of the coccus bacterium *Staphylococcus aureus* (see Figure 1.5C, p. 12) is resistant to all antibiotics, except one with side-effects that can be lethal to the patient! These examples highlight the importance of the measures listed above, which aim to guard against the development of resistance.

SUMMARY

The body's natural defences against disease are supported by advances in modern medicine. Disinfectants and antiseptics help to prevent the spread of infection. Antibiotic drugs, in particular, have had a major impact on treatment. Diseases which previously killed many people can now be cured. However, some bacteria have developed resistance to antibiotics. Sensible use of drugs and the discovery of new ones together help to combat the problem.

CHECKPOINT

❶ Discuss how the body's natural defences keep most of us healthy for most of our lives.

❷ Distinguish between:

antiseptics	disinfectants
bactericides	bacteristats
antisepsis	asepsis

❸ Discuss how different antibiotic drugs attack bacteria.

❹ How do bacteria become resistant to antibiotic drugs? Discuss the precautions which can be taken to slow the development of resistance.

? THEME QUESTIONS

● **Topic 11**

1 The table below gives the approximate amounts of energy required by different people.

Energy (kJ/day)		Energy (kJ/day)	
Newborn baby	1890	Office worker	11 327
Child 1 year	3347	Factory worker	12 575
Child 3 years	5852	Coal miner	15 075
Child 5 years	7621	Pregnant woman	10 065
Girl 12–15 years	9705	Breast-feeding	11 325
Boy 12–15 years	11 753		

(a) Explain the increase in energy requirement from a newborn baby to a 12–15 year old.
(b) Why do the following require more energy than an office worker:
    (i) a 12–15 year old boy (ii) a coal miner?
(c) Why does a breast-feeding woman require more energy than a pregnant woman?

2

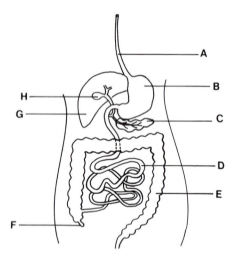

(a) Copy and label the parts of the diagram of the human alimentary canal.
(b) Which organs in the diagram (i) produce bile (ii) store bile (iii) have acidic conditions (iv) reabsorb a lot of water from food?
(c) What are the two major functions of the gland C?
(d) The organ D is very long. How does this help it to carry out its function?
(e) How would organ F be different in the alimentary canal of a rabbit? How is this connected with the rabbit's diet?

3 A scientist carried out an experiment growing cereal plants in a field. He took a sample of the plants every four hours and measured how much sugar had been made in their leaves. The sugar concentrations, expressed as a percentage of the dry mass of the leaves, are given in the following table.

Time of day	Sugar concentration
4.00 a.m.	0.45
8.00 a.m.	0.60
12 noon	1.75
4.00 p.m.	2.00
8.00 p.m.	1.45
12 midnight	0
4.00 a.m.	0.45

(a) Plot the data on graph paper, with time of day on the horizontal axis.
(b) From the graph, suggest the likely sugar concentration at (i) 10 a.m. and (ii) 2 a.m.
(c) At what time of day is the sugar concentration likely to be at a maximum?
(d) Look at the graph and try to explain the variations that occur in the sugar concentration over 24 hours.

● **Topic 12**

4 Look carefully at the diagram of the alveoli and blood capillary in a lung on the next page.
(a) Which arrow represents (i) deoxygenated blood (ii) carbon dioxide?
(b) (i) By what process is oxygen able to pass from the air in the alveoli into the red blood cells in the capillary?

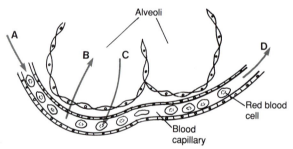

(ii) Through how many layers of cells does the oxygen have to pass?
(c) Give two features of the alveoli that help this passage of gas.
(d) Which structures allow gas exchange to take place in (i) a leaf (ii) a fish?

5 The respirator was set up as shown in the diagram and after a short interval of time the coloured liquid rose in arm X of the capillary tubing.

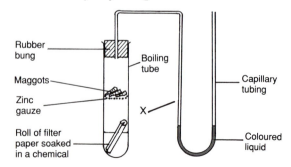

(a) (i) Name the chemical in the boiling tube.
   (ii) Which gas does it absorb?
   (iii) What is the function of the roll of filter paper?

(b) Explain why the coloured liquid rose up arm X.

(c) Suggest two possible sources of error in this experiment.

(d) If germinating green peas are used instead of maggots, the boiling tube has to be blacked out. Why is this so?

● Topic 13

6 The diagram shows a kidney dialysis machine in simplified form. The patient's blood flows on one side of a selectively permeable membrane (like visking tubing). The dialysing fluid, which has a composition similar to that of human plasma, flows on the other side of the membrane.

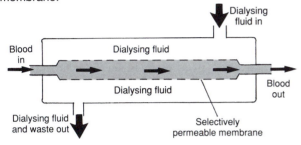

(a) By what processes does (i) excess water and (ii) excretory products pass from the blood into the dialysing fluid?

(b) (i) Name the excretory products that would pass out of the blood.
   (ii) Why should protein not pass out of the blood?

(c) (i) What structures in the kidney would act like the selectively permeable membrane?
   (ii) Suggest why the presence of protein in the urine could be an indication of kidney damage.

(d) In the cases of permanent kidney failure it is more convenient and more economical for a patient to be given a kidney transplant rather than treatment with an artificial kidney machine. Explain why this is so.

● Topic 14

7 The figure below is a histogram showing the rate of blood flow to various organs.

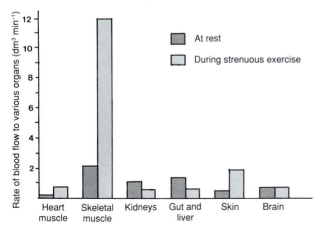

(a) Copy the following table and record the volume of blood flowing per minute to each of the organs when the body is at rest, and when the body is undergoing strenuous exercise.

	Volume of blood (dm³/min)	
	At rest	*During strenuous exercise*
Heart muscle		
Skeletal muscle		
Kidneys		
Gut and liver		
Skin		
Brain		

(b) What is the total volume of blood per minute being pumped by the left ventricle to all these organs when the body is (i) at rest (ii) undergoing strenuous exercise?

(c) If the pulse rate when the body is at rest is 70 beats per minute, what volume of blood is pumped out by the left ventricle at each heart beat?

(d) If the pulse rate during strenuous exercise is 160 beats per minute, what volume of blood is pumped out by the left ventricle at each heart beat?

(e) Account for the changes in the rate of blood flow to the following organs as a result of undergoing strenuous exercise.
   (i) Heart muscle
   (ii) Skeletal muscle
   (iii) Brain
   (iv) Gut and liver   (MEG)

● Topic 15

8 (a) Describe the events which occur after the release of sperms into the vagina of a human female until implantation of the embryo.

(b) The chart below shows how the thickness of the uterus lining and the levels of two hormones A and B, made in the ovaries, vary during the menstrual cycle. The first month is a normal menstrual cycle but fertilisation occurs during the second month.

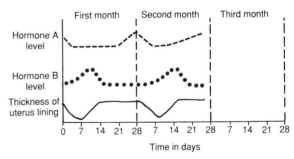

(i) Say when was the most likely time of ovulation in the first month.
(ii) During which period in the second month is fertilisation likely to have occurred?

(c) Copy and complete the chart above to show what happens, following fertilisation, to:
(i) the uterus lining in the third month,
(ii) the levels of hormones A and B in the third month.

(d) One type of contraceptive pill contains a mixture of hormones A and B.
(i) Explain briefly how this pill works as a contraceptive.

(ii) If she is using the contraceptive pill, it is usual for a woman to take a hormone pill each day for 21 days and then to take a pill without any hormones for the next 7 days. What is the advantage of taking the hormones for 21 days only?

(NEA)

● *Topic 16*

9 A young plant is grown in a pot as shown in diagram A. It was then turned on its side as in B, and the result after several days is shown in diagram C. The pot is shown in section.

(a) Name **two** factors which could be responsible for the growth of the stem as shown in C.

(b) Name **one** factor which could be responsible for the growth of the roots in the direction shown in C.

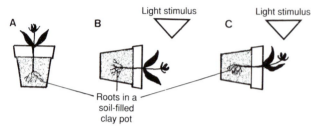

(c) To make the responses shown in the diagram possible, the growing points produce a chemical substance. Name this substance.

(d) State the advantages of these responses in (i) the stem and (ii) the root, to the plant when growing in its natural habitat.

(WJEC)

● *Topic 17*

10 The diagram below shows a section through the ball and socket joint in the human hip.

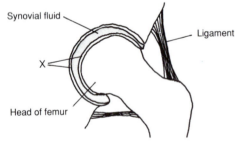

(a) Name the structure X.

(b) The ligament is playing a part in holding the bones together. As well as being strong, suggest another important property of the ligament.

(c) What is the function of the synovial fluid in the joint?

(d) The diagram below shows the positions of the bones and main muscles of the legs of a human when running. (For clarity, each muscle is shown on one leg only.)

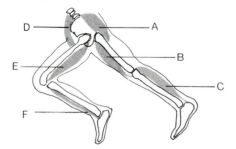

(i) Explain the term 'antagonistic muscles', illustrating your answer by reference to **two** sets of antagonistic muscles from the muscles labelled A to F on the diagram.

(ii) Suggest why muscle C is more powerfully developed than muscle F.

(NEA)

● *Topic 18*

11 An adult blowfly lands on the surface of an open meat sandwich on a hot summer's day and lays eggs.

(a) Name **one** type of microorganism which could be transferred from the blowfly to the meat.

(b) Explain why the growth of some microorganisms on the meat might have been reduced if a large growth of penicillium mould was present on the bread on which the meat was placed.

(c) Describe how the blowfly eggs will eventually develop into adult blowflies.

(d) Explain why microorganisms would be less likely to grow successfully if table salt had been sprinkled over the meat when the sandwich was made.

(MEG)

12 The diagram shows apparatus used in an experiment to investigate the conditions which cause milk to decay.

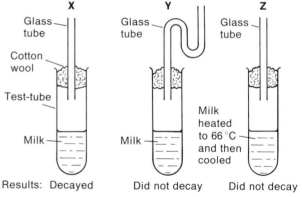

Results: Decayed   Did not decay   Did not decay

(a) (i) Name a type of organism which causes food to decay.
(ii) Explain why the milk in test tube **Y** did **not** decay though the glass tube was open to the air.

(b) (i) Explain why the milk in test tube **Z** did not decay for several days.
(ii) Milk is often preserved using the type of practice shown in **Z**. State **one** advantage of preserving milk in this way rather than boiling the milk.

(c) State **two** other methods of food preservation (**not** connected with this experiment).

(d) Name the parasite responsible for AIDS.

(e) The AIDS parasite can be passed to others by sexual intercourse. Briefly describe **three** non-sexual ways in which the parasite might be passed on.

(f) Describe **three** methods being used to control the spread of AIDS.

(g) AIDS creates social and medical problems. Briefly describe **three** AIDS-related problems.

(MEG)

13 'I think that great things are coming to pass. Joseph Meister has just left the laboratory. The last three inoculations have left some pink marks under the skin … The lad is very well this morning … he has a good appetite and no feverishness'

That was Louis Pasteur writing about the first human test of his anti-rabies treatment. He used it in 1885 to treat a small boy who had been bitten by a dog with rabies.

(a) Vaccines can be made by making a weaker and weaker strain of the virus which causes the disease. What do the white blood cells make in response to the weak strains of the virus getting into the body when the person is vaccinated?

(b) What is **one** disadvantage of using weakened forms of the virus as a vaccine?

(c) What technology is being used to make safer vaccines?

(d) Use these shapes to draw a diagram of a antibody–antigen complex.

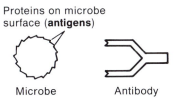

Proteins on microbe surface (**antigens**)

Microbe   Antibody

(e) What is the name for antibodies which are all of the same type?

(f) Monoclonal antibodies attached to radioactive materials can be used to locate cancerous cells in the body.
(i) How could you find the antibodies when they have found the cancerous cells?
(ii) Once the cancerous cells have been found they could be destroyed by drugs attached to monoclonal antibodies, without the drugs damaging any cells which are not cancerous. Explain how this could happen.

(MEG)

14 The heart muscle has its own supply of blood from blood vessels that run all over its surface. In some people these vessels can become blocked with a fatty substance called **cholesterol**. The diagram shows a human heart in which a clot has blocked the right coronary artery.

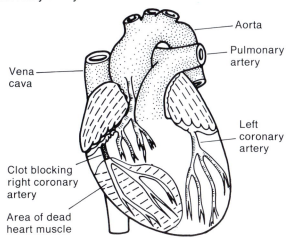

Aorta

Pulmonary artery

Vena cava

Left coronary artery

Clot blocking right coronary artery

Area of dead heart muscle

(i) Look at the diagram and state the name of the major blood vessel which supplies blood to the right coronary artery.
(ii) Look at the diagram and explain why the area of heart muscle has died.
(iii) Describe the sort of people whose blood vessels are most likely to be blocked with cholesterol.

(MEG)

15 (a) What type of drug can be used to fight an infection caused by bacteria?

(b) Who made the observation that a mould *Penicillium* stopped bacteria from growing?

(c) Before penicillin was made in large amounts it was so precious that it was recycled from patient to patient.
How might penicillin be recycled?

(d) Alexander has a cold. His doctor says that an antibiotic would be of no use in curing a cold. Why is the doctor right?

(e) The Health Centre has prescribed tetracycline for Theresa's infected cut on her finger. After Theresa has taken half the pills her finger looks better. Why must she make sure that she finishes **all** the pills?

(f) People who are taking tetracycline pills often eat unpasteurised live yoghurt during their course of treatment. Why do they do this?

(MEG)

16 Darren has broken his arm. To help healing, metal plates were fixed to the bones in Darren's arm. The diagram below shows this.

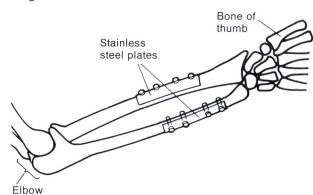

Bone of thumb

Stainless steel plates

Elbow

After a few weeks, it was found that healing had not started.
(a) What technique could be used to show this?
The healing of a broken bone is controlled by hormones. When Darren's arm failed to heal, the doctors decided to take several small pieces of bone from his hip and place them near the ends of the broken bones. As a result, Darren's arm began to heal rapidly.
(b) (i) What is meant by a hormone?
(ii) Suggest why the break failed to heal until pieces of hip bone were placed near the broken bones.

(MEG)

17 The diagram below shows a potato plant. The tubers are the part that people eat.

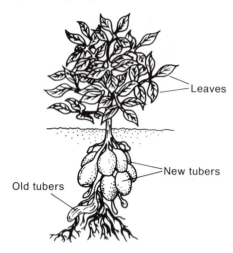

Leaves

New tubers

Old tubers

Read the following passage about disease in potatoes.

1 A disease of potatoes is caused by a fungus. The fungus attacks the leaves and the stems of the potato plant. The fungus damages the 'tubes' in the plant along which food passes from the leaves to
5 the tubers in the soil. The fungus feeds on the potato plant cells by external digestion.

The fungus is mainly spread by spores in warm, damp weather. Potato crops can be protected from the fungus by spraying with chemicals. The spread
10 of the fungus can also be reduced by planting potato tubers which are not infected.

Some types of potato plant are better able than others to resist the disease. The variety 'Majestic' is more resistant to the disease than the variety
15 'King Edward'.

Use the information in the passage and your knowledge to answer the following questions.

(a) (i) Name a food produced in the leaves (line 4).
(ii) Which 'tubes', xylem and phloem, carry food from the leaves to the tubers (line 5)?
(iii) Energy is needed to move the food inside a plant. Which process releases this energy?

(b) Describe how the fungus feeds by external digestion (line 6).
Use the following words in your answer:
enzymes   absorb
(c) (i) Which structures of the fungus spread the disease (line 7)?
(ii) Under what conditions does the fungus spread rapidly (lines 7–8)?
(d) Give **two** ways by which farmers can reduce the spread of the disease (lines 9–15).
(e) The graph below shows potato production in Britain between 1965 and 1993. The potato production is given in tonnes per hectare.

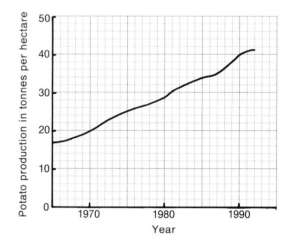

(i) How many tonnes per hectare were produced in 1970 and how many in 1985?
(ii) By how much did potato production per hectare increase between 1970 and 1985?
(f) If there is no disease, more potatoes can be produced. Farmers can increase potato production in a field by other methods. Suggest **two** ways by which this can be done.

(LEAG)

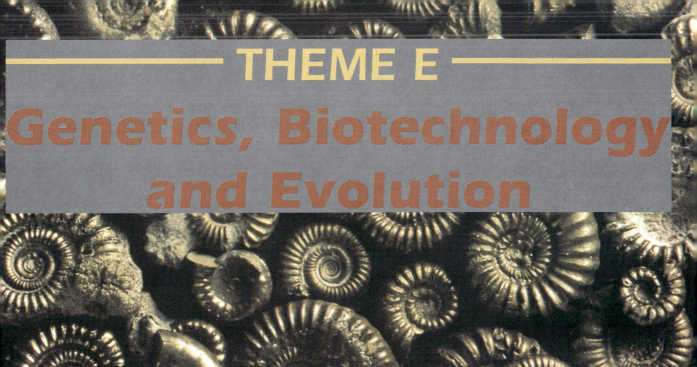

THEME E

Genetics, Biotechnology and Evolution

Geneticists help our understanding of heredity by investigating how offspring inherit characteristics from their parents. Biotechnologists turn our understanding to good use. New techniques such as the use of monoclonal antibodies and genetic engineering help fight diseases, make new products, improve food production and protect the environment. Genetics also helps our understanding of evolution: the process whereby living things become different from their ancestors through thousands of generations of change due to natural selection. The variety of life today is a result of evolution from the origins of life on Earth around 4000 million years ago.

INHERITANCE AND GENETICS

19.1 From generation to generation

Glance around your classmates. Almost certainly you look different from one another. However, you each look something like your parents or even grandparents because you inherited their DNA and, therefore, the characteristics that DNA controls. The study of the patterns of heredity (the ways in which offspring resemble or differ from their parents) is called genetics.

Blending inheritance

Figure 19.1A ● Three generations of a family. *What characteristics do you think they have in common?*

IT'S A FACT

People used to think that blood was responsible for offspring inheriting a mix of their parents' characteristics. Even today we speak of 'blood relation', 'royal blood' and 'blood line'.

Early on in our history parents must have noticed that their children did not look exactly like them. Each child shows a mixture of the mother's and father's characteristics. The idea grew up that the parents' characteristics are somehow blended in their children. Differences between a child and his or her parents were supposed to be due to the mixing effect. 'Stronger' characteristics in one of the parents were thought to account for differences between brothers and sisters.

If parents' characteristics really do blend in their offspring what colour flowers would you expect in the offspring of red and white flowers? What colour flowers would you expect in the offspring of the offspring? Would the red and the whites of the original parents ever be seen again? Figure 19.1B shows you what does happen when the original parents are pure-breeding red and white roses. *Does Figure 19.1B support or contradict the ideas of blending inheritance?*

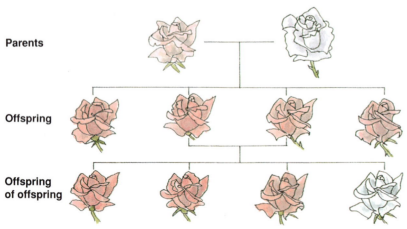

Parents

Offspring

Offspring of offspring

Figure 19.1B ● Inheritance of flower colour in pure-breeding roses

19.2 **Enter Gregor Mendel**

Like any other branch of science, genetics has its own vocabulary. Before you start reading about Gregor Mendel's experiments you need to know the meaning of a few words. Checking the genetics glossary will help you.

Genetics glossary

- **pure-breeding**   characteristics that 'breed true' appearing unchanged generation after generation
- **parental generation** (symbol **P**)   individuals that are pure-breeding for a characteristic
- **first filial generation** (symbol F_1)   the offspring produced by the parental generation
- **second filial generation** (symbol F_2)   the offspring of the first filial generation
- **dominant characteristic**   any characteristic that appears in the F_1 offspring of a cross between parents with contrasting characteristics such as tallness and shortness in pea plants
- **recessive characteristic**   any characteristic present in the parental generation that misses the F_1 generation but reappears in the F_2 generation

Mendel's experiments

The work of Gregor Mendel marks the beginning of modern genetics. At the monastery where Mendel was a monk, he kept a small garden plot where he experimented with breeding the garden pea (*Pisum sativum*). He observed the way in which its characteristics were inherited from one generation to the next.

Mendel's choice of the garden pea for his experiments was fortunate:
- Pea plants are easy to grow.
- Pea plants have different characteristics which breed true, appearing unchanged generation after generation (for example height of plant, colour of seed).
- The petals of the pea flower enclose the stamens and the carpels. Cross pollination, therefore, does not usually occur. If pea plants could naturally cross pollinate, Mendel's experiments would have been unsuccessful. *Can you think why?*

Who's behind the science

Gregor Mendel was born in 1822. He was the son of an Austrian farmer. As a young man he entered the Augustinian monastery in the town of Brünn (now Brno in Czechoslovakia) and was ordained as a priest at the age of 25. Mendel trained in mathematics and natural history at the university of Vienna and then taught in the high school at Brünn.

Figure 19.2A ● The pea plant and its flower showing how the petals enclose the stamens and carpels

LOOK AT LINKS
for **pollination**
See Topic 15.2.

Mendel's success also lay in the methodical way he planned his work. He only studied one characteristic at a time in thousands of pea plants and his mathematical training allowed him to analyse his results.

To prevent self-pollination in the plants in his experiments, Mendel prised open the flower buds before the pollen matured and removed the anthers. He then pollinated the flowers with pollen from other mature plants and tied small, muslin bags over the pollinated flowers to prevent any stray pollen settling on the flowers.

Mendel took pollen from **pure-breeding** short plants and dusted it onto the stigmas of **pure-breeding** tall plants and vice versa. These plants were the **parental generation**. He collected and grew the seeds they produced. (See Figure 19.2B.)

It did not matter if Mendel took pollen from tall or short parent plants, the **F_1 generation** were always tall. He called 'tallness' a **dominant** characteristic.

Mendel then let the F_1 generation self-pollinate. He collected and grew their seeds. The 'shortness' characteristic that skipped the F_1 generation re-appeared in the **F_2 generation**. Mendel called 'shortness' a **recessive** characteristic. (See Figure 19.2C.)

How characteristics are inherited

Introductory Genetics
(program)

This set of three programs is designed to clarify your ideas about genetics and inheritance. Test your understanding with the questions following each program.

From his results Mendel realised that parents passed 'something' on to their offspring which made them look like their parents. When these offspring became parents they passed on the 'something' to their offspring, and so on from generation to generation. Today, we call the 'something' which parents pass to offspring **genes**.

Mendel reasoned that sexually reproduced offspring receive the same number of genes from each parent and that any particular characteristic, therefore, must be controlled by a pair of genes. Paired genes controlling a particular characteristic are called **alleles**. They may be identical to one another or different. An individual with identical alleles controlling a particular characteristic is called a **homozygote** (*homo-* means the same); an individual with different alleles controlling a characteristic is called a **heterozygote** (*hetero-* means different).

Mendel concluded that alleles must separate when **gametes** form. We know that the separation of alleles occurs at **meiosis** and that only one allele goes to each gamete. Mendel, however did not know this. Nearly 30 years went by after his experiments with peas before meiosis was discovered.

LOOK AT LINKS
for **gametes** and **meiosis**
See Topic 15.1.

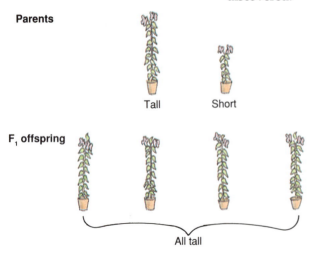

Parents

Tall   Short

F_1 offspring

All tall

Figure 19.2B ● The F_1 offspring of tall and short parents are all tall

F_1 parents

Tall   Tall

F_2 offspring

Tall   Short

Figure 19.2C ● F_1 parents produce a mixture of tall and short F_2 offspring

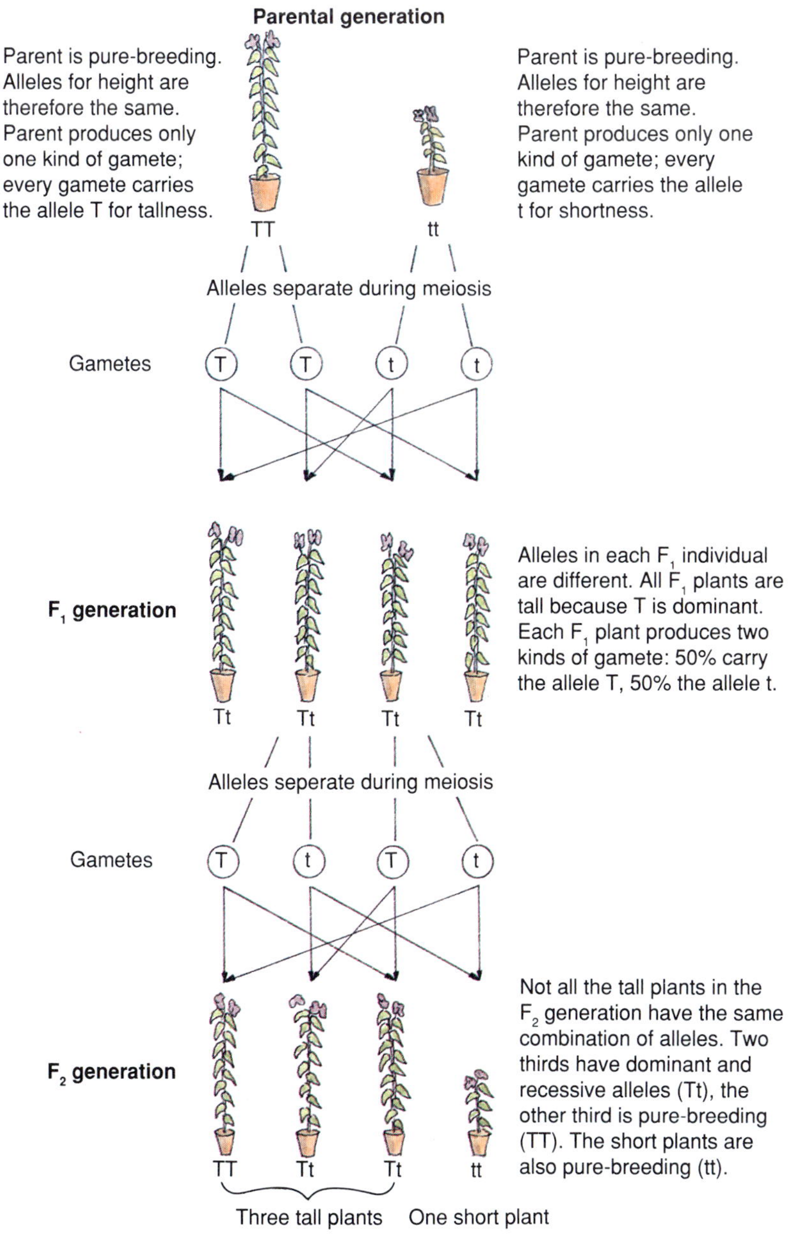

Parental generation

Parent is pure-breeding. Alleles for height are therefore the same. Parent produces only one kind of gamete; every gamete carries the allele T for tallness.

Parent is pure-breeding. Alleles for height are therefore the same. Parent produces only one kind of gamete; every gamete carries the allele t for shortness.

TT   tt

Alleles separate during meiosis

Gametes   (T) (T)   (t) (t)

F₁ generation

Tt   Tt   Tt   Tt

Alleles in each F₁ individual are different. All F₁ plants are tall because T is dominant. Each F₁ plant produces two kinds of gamete: 50% carry the allele T, 50% the allele t.

Alleles seperate during meiosis

Gametes   (T) (t)   (T) (t)

F₂ generation

TT   Tt   Tt   tt

Not all the tall plants in the F₂ generation have the same combination of alleles. Two thirds have dominant and recessive alleles (Tt), the other third is pure-breeding (TT). The short plants are also pure-breeding (tt).

Three tall plants   One short plant

Figure 19.2D ● How alleles controlling a characteristic pass from one generation to the next. *Which plants are homozygotes and which are heterozygotes? Why are only homozygotes pure-breeding?*

Mendel used letter symbols to simplify his observations. Capital letters symbolised alleles for dominant characteristics and small letters symbolised alleles for recessive characteristics. For example, he used **T** for the allele which produced tallness in pea plants and **t** for the allele which produced shortness, and set out the crosses between plants diagrammatically (see Figure 19.2D).

Mendel repeated his experiments using other pairs of characteristics. For example, he crossed purple flowered plants with white flowered plants. He always found that the flowers in the F₁ generation were purple. In other words, purple was a dominant characteristic; white was a recessive characteristic. Table 19.1 summarises the different pairs of dominant and recessive characteristics studied by Mendel in his experiments.

Table 19.1 ● Dominant and recessive characteristics in pea plants. (Note: axial flowers sprout along the stem; terminal flowers at the end.)

Characteristic	Dominant	Recessive
Seed shape	Round seed	Wrinkled seed
Seed colour	Yellow seed	Green seed
Seed coat colour	Coloured seed coat	White seed coat
Pod shape	Smooth pod	Wrinkled pod
Pod colour	Green pod	Yellow pod
Flower position	Axial flowers	Terminal flowers
Plant height	Tall stem	Short stem

1 The table shows the results of breeding experiments with pea plants beginning with parents pure-breeding for tallness and shortness.

Cross	Original parental cross	F_1 plants from parental cross	F_2 plant from F_1 cross
Height	Tall x short	All tall	779 tall: 268 short

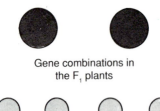

Gene combinations in the F_1 plants

Gametes

(a) Explain how you can tell from the results that tallness is dominant and shortness is recessive.
(b) To the nearest whole number, what is the ratio of tall to short plants in the F_2 generation?
(c) All the F_1 plants are tall. Explain how the combination of their alleles is different from that of the parent tall plant.
(d) Copy and complete the diagram opposite to explain the F_2 results obtained from crossing within the F_1 generation.

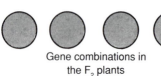

Gene combinations in the F_2 plants

2 What is the distinction between genes and alleles?

3 Match the terms in Column A with their definitions in Column B.

Column A	Column B
Gene	A gene which can express itself despite the presence of another version of the gene
Recessive gene	Portion of DNA molecule which controls a specific characteristic such as eye colour in humans and height in pea plants
Dominant gene	A plant which will always breed true for a particular characteristic
Pure-breeding plant	A gene not expressed in the presence of another version of the gene

19.3 Genes at work

In this section you will find out about some interesting examples of human genetics.

LOOK AT LINKS
for **blood groups**
See Topic 14.2.

Genetics of human blood groups

Sometimes a characteristic is controlled by more than two alleles. Human blood groups, for example, are controlled by three, A, B and O. An individual has two out of the three alleles.

The A and B alleles control production of the antigens which determine a person's blood group. The A and B alleles are dominant to the O allele but not to each other. So, if the A and B alleles are both present, then the person's blood group is AB. If neither the A nor B alleles are present, then the person's blood group is O. (See Table 19.2.)

Table 19.2 ● The genetics of blood groups (Notice that blood groups A and B each have two possible combinations of alleles)

Alleles	Antigen on red blood cells	Blood group
AA or AO	A	A
BB or BO	B	B
AB	A and B	AB
OO	None	O

RESOURCE ACTIVITY PACK

LOOK AT LINKS
for **homologous pairs of chromosomes**
See Topic 15.1.

Inheritance of sex

Figure 19.3A shows the chromosomes of a man and a woman. In each case a photograph of all the chromosomes from the nucleus of a body cell has been cut up and the chromosomes arranged into **homologous pairs** and in order of size.

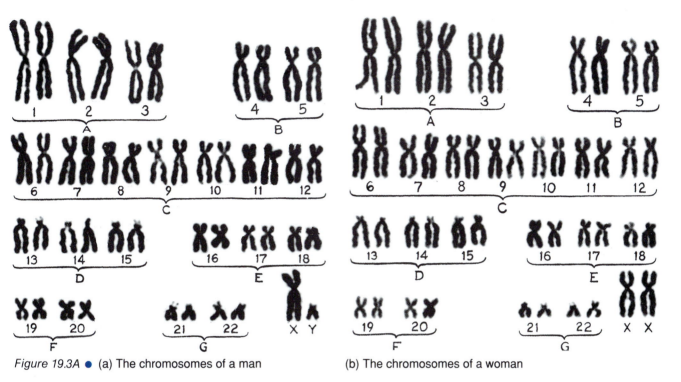

Figure 19.3A ● (a) The chromosomes of a man   (b) The chromosomes of a woman

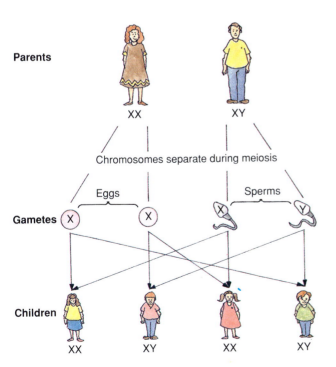

Figure 19.3B ● Inheritance of sex in humans

Of the 23 pairs of chromosomes in each photograph, the chromosomes in each of 22 pairs are similar in size and shape in both the man and the woman. Notice, however, that the chromosomes of the 23rd pair in the man are different from the 23rd pair in the woman. These are the sex chromosomes. The larger chromosomes are called the X chromosomes; the smaller chromosome is called the Y chromosome.

Since the body cells of a woman each carry two X chromosomes, meiosis can only produce eggs containing an X chromosome. Each body cell of a man, however, carries an X chromosome and a Y chromosome, so meiosis produces two types of sperm. Of the sperms produced, 50% carry an X chromosome and 50% carry a Y chromosome. A baby's sex depends on whether the egg is fertilised by a sperm carrying an X chromosome or one carrying a Y chromosome (see Figure 19.3B). The birth of almost equal numbers of girls and boys is governed by the production of equal numbers of X and Y sperms at meiosis.

Human genetic diseases

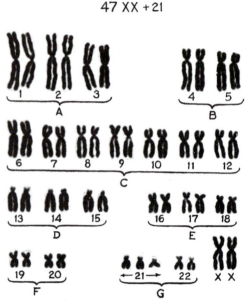

47 XX + 21

Figure 19.3C ● (a) A child with a variety of Down's syndrome working alongside pupils in a normal school class

(b) The chromosomes of a person with Down's syndrome

Sex Determination
(program)

What are the chances of an offspring being male or female? How does this affect the population overall? Find out, using the program.

Down's syndrome is one of the most familiar conditions resulting from a genetic abnormality. It is called a syndrome because it involves a number of disorders. People with Down's syndrome are usually short and stocky. They are more likely to suffer from infectious diseases, find speech difficult and have abnormalities of the heart and other organs. Most are also slightly mentally handicapped, which often improves with treatment. Some are above average in intelligence.

Figure 19.3C shows the chromosomes of a person with Down's syndrome. Carefully examine the 21st set of chromosomes and compare them with the 21st set in Figure 19.3A. *Can you see the extra one making 47 chromosomes in all?* A defect occurring during cell division produces the extra copy of chromosome 21. It is the presence of the extra chromosome that causes Down's syndrome.

Defects occurring during cell division can also affect the numbers of sex chromosomes (set 23). In humans a combination of an X and a Y chromosome (XY) produces male children. However, XXY, XXXY and XXXXY combinations also produce males. The combinations XXX and X0 (only one X chromosome present) as well as the normal combination XX produce female children. Men and woman with these unusual combinations of sex chromosomes are usually sterile and suffer from other abnormalities.

● *Sex linkage*

Haemophilia is a genetic condition in which the blood does not clot properly after an injury. In its most common form the person fails to produce the clotting agent factor VIII. The allele responsible for haemophilia is recessive and is located on the X chromosome. Therefore, haemophilia is a **sex-linked** disease.

Although women may carry the defective allele on one of the X chromosomes, they do not usually suffer from haemophilia. This is

LOOK AT LINKS
for **factor VIII** and **haemophilia**
See Topic 14.2.

because the normal allele on the other X chromosome is dominant. The dominant allele masks the effect of its recessive partner and ensures enough factor VIII is made for normal blood clotting to take place. A woman who carries the recessive allele on one of the X chromosomes is called a **carrier**. Although she does not suffer from the disease she is able to pass it on to her children. For a woman to have haemophilia she would have had to receive the recessive allele for the characteristic from both her mother and her father. Since the recessive allele is rare, this only happens very occasionally.

Men have one X chromosome and one Y chromosome. The Y chromosome does not carry as many genes as the X chromosome. If a man inherits the recessive allele for haemophilia on the X chromosome, there is no dominant allele on the Y chromosome to mask the effect of the recessive allele. No factor VIII is produced and the man suffers from haemophilia.

Haemophilia is a genetic disease with a royal connection. Queen Victoria was a carrier of the haemophilia allele. One of her sons and two of her daughters inherited the defective allele. Although haemophilia is usually rare, various intermarriages caused the defective allele to spread among the royal families of Europe.

Figure 19.3D is a **pedigree chart** of Queen Victoria's family. A pedigree chart shows genetic data about related individuals through a number of generations.

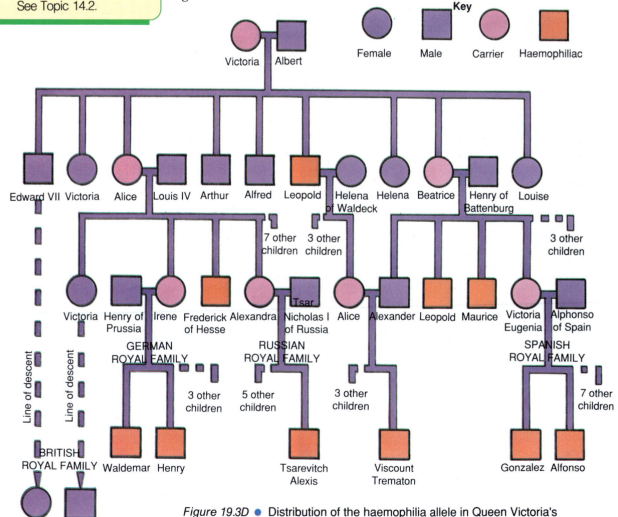

Figure 19.3D ● Distribution of the haemophilia allele in Queen Victoria's descendants. (Notice that there are no female haemophiliacs and that the British Royal Family escaped the disease because Edward VII did not inherit the defective allele.)

Red–green colour-blindness is another sex-linked disorder caused by a recessive allele on the X chromosome. As with haemophilia, women can be carriers but rarely suffer from the disorder. Red–green colour-blindness occurs in 8% of men but only 0.04% of women.

LOOK AT LINKS
for **base pairs** and **DNA**
See Topic 10.5.

LOOK AT LINKS
for **amniotic fluid**
See Topic 15.5.

LOOK AT LINKS
for **ultrasound scanners**
See *KS: Physics*, Topic 12.2.

Shall we start a family?

A couple thinking of starting a family can be especially anxious that any children they have are fit and healthy. If either partner has a family history of genetic disease, the couple may want to know what the chances are of handing on the problem to any children they have. A pregnant woman may want to know if the baby she is carrying has inherited a particular disease. An older woman may wish to evaluate the risk of having a baby with Down's syndrome (the likelihood of having a child with Down's syndrome increases with the mother's age). These are just some of the worries that people can have about starting a family.

Scientists and doctors have developed techniques which allow them to diagnose genetic disorders in prospective parents and in babies developing in the uterus.

- **Genetic probes**. For some types of genetic disorder, differences between the defective allele responsible for the disorder and its normal partner can be detected with genetic probes. The probes detect differences in the sequences of **base pairs** on the **DNA**. Genetic probes give very accurate results. (Figure 19.3E(a).)
- **Amniocentesis**. **Amniotic fluid** contains living cells from the foetus. These cells, once removed, can be grown in the laboratory and examined for genetic disorders. A thin needle is used to withdraw fluid from the amniotic cavity. Doctors work out where the foetus is in the uterus by using an **ultrasound scanner**. (Figure 19.3E(b).)
- **Statistical evidence**. The risk of having a child with a genetic disorder is well documented for some disorders such as Down's syndrome. (Figure 19.3E(c).)

Figure 19.3E ● (a) A genetics laboratory where the results from genetic probes are analysed

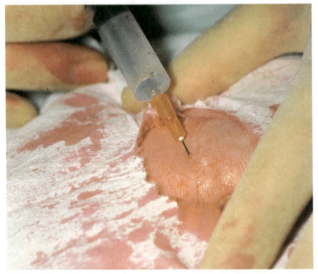

(b) Amniocentesis in progress – amniotic fluid is being withdrawn through a thin needle piercing the abdomen of a pregnant woman

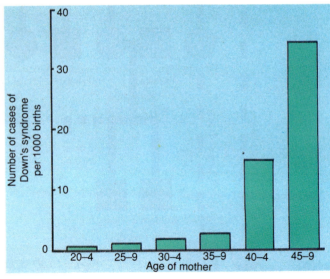

(c) Statistical evidence – the risk of having a baby with Down's syndrome

Any prospective or expectant parents who have reason to be worried about genetic disorders can attend **genetic counselling**. During genetic counselling, statistical evidence and results from amniocentesis and genetic probes are discussed by the couple and a specially trained genetic counsellor. The counselling helps couples to make informed choices about whether to start a family if there is a chance of having children with genetic disorders. It also helps expectant mothers and their partners to decide whether to continue with a pregnancy when they know their baby has a genetic disorder.

CHECKPOINT

❶ Defective alleles on the X chromosome are the cause of red–green colour-blindness. How do you account for the difference in the occurrence of red–green colour-blindness between men and woman?

❷ Explain why the sex of a baby is determined by the father and not the mother. How does this account for the birth of almost equal numbers of boys and girls?

❸ Women carry the defective allele for haemophilia but do not usually suffer from the disease. Explain why this is so.

19.4 Variation: genes and the environment

Look at your classmates, the members of your family and your friends. They have different coloured hair, different coloured eyes, different shaped faces. They all show variations in the different characteristics that make up physical appearance.

Figure 19.4A shows another human characteristic that varies: height. The average height of the adult population lies at the centre of the curve. Most people have heights close to the average. Only a few individuals are really short or really tall compared with the majority.

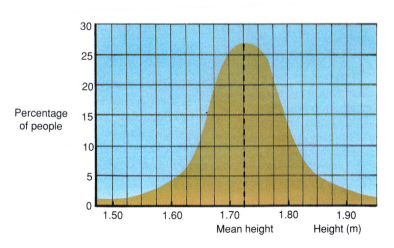

Figure 19.4A ● Variation in height of the adult human population

Figure 19.4B ● (a) Special training and a high protein diet have produced this weight-lifter's bulging muscles. Compare him with the other person who has a less strenuous life style.

(b) High winds and salt spray have made this tree lop-sided. The other tree, planted in a sheltered site, is unaffected by prevailing winds and grows vertically.

The outward appearance of an organism is called its **phenotype**; the genes contained in its cells form its **genotype**. Our appearances (our phenotype) vary because each of us inherits a different combination of alleles (our genotype). The same is true of all the members of all the different species of organism living now and in the past. Appearances, however, are not solely determined by the alleles that living things inherit. Figure 19.4B shows how the environment plays a part as well. The characteristics produced by the environment are said to be **acquired**. Acquired characteristics are not inherited.

CHECKPOINT

1 Inherited characteristics and acquired characteristics are sources of variation in species.
   (a) Distinguish between inherited and acquired characteristics.
   (b) Which type of characteristic is important for the evolution of new species? Try to give reasons for your answer.

FIRST THOUGHTS

19.5 The growth of modern genetics

Developments in cell biology and molecular biology have helped to explain the results of Mendel's breeding experiments. Discovering the structure of DNA and the nature of genes has opened up the possibility of manufacturing processes based on biotechnology. This section tells you about these exciting prospects.

Mendel reported the results of his experiments in 1865 at a meeting of the Brünn Natural History Society. Nobody really understood what he was talking about. His paper was published the following year but was ignored (Figure 19.5A). Mendel continued his breeding experiments with pea plants until 1871 when he was made Abbot of the monastery. His new duties gave him little time for further research. He died in 1884.

There matters rested until 1900 when Mendel's work was re-discovered by three biologists working independently of one another. The growth of modern genetics dates from their recognition of the importance of Mendel's work.

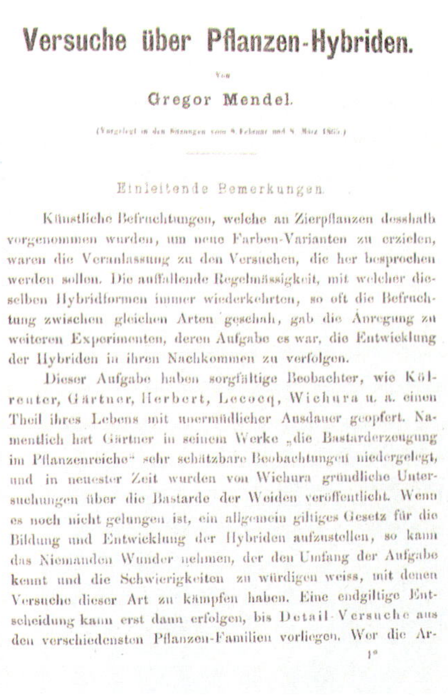

Versuche über Pflanzen-Hybriden.

Von

Gregor Mendel.

(Vorgelegt in den Sitzungen vom 8. Februar und 8. März 1865.)

Einleitende Bemerkungen.

Künstliche Befruchtungen, welche an Zierpflanzen desshalb vorgenommen wurden, um neue Farben-Varianten zu erzielen, waren die Veranlassung zu den Versuchen, die hier besprochen werden sollen. Die auffallende Regelmässigkeit, mit welcher dieselben Hybridformen immer wiederkehrten, so oft die Befruchtung zwischen gleichen Arten geschah, gab die Anregung zu weiteren Experimenten, deren Aufgabe es war, die Entwicklung der Hybriden in ihren Nachkommen zu verfolgen.

Dieser Aufgabe haben sorgfältige Beobachter, wie Kölreuter, Gärtner, Herbert, Lecocq, Wichura u. a. einen Theil ihres Lebens mit unermüdlicher Ausdauer geopfert. Namentlich hat Gärtner in seinem Werke „die Bastarderzeugung im Pflanzenreiche" sehr schätzbare Beobachtungen niedergelegt, und in neuester Zeit wurden von Wichura gründliche Untersuchungen über die Bastarde der Weiden veröffentlicht. Wenn es noch nicht gelungen ist, ein allgemein giltiges Gesetz für die Bildung und Entwicklung der Hybriden aufzustellen, so kann das Niemanden Wunder nehmen, der den Umfang der Aufgabe kennt und die Schwierigkeiten zu würdigen weiss, mit denen Versuche dieser Art zu kämpfen haben. Eine endgiltige Entscheidung kann erst dann erfolgen, bis Detail-Versuche aus den verschiedensten Pflanzen-Familien vorliegen. Wer die Ar-

Figure 19.5A ● The title page of Mendel's paper. Almost all of the notebooks containing the results of his experiments were destroyed soon after his death.

IT'S A FACT

The fruit fly *Drosophila* is a small fly which feeds on the sugar it finds in rotting fruit. The fly has been studied intensively and today more is known about its genetics than any other animal. It is ideal for genetic experiments because:

• It reproduces quickly (a generation every two weeks),
• It is easy to keep,
• Its cells each have only four pairs of chromosomes in the nucleus (the fewer chromosomes the better for genetic experiments),
• It has giant chromosomes in the salivary glands (giant chromosomes are easy to see under the microscope).

In the interval between the publication of Mendel's paper and its rediscovery, important advances had been made in cell biology. New ways of staining cells showed that the nucleus contained strands of material that took up dyes very strongly. They were called chromosomes, which literally means 'coloured bodies'. Mitosis and meiosis were seen for the first time.

Early work on genetics depended on breeding experiments to establish the effect of genes on the appearances of organisms. Questions about what genes are and how they produce their effects were not tackled until new techniques in molecular biology were developed. Figure 19.5B shows how developments in genetics, cell biology and molecular biology were brought together to give a full explanation of Mendel's observations. Today the techniques of cell biology and molecular biology have revolutionised genetics. New discoveries are stimulating the growth of biotechnology bringing benefits to agriculture, industry and medicine.

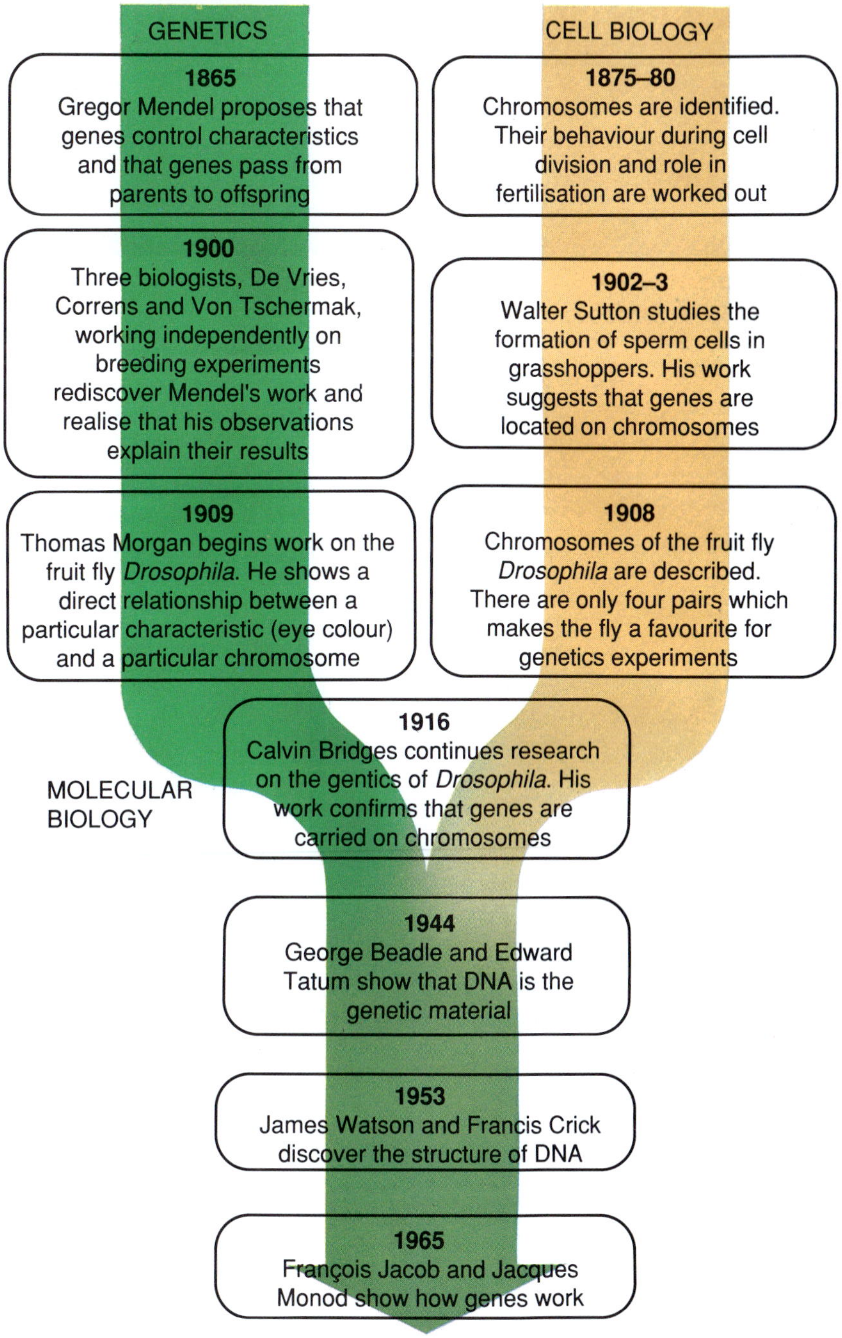

GENETICS

CELL BIOLOGY

1865
Gregor Mendel proposes that genes control characteristics and that genes pass from parents to offspring

1875–80
Chromosomes are identified. Their behaviour during cell division and role in fertilisation are worked out

1900
Three biologists, De Vries, Correns and Von Tschermak, working independently on breeding experiments rediscover Mendel's work and realise that his observations explain their results

1902–3
Walter Sutton studies the formation of sperm cells in grasshoppers. His work suggests that genes are located on chromosomes

1909
Thomas Morgan begins work on the fruit fly *Drosophila*. He shows a direct relationship between a particular characteristic (eye colour) and a particular chromosome

1908
Chromosomes of the fruit fly *Drosophila* are described. There are only four pairs which makes the fly a favourite for genetics experiments

MOLECULAR BIOLOGY

1916
Calvin Bridges continues research on the gentics of *Drosophila*. His work confirms that genes are carried on chromosomes

1944
George Beadle and Edward Tatum show that DNA is the genetic material

1953
James Watson and Francis Crick discover the structure of DNA

1965
François Jacob and Jacques Monod show how genes work

Figure 19.5B ● The growth of modern genetics

Improving the oil palm

We use vegetable oil in the home for cooking and industry uses huge quantities for making margarine, detergents and soaps. *Where does the oil come from?* Rapeseed is a major source but the fruits of the oil palm tree and sunflower seeds are even more important (see Figure 19.5C). Oil palm oil and sunflower oil provide 30% of the world's supply of vegetable oil.

(a) Oil palm

(b) Sunflower

(c) Rape

Figure 19.5C ● Major sources of vegetable oil

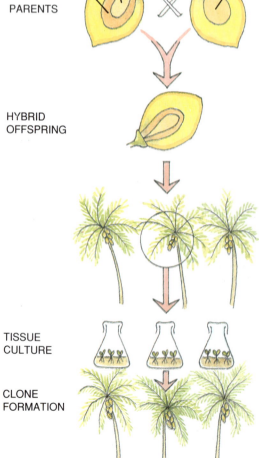

PARENTS

Shell Seed Seed

Trees producing fruit with a thick shell around the seed are crossed with trees producing fruit without a shell around the seed

HYBRID OFFSPRING

The thin-shelled fruit of some of the hybrid offspring produce a lot of oil

It is only possible to find out which offspring will have fruit that produces a lot of oil by growing plants from the hybrid seeds. This takes about three years

A tree is selected that has fruit which produce a lot of oil. Root tissue from the selected tree is placed in a medium which contains all the nutrients needed for growth. Auxin is used to stimulate growth

TISSUE CULTURE

CLONE FORMATION

Plants develop from the root tissue. They are genetically the same (clones) since they come from the same parent. All the clones have fruit which produce a lot of oil

Oil palm trees grow in the tropics. Each year trees produce about one tonne of oil per hectare. Some trees, however, produce much more. Breeding new stocks of trees from these high yielding varieties seems an obvious way of improving oil production. Unfortunately breeding oil palms is a slow, unreliable process. Trees vary greatly in oil content even when bred from good oil producing parents. Scientists have turned to **cloning** techniques to produce a consistent supply of high yielding trees (see Figure 19.5D).

IT **Genetics Maize** (program)

Use the program to investigate monohybrid inheritance. The program looks at the grain colour of corn cobs.

Figure 19.5D ● Breeding programme for the oil palm. Oil palms are produced which yield up to six times more oil than oil palms bred by traditional methods (Note: a hybrid is the offspring of a cross between two genetically unlike individuals)

TOPIC 20 BIOTECHNOLOGY

  Introducing biotechnology

 FIRST THOUGHTS

Biotechnology has a long history. This section explains the modern industrial process and the biology behind some traditional products of biotechnology.

 LOOK AT LINKS
for more about
microorganisms
See Topic 1.5.

 **RESOURCE ACTIVITY PACK**

The word **biotechnology** describes the way we use plant cells, animal cells and microorganisms to produce substances that are useful to us. Although the word biotechnology is new, the processes of biotechnology have a long history. The use of moulds to make cheese, and bacteria to make vinegar are early examples. For thousands of years we have exploited **yeast** to make wine, beer and bread. The ancient civilisations of Egypt and China knew that lactic acid bacteria preserved milk by turning it into yogurt. Figure 20.1A shows traditional products of biotechnology.

Figure 20.1A ● Biotechnology has a long history

 TRY THIS

Microorganisms make food. Imagine that you are a person ambitious to set up a business using traditional biotechnology to produce traditional foods. First you must find out how microorganisms work on:
- flour and fat to make bread
- milk to make cheese and yogurt
- grapes to make wine
- beer, wine and cider to make vinegar
1. Identify the species of fungi or bacteria responsible for producing each type of food listed above.
2. In each case map out a detailed flow chart showing the production process from starter material(s) to food product.
3. Why must hygiene be strictly controlled for each process? Summarise the precautions taken.

Production techniques

Does your family make wine at home? If so, you may keep the glass containers (called demi-johns) full of grape-juice, water and yeast next to a radiator or in the airing cupboard.

Figure 20.1B ● Demi-johns of wine – each demi-john has a volume of about 10 litres. The cork is protected by a trap which allows carbon dioxide gas produced during fermentation to escape but prevents unwanted air-borne microorganisms from contaminating the mixture. The demi-johns and traps are sterilised before pouring in the mixture of grape-juice, water and yeast

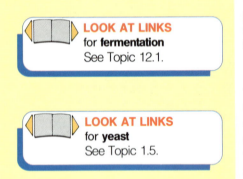

LOOK AT LINKS
for **fermentation**
See Topic 12.1.

LOOK AT LINKS
for **yeast**
See Topic 1.5.

Figure 20.1B shows the set up. You may not realise it, but making wine at home illustrates the principles behind biotechnology: demi-johns are **fermenters**, grape-juice is a **nutrient solution** and yeast is the **cell culture** that ferments the sugar in grape-juice into ethanol (alcohol) – the desired **product**. Warmth from the radiator provides the best temperature for the fermentation reactions to take place. Fermenter, nutrient, cell culture and product come together in an industrial process which is illustrated in Figure 20.1C.

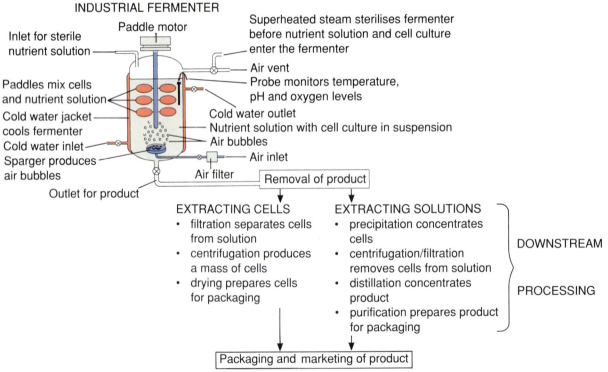

INDUSTRIAL FERMENTER

Inlet for sterile nutrient solution

Paddle motor

Superheated steam sterilises fermenter before nutrient solution and cell culture enter the fermenter

Air vent

Probe monitors temperature, pH and oxygen levels

Paddles mix cells and nutrient solution

Cold water jacket cools fermenter

Cold water inlet

Sparger produces air bubbles

Outlet for product

Cold water outlet

Nutrient solution with cell culture in suspension

Air bubbles

Air inlet

Air filter

Removal of product

EXTRACTING CELLS
- filtration separates cells from solution
- centrifugation produces a mass of cells
- drying prepares cells for packaging

EXTRACTING SOLUTIONS
- precipitation concentrates cells
- centrifugation/filtration removes cells from solution
- distillation concentrates product
- purification prepares product for packaging

DOWNSTREAM

PROCESSING

Packaging and marketing of product

Figure 20.1C ● Industrial biotechnology

Louis Pasteur was born in 1822 in the tiny French town of Dôle near the Swiss border. He became a famous scientist. To begin with, Pasteur worked on the chemistry of crystals at the University of Strasbourg. In 1854 a move to Lille established his work with microorganisms. He proved that bacteria made milk sour and turned ethanol into the ethanoic acid of vinegar. What Pasteur did not know (nor did anyone) was where the bacteria came from. Did 'bad' food make microorganisms or did microorganisms contaminate food and turn it 'bad'?

By now Pasteur had moved to Paris. There he investigated the question of where microorganisms came from. The diagrams show his experiments with swan-necked flasks. Study them carefully. *From the evidence, what do you think Pasteur's answer to the question was?*

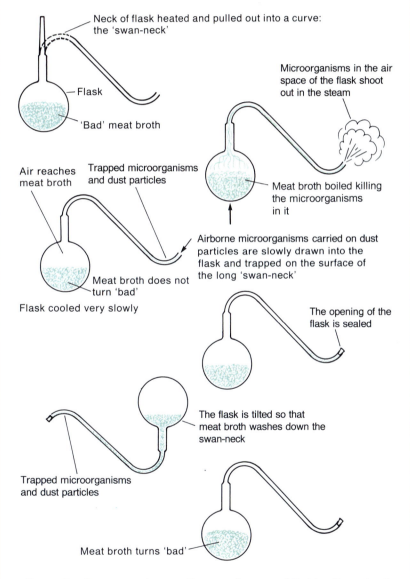

Neck of flask heated and pulled out into a curve: the 'swan-neck'

Flask

'Bad' meat broth

Microorganisms in the air space of the flask shoot out in the steam

Meat broth boiled killing the microorganisms in it

Air reaches meat broth

Trapped microorganisms and dust particles

Meat broth does not turn 'bad'

Flask cooled very slowly

Airborne microorganisms carried on dust particles are slowly drawn into the flask and trapped on the surface of the long 'swan-neck'

The opening of the flask is sealed

The flask is tilted so that meat broth washes down the swan-neck

Trapped microorganisms and dust particles

Meat broth turns 'bad'

Soon after these experiments, Pasteur discovered that heating wine to 55 °C killed the microorganisms that turned it into vinegar and yet did not affect the taste. Pasteur's process is called **pasteurisation**. It is used to make milk safe to drink (see p. 197). Pasteur went on to make many important discoveries about microorganisms that cause disease. For example he worked on the bacteria that cause anthrax and rabies. Pasteur died in 1895 after a life-time devoted to science.

Scaling up

There is a big jump from home wine-making to industrial biotechnology. Fermentation in a 10-litre demi-john usually looks after itself, with perhaps an occasional shake to help air to circulate. Scaling up to an industrial fermenter perhaps 25 000 times the size creates considerable problems:

* **Sterile** conditions in the fermenter and in the pipelines leading to and from the fermenter must be maintained. **Superheated** steam at around 120 °C is pumped through the system to kill unwanted micro-organisms.
* **Oxygen** levels at the centre of the fermenter may drop, slowing the growth of the cell culture. Sterile air is fed into the bottom of the fermenter through a perforated metal ring or disc called a **sparger**. Air bubbles rise through the liquid. Motorised paddles stir the mixture.
* **Heat** generated by the fermentation reactions and motorised paddles can quickly raise the temperature inside the fermenter. Temperatures of more than 60 °C would kill the cell culture and damage the product. A cooling system, therefore, is vital.

Notice the label 'downstream processing' in Figure 20.1C. When fermentation is complete, downstreaming is the process that collects and purifies the product, and packages it. The product may be cells, substances that cells produce or substances that cells ferment from nutrient solutions. Some of the products and uses of biotechnololgy are investigated in the following sections.

SUMMARY

Ancient civilisations used the process of biotechnology to produce food. Modern biotechnology produces many useful substances and operates on an industrial scale.

CHECKPOINT

❶ Describe the mode of action of **three** types of microorganism used to produce food (see *Resource Bank*).

❷ A demi-john is a glass container used for making home-made wine. 'In principle an industrial fermenter is a large version of a demi-john'. Do you think this statement is correct? Give reasons for your answer.

❸ What is 'downstream processing'?

20.2 Using biotechnology

FIRST THOUGHTS

Biotechnology is creating new industries which earn millions of pounds annually. This section tells you about some of the developments.

Moths, Mice and Other Beasties (programs)

This is a set of five programs on genetics. You can use this, and the worksheets that go with it, to go over the work you have done on genetics (see Topic 19).

Before the First World War, glycerol and propanone (which are used in the manufacture of explosives) were made by using bacteria. After the war, the petrochemical industry expanded and the manufacture of glycerol and propanone from oil replaced these early examples of biotechnology. However, rising oil costs have revived interest in the production of these chemicals by biotechnology.

In 1928, Alexander Fleming reported that the mould *Penicillin notatum* made a substance that killed bacteria. The substance was called penicillin: the first of a family of antibiotic drugs made by biotechnology.

Modern biotechnology is branching out in new and exciting ways. Central to success is our understanding that the products of biotechnology are the result of the action of genes. We now know what genes are, how they work and how to manipulate them to our advantage.

DNA technology

A person's DNA is as unique as their fingerprints. DNA 'fingerprinting' can help to identify criminals in cases where the criminal's body cells are found at the scene of the crime. Except in the case of identical twins, the chances of two people having the same DNA 'fingerprint' are millions to one against.

Figure 20.2A shows how DNA 'fingerprints' can help to identify criminals. The X-ray film shows the DNA maps made from a semen stain found at the scene of a rape, the body cells of a suspect and the body cells of the victim. *Do you think the police have caught the right man?*

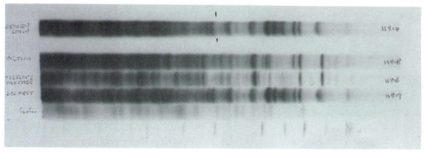

Figure 20.2A ● DNA fingerprinting

Biosensors

An **antibody** will attach itself only to a particular antigen; an enzyme will catalyse only a particular reaction or group of reactions. Biosensors use the sensitivity of these reactions and microelectronic circuits to detect minute amounts of chemicals. In the future, biosensors will help scientists to diagnose disease and to monitor pollution in the environment.

Figure 20.2B shows a biosensor that is able to detect glucose levels in the blood. If this is coupled to a portable insulin pump, the glucose level in the blood of a **diabetic** can be continuously monitored, and the correct level of insulin maintained without the need for daily injection.

LOOK AT LINKS
for **diabetes** and **insulin deficiency**
See Topic 16.5.

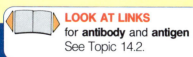

LOOK AT LINKS
for **antibody** and **antigen**
See Topic 14.2.

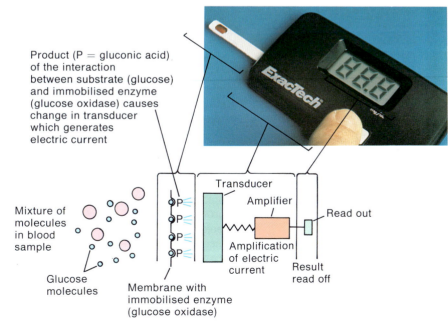

Product (P = gluconic acid) of the interaction between substrate (glucose) and immobilised enzyme (glucose oxidase) causes change in transducer which generates electric current

Mixture of molecules in blood sample

Glucose molecules

Membrane with immobilised enzyme (glucose oxidase)

Transducer

Amplifier

Amplification of electric current

Read out

Result read off

Figure 20.2B ● How a biosensor works

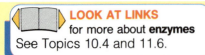

LOOK AT LINKS
for more about **enzymes**
See Topics 10.4 and 11.6.

Enzymes

Whoever wore the clothes about to go into the washing machine shown in Figure 20.2C is a careless eater! Notice the egg stains, grease marks and sticky jam. Also notice that the washing powder to be used is biological. It contains enzymes! The enzymes are sealed in capsules to protect them from the action of the detergent powder. Encapsulation also protects the people making the washing powder from possible harmful side-effects caused by the enzymes. *What do you think such side-effects might be? Give reasons for your answer. What are enzymes doing in washing powder? Which types of enzyme are best for washing food-stained clothing? Table 11.8 on p. 203 will help you answer these questions.*

Enzymes are **protein catalysts** which speed up chemical reactions taking place inside cells. Biotechnology depends on them! Most industrial enzymes come from microorganisms. Fungi and bacteria are grown in a nutrient solution inside large fermenters. Enzymes are secreted into the nutrient solution, then filtered off, concentrated and packaged for sale either as liquids or as powders.

Enzymes are very useful industrial catalysts because:
* only a particular reaction is catalysed, making it easier to collect and purify the product
* activity is high at moderate temperature and pH
* only small amounts are required
* the enzyme is not used up in the reaction – at the end of the reaction it is as good as new and can be used again.

Table 20.1 summarises the sources and uses of some important industrial enzymes.

Figure 20.2C ● Wash time with biological washing powder

Table 20.1 ● Some important industrial enzymes (B) = Bacterium, (F) = Fungus. Notice that carbonic anhydrase is derived from red blood cells

Use	Enzyme	Source
Industrial		
Food production, leather making, brewing and washing powder manufacture	*Amylase:*	
	● converts starch into glucose producing syrup for the food industry	*Bacillus subtilis* (B)
	● used in biological washing powders	*Aspergillus oryzae* (F)
	Glucose isomerase:	
	● converts glucose into fructose, which is used as a sweetener in foods	*Bacillus coagulans* (B)
	Proteases:	
	● 'clears' beer of yeast haze	*Streptomyces spp* (B)
	● coagulates milk for cheesemaking	
	● removes hairs from skins used to make leather	*Bacillus subtilis* (B)
Medical		
Diagnosis and treatment	*Glucose oxidase:*	
	● indicates glucose levels in blood	*Aspergillus niger* (F)
	L-asparaginase:	
	● breaks down the amino acid l-asparagine required for the growth of cancers	*Escherichia coli* (B)
	Streptokinase:	
	● dissolves blood clots and cleans wounds	*Streptomyces* spp (B)
Analysis		
Environmental pollution and crime detection	*Carbonic anhydrase:*	
	● detects insecticides	Red blood cells
	Urease:	
	● indicates amount of urea in urine and blood	*Bacillus pasteurii* (B)

LOOK AT LINKS
for more about **catalysts**
See *KS: Chemistry*, Topic 23.7.

The enzyme added to the substances whose reaction it catalyses may be lost when the product is collected. Dilution also occurs, making it difficult to recover the enzyme when the extraction is complete. Recent developments in the manufacture of enzymes have overcome these problems. Various insoluble materials are used to bond to the enzyme, so that the enzyme can still catalyse reactions but remains attached to the insoluble support and is not lost when the products are collected. The enzyme is said to be **immobilised**.

Immobilised enzymes are:

- easily recovered and can be used again and again
- often active at temperatures that would destroy the unprotected enzyme. For example, immobilised *glucose isomerase* (see Table 20.1) is stable at 65 °C whereas unprotected *glucose isomerase* is soon destroyed at 45 °C
- not diluted and therefore do not contaminate the product.

Immobilised enzymes are the biological heart of different types of biosensor. For example immobilised *glucose oxidase* is a vital part of the biosensor shown in Figure 20.2B. The enzyme catalyses the reaction:

$$\text{Glucose} + \text{O}_2 + \text{H}_2\text{O} \xrightarrow{\text{\textit{Glucose oxidase}}} \text{Gluconic acid} + \text{H}_2\text{O}_2$$

Glucose + Oxygen + Water → Gluconic acid + Hydrogen peroxide

The acid produced conducts an electric current which is proportional to the amount of glucose in solution – an important measurement for people suffering from diabetes.

Biofuel

The Sun floods the Earth with light energy. Some of the light energy is absorbed by chlorophyll (the green pigment in plant cells) and used in the chemical reactions of photosynthesis. The sugars formed are a source of energy which plants use to live and grow. New plant material, therefore, represents a store of energy which biotechnology converts into fuels such as ethanol.

Brazil's cars run on ethanol produced by biotechnology from sunshine and sugar cane! During the 1970s oil prices rose steeply. Brazil's lack of oil resources led to the **'gasohol programme'** with the aim of making the country self-sufficient in energy by the year 2000. Early problems with corrosion of carburettors and fuel pumps have been overcome, and engines in Brazilian cars can now run on ethanol or an ethanol–petrol mixture.

Brazil is an ideal environment for growing lots of sugar cane. Other sources of carbohydrate such as cassava roots, straw and waste from the timber industry are also used in the manufacture of gasohol.

Most yeasts that are used to ferment sugar die when the concentration of ethanol is more than 15%. Gasohol fuel is 96% pure ethanol. It is uneconomic to burn fossil fuel to drive off (distil) the excess water (the final cost of the product would be much more than that of petrol). Instead the waste (called **bagasse**) left over after cane has been crushed to extract its juice is used.

IT'S A FACT

Each day energy from the Sun equivalent to one-fifth of the known reserves of fossil fuels (coal, gas and oil) reaches the Earth's surface.

LOOK AT LINKS
for more about **gasohol**
See *KS: Chemistry*, Topic 31.1.

SUMMARY

Biotechnology is creating products for industrial, medical and analytical purposes. The new processes depend on the activities of microorganisms.

CHECKPOINT

❶ How does a biosensor work? Explain why biosensors help scientists diagnose disease and monitor pollution in the environment.

❷ Why are enzymes useful industrial catalysts?

❸ What is an immobilised enzyme?

❹ 'Plants represent a store of energy'. Explain this statement and describe how biotechnology converts the stored energy in plants into fuel.

20.3 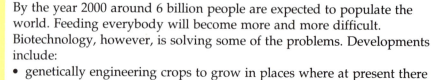 Biotechnology on the farm

FIRST THOUGHTS

Strawberries growing in winter, kid goats born to sheep, potato and tomato plants joined to form a 'pomato' – what other changes are taking place on the farm because of biotechnology? Read this section and find out.

LOOK AT LINKS
for more about the **human population**
See Topic 3.4;
for **plants fixing nitrogen**
See Topic 2.6.

By the year 2000 around 6 billion people are expected to populate the world. Feeding everybody will become more and more difficult. Biotechnology, however, is solving some of the problems. Developments include:

- genetically engineering crops to grow in places where at present there is little chance of success
- altering nitrogen-fixing bacteria so that they can live in the roots of cereal crops
- designing insecticides, produced by bacteria, which are selective for particular insect pests
- producing plants resistant to disease
- developing livestock to produce more and better quality meat and milk.

Can we look forward to a future where crops tolerate cold, flourish in drought conditions and resist insects and disease? Think carefully about your vision of what is possible. Make a list of your ideas.

Genetic engineering and *Agrobacterium tumefaciens*

LOOK AT LINKS
for more about **DNA**
See Topic 10.5.

In the early 1970s a type of enzyme (called a **restriction enzyme**) which cuts DNA into pieces was discovered in several species of bacteria. It made it possible to isolate desirable genes from one type of cell and transfer them to another type of cell (the host cell). Another type of enzyme (called **ligase**) enables the desirable genes to combine with the genetic material of the host cell. These discoveries marked the beginning of **genetic engineering**.

LOOK AT LINKS
for more about **clones**
See Topic 15.1.

Scientists use genetic engineering to isolate desirable genes and insert them into crop plants. The bacterium *Agrobacterium tumefaciens* is an ideal cell for introducing desirable genes into host cells. Figure 20.3A shows the technique.

Notice that cells of *Agrobacterium* each have a loop of DNA called the *Ti* **plasmid** Also notice that a desirable gene has been inserted into the *Ti* plasmid. The plant is infected with engineered *Agrobacterium* and, because of the *Ti* plasmid, produces a cancerous growth (**tumour**) called a **crown gall**. Cells in the gall each contain the *Ti* plasmid with the desirable gene in place. Plantlets can be cultured from small pieces of tissue cut out of the gall. The plantlets are genetically identical, forming a **clone**, each one carrying the *Ti* plasmid with its desirable gene. The plantlets are transferred to soil where they grow into a mature crop, helped by the characteristic which the gene inserted into the *Ti* plasmid controls.

LOOK AT LINKS
for more about **cancer**
See Topic 9.4.

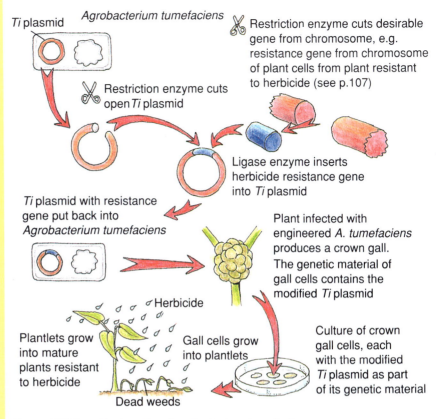

Figure 20.3A ● Engineering *Agrobacterium tumefaciens*

Nitrogen fixation

Nitrogen (N_2) is an essential element of proteins, DNA and RNA. Crop growth depends on the availability of nitrogen. Different species of the bacterium *Rhizobium* live in swellings (called **nodules**) on the roots of **leguminous** plants (beans, peas, soybeans, clover). Each nodule contains millions of bacteria which convert nitrogen into nitrates. This process is called **nitrogen fixation**. The host plant uses the nitrates (as a source of nitrogen) and provides the bacteria with sugar.

Cereal crops account for 50% of food supplies world-wide. Unfortunately they do not have nitrogen-fixing bacteria. As a result, huge amounts of nitrogen fertiliser (one hundred million tonnes per year) are applied to crop plants to boost growth and productivity. Surplus fertiliser runs off the land and pollutes drinking water, with serious consequences for the environment.

Biotechnology offers alternative strategies. For example, increasing the range of plants which associate with *Rhizobium* is one idea under investigation. Success depends on understanding the mechanism by which *Rhizobium* recognises legume root tissue and using this knowledge to achieve the same result with non-leguminous plants like cereals.

Other ideas involve manipulating the set of about 12 genes (called the *nif* **genes**) which control nitrogen fixation in *Rhizobium*. The options include inserting the *nif* genes into:

- suitable cereal plants
- bacteria which already live in association with cereal plants
- the *Ti* plasmid of *Agrobacterium tumefaciens*, and then infecting suitable plants with the engineered bacterium, causing crown gall formation.

Looking at Figure 20.3A once more will remind you of the idea.

How do you think food production and the environment would benefit from this type of technology? Make a list of your ideas.

LOOK AT LINKS
for more about **nitrogen**
See *KS: Chemistry*, Topics 14.4 and 14.5;
for more about **proteins**
See Topic 10.4.

LOOK AT LINKS
The relationship between different species where both partners benefit is called **mutualism**
See Topic 3.7.

IT'S A FACT

Cereal crops are a vital source of carbohydrate and protein. We depend on eight species world-wide: wheat, barley, rice, oats, rye, maize, sorghum and millet.

LOOK AT LINKS
for more about **pollution**
See Theme B;
for more about **fertilisers**
See Topic 6.6;
for more about **resistance**
See Topic 21.4.

LOOK AT LINKS
for more about **insecticides**
See Topic 6.8.

IT'S A FACT

The Red Queen Effect
The Red Queen from *Through the Looking Glass* says to Alice, 'You have to run faster than that to stay in the same place'. Scientists developing insecticides to combat pests and drugs to fight disease would probably agree with the Red Queen. *Briefly explain why scientists might think this.*

Bacterial insecticides

Insects quickly develop resistance to conventional insecticides. More and more insecticide has to be used until its inefficiency and expense forces the farmer to switch to another compound. Resistance to the new insecticide soon appears and the cycle starts again. Scientists are in a race between resistance developing and finding new chemicals with which to control insect pests. Even so, insect damage to crops costs around 1.5 billion pounds per year worldwide.

Bacterial insecticides are safer to use and environmentally more acceptable than conventional insecticides. However, only about 20 such products are registered for use. Conventional insecticides, therefore, are likely to remain an important weapon against insect pests for some time.

The bacterium *Bacillus thuringiensis* kills leaf-eating caterpillars and the larvae of flies and mosquitoes. For example, it kills the codling moth, a pest of applies and pears, and the cabbage looper moth, a pest of lettuces, broccoli, cabbages and potatoes. The damage is caused by a poison called **insecticidal crystal protein (ICP)**, which is produced by the bacterium. ICP attacks the gut lining, the caterpillar stops feeding and eventually dies.

Batches of *Bacterium thuringiensis* are fermented in large quantities. The product is drawn off, mixed with a sticky substance and sprayed on to crops as a mixture of bacterium and poisonous ICP.

Unfortunately *Bacterium thuringiensis* is a poor survivor when exposed to the weather. As an alternative to spraying, therefore, plants are engineered to carry the gene for the ICP poison. The gene is inserted into the *Ti* plasmid (see p. 365). *Draw a flow diagram to show the method. Look once more at Figure 20.3A to help you with your task.*

Cold tolerance

Late frosts in spring-time often damage crops. The bacterium *Pseudomonas syringae* is nearly always found on the surfaces of leaves. It feeds on frost-affected plants and damages them. It contains a gene which produces a protein around which ice crystals form at temperatures between 0 °C and −7 °C.

If the bacterium is not present the plants are unaffected by ice crystals unless the temperature drops below −7 °C. Scientists have produced a strain of *P. syringae* without its 'ice-protein' gene. This strain competes and replaces wild-type 'ice' bacteria when it is sprayed over crops. Frost-sensitive crops like strawberries are less vulnerable to damage after the treatment and can be grown at a time earlier in the year when frosts would kill untreated plants.

Leaf tissue stripped from leaf surface

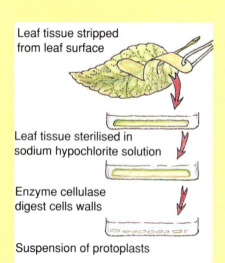

Leaf tissue sterilised in sodium hypochlorite solution

Enzyme cellulase digest cells walls

Suspension of protoplasts

Figure 20.3B ● Preparing protoplasts

Protoplasts

Protoplasts are plant cells with their walls removed. Figure 20.3B shows the idea.

Protoplasts will take up substances added to the medium in which they are growing. For example, scientists have incorporated genes into protoplast DNA by adding genes to the growth medium. The protoplasts grow into complete cells which then divide. Treatment with **auxins** stimulates shoot and root development and whole plants can be raised carrying the genes engineered into the original protoplasts. *Think of some of the benefits of this technique. Make a list of your ideas.*

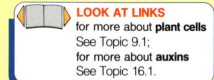
LOOK AT LINKS
for more about **plant cells**
See Topic 9.1;
for more about **auxins**
See Topic 16.1.

LOOK AT LINKS
for **hybrids**
See Topic 1.3.

Protoplasts can be fused together (Figure 20.3C) and grown to produce hybrid plants with new mixtures of characteristics. The **pomato** mentioned in *FIRST THOUGHTS* was produced in the late 1970s as an early demonstration of the technique. The possible benefits of protoplast fusion include overcoming diseases, such as potato leaf roll disease, shown in Figure 20.3D.

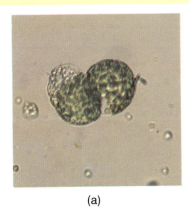

(a)

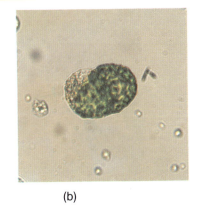

(b)

Figure 20.3C ● Fusing protoplasts: (a) Early stage (b) Late stage

Figure 20.3D ● Potato leaf roll disease

LOOK AT LINKS
for more about **Factor 8**
and **haemophilia**
See Topic 14.2.

LOOK AT LINKS
for **AIDS**
See Topic 14.2.

Improving animals

The sheep shown in Figure 20.3E are transgenic. Genetic engineering has inserted the human gene for **Factor 8** (an essential blood-clotting protein) into their DNA. The sheep will make Factor 8 which can be drawn off in the milk, purified and used to treat **haemophilia**. *Why are new sources of blood products important? Think about blood-borne diseases such as **AIDS** before making a list of your answers.*

New techniques are also helping to conserve rare animal species. Figure 20.3F shows a zebra foal born to a mare. Fertilised zebra eggs were placed inside the mother animal who is called a **surrogate**. Such techniques are useful because:

- a mother of a common species can give birth to a youngster of a rare species
- fertilised eggs of rare species can be cooled and preserved for years after the original parents have died. When a suitable surrogate is available the fertile eggs can be transplanted.

Figure 20.3E ● Transgenic sheep

Figure 20.3F ● A surrogate mother with her foal

LOOK FOR LINKS
for more about
conservation
See Topic 5.4.

These techniques are an increasingly important source of **rare** species. Zoos are developing the ideas to help **conservation**. Their breeding programmes build up numbers of rare stock and arrange for their release into the wild. Figure 20.3G gives you the idea.

(b) Arabian oryx in half-way station, before release into the wild

Figure 20.3G ● Arabian oryx conservation programme

(a) Arabian oryx in zoo

(c) Arabian oryx after release into the wild in Oman

Biotechnology is also producing healthier farm animals which produce more meat, milk and fibre for clothes manufacture. For example, the gene for **bovine somatotropin (BST)** has been genetically engineered into bacteria. As a result large amounts of this hormone are available for injection into cows.

BST stimulates growth and increases milk production. The meat from treated cows carries less fat, but there are worries about the long-term effects on human health. Analysis shows traces of BST in the meat and milk. Although we break down BST in the gut and the hormone seems to be active only in cattle, public concern remains. *Where do you think the balance lies? Even though there is no evidence of ill-effects, do you think people's worries are sufficient reason for abandoning developments? Do you think it cruel to treat animals in this way? Carefully consider the arguments before listing your answers.*

CHECKPOINT

❶ How is biotechnology helping to solve some of the problems of feeding the world's growing human population?

❷ Describe the discoveries made in the early 1970s which marked the beginning of genetic engineering.

❸ Why is the bacterium *Agrobacterium tumefaciens* an ideal cell for introducing desirable genes into host cells?

❹ How does the bacterium *Bacillus thuringiensis* kill insect pests?

❺ What is the advantage of developing crops that are able to tolerate cold weather?

❻ Describe how biotechnology is helping zoos to conserve rare animal species.

FIRST THOUGHTS

Eating microorganisms – bacteria, algae and fungi – may seem odd, but it could help to feed the world's growing population. Biotechnology is the key to new types of food based on micro-organisms.

Food! We all need it, but how can we produce enough food to feed the world's growing population? Farmers try to produce as much food as possible by farming land intensively. Eating microorganisms produced by biotechnology is another option.

Think about these facts:

- microorganisms double their mass within hours. Plants and animals may take weeks.
- microbial mass is at least 40% protein and has a high vitamin and mineral content.

Eating microorganisms may not seem so odd after all! In the light of the facts, do you agree with this statement? Give reasons for your answer.

High-protein food produced from microorganisms is called **single-cell protein (SCP)**. The idea is not new. During the First World War, food shortages in Germany were overcome with the help of scientists who discovered how to grow large amounts of yeast to fill out sausages and soups. 'Marmite' and similar spreads made from yeast are firm favourites at tea-time. Yeast left over from brewing beer is turned into animal feed for livestock on the farm.

LOOK AT LINKS

for **bacteria**, **algae** and **fungi**
See Topic 1.5;
for **intensive farming**
See Topic 6.1;
for the meaning of a 'balanced diet'
See Topic 11.3.

SCP production

In the 1960s, oil was used as a nutrient for growing SCP micro-organisms. However, massive increases in the oil prices in the 1970s made SCP production based on oil uneconomical. To-day glucose syrup, fruit pulp, waste from paper making, agricultural waste and sewage are just some examples of the wide range of nutrients used to grow SCP microorganisms.

Large fermenters capable of producing thousands of tonnes of SCP per year under sterile conditions are run continuously for months at a time. Nutrients are replaced as they are used up, and the temperature and pH carefully controlled. Microorganisms are harvested at regular intervals and processed.

IT'S A FACT

In the early 1500s, the Aztec people of Mexico made cakes from the bacterium *Spirulina maxima*, which grows in shallow lakes. Renewed interest may mean that *Spirulina* becomes an important source of SCP.

Quorn – a case study

Figure 20.4A shows the mould *Fusarium graminearum*. With 45% protein content and 13% fat content, *Fusarium* is as nutritious as meat with the added advantage that, unlike meat, it is high in fibre and cholesterol-free. Hyphae of the mould *Fusarium* form a highly nutritious food called **mycoprotein**.

LOOK AT LINKS

for the importance of fibre in the diet
See Topic 11.3;
for **cholesterol** and **heart disease**
See Topic 14.5;
for **world hunger** and **the food crisis**
See Topic 6.1.

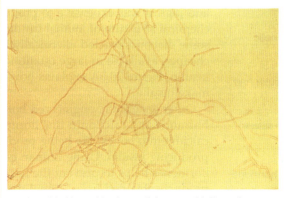

Figure 20.4A ● Hyphae of the mould *Fusarium graminearum*

Mycoprotein is marketed as 'Quorn'. The product looks like sheets of moist uncooked pastry. Careful control of hyphal length during fermentation and the addition of colours and flavours results in a range of food including biscuits, soups and drinks, as well as substitutes for ham and chicken (Figure 20.4B). Quorn also blends well with meat products such as sausages.

Figure 20.4B ● Quorn, pineapple and coconut curry. The chunks which look like meat are Quorn

SUMMARY

SCP is a nutritious food obtained from microorganisms. Despite early failures, products based on SCP are now available in shops. Wider acceptance of SCP as a food depends on careful planning and marketing of the products.

Is SCP a sign of the future? Food can be grown in large quantities in fermenters which take up little space and in a controlled environment independent of the weather. The technology is established and can be adapted for developing countries that experience food shortages. *Do you think people might be reluctant to eat food made from microorganisms? Devise ways of making SCP acceptable. Discuss the contribution that SCP production could make to solving the world's food crisis.*

20.5 Biotechnology and medicine

FIRST THOUGHTS

New drugs, new vaccines, prevention of diseases and better diagnosis are hopes for the future all because of biotechnology. This section highlights some important advances.

Biotechnology has enormous potential in the field of medicine. Cancer and heart disease, for example, are responsible for more than 50% of deaths in the developed countries of the world. Biotechnology will play a key role in the prevention, diagnosis and treatment of these and other diseases.

The human hormone **insulin** was the first substance made by genetic engineering to be given to humans. In 1980, volunteer diabetics successfully tried out genetically engineered insulin and by 1982 it was in general use. Before then, insulin was obtained from slaughtered cattle and pigs. It was expensive to produce and in limited supply. The chemical structure of animal insulin is different from human insulin; some diabetics reacted allergically to it. Genetically engineered insulin is cheaper, available in large quantities and chemically the same as human insulin. Figure 20.5A explains how scientists use genetic engineering to make insulin.

Other genetically engineered hormones including **human growth hormone** and **calcitonin** (the hormone that controls the absorption of calcium into bones) are being produced by methods similar to those for producing insulin. Using human genes to produce substances like hormones helps to prevent the harmful side-effects that can come from products obtained from animal tissues. It also reduces the use of animals for medical research.

LOOK AT LINKS
for more about **genetic engineering**
See Topic 20.3;
for more about **insulin** and **diabetes**
See Topic 16.5.

LOOK AT LINKS
for **human growth hormone**
See Topic 16.5.

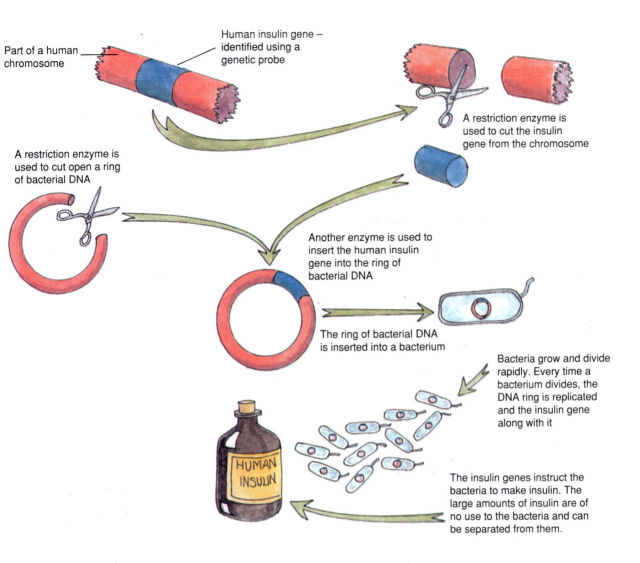

Part of a human chromosome

Human insulin gene – identified using a genetic probe

A restriction enzyme is used to cut the insulin gene from the chromosome

A restriction enzyme is used to cut open a ring of bacterial DNA

Another enzyme is used to insert the human insulin gene into the ring of bacterial DNA

The ring of bacterial DNA is inserted into a bacterium

Bacteria grow and divide rapidly. Every time a bacterium divides, the DNA ring is replicated and the insulin gene along with it

HUMAN INSULIN

The insulin genes instruct the bacteria to make insulin. The large amounts of insulin are of no use to the bacteria and can be separated from them.

Figure 20.5A ● Making genetically engineered insulin

Monoclonal antibodies

White blood cells produce millions of antibodies to defend the body from attack by bacteria, viruses, fungi and other potentially dangerous antigens. It is difficult to separate different antibodies into pure samples of the antibodies required to fight specific antigens. The problems can be overcome by fusing white blood cells that produce a particular antibody with a type of rapidly dividing cancer cell. The fused cells only produce the antibody required. Pure samples of antibodies made in this way are called **monoclonal antibodies**.

Monoclonal antibodies have a wide range of uses. Scientists hope to be able to use monoclonal antibodies to treat cancer. Some types of cancer cell make proteins (antigens) that are different from the proteins made by healthy cells. If monoclonal antibodies that attach to only the abnormal proteins can be made, then it should be possible to target the cancer cells with drugs without affecting the healthy cells (Figure 20.5B).

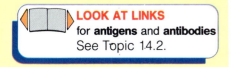

LOOK AT LINKS
for **antigens** and **antibodies**
See Topic 14.2.

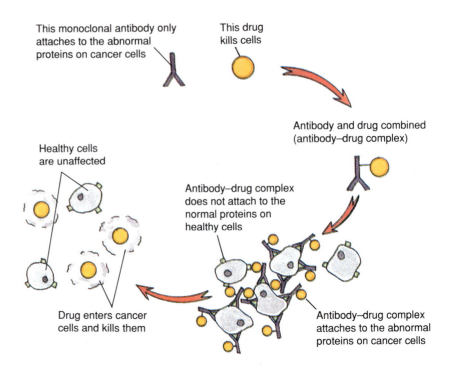

This monoclonal antibody only attaches to the abnormal proteins on cancer cells

This drug kills cells

Antibody and drug combined (antibody–drug complex)

Healthy cells are unaffected

Antibody–drug complex does not attach to the normal proteins on healthy cells

Drug enters cancer cells and kills them

Antibody–drug complex attaches to the abnormal proteins on cancer cells

Figure 20.5B ● How monoclonal antibodies could be used to treat cancer.

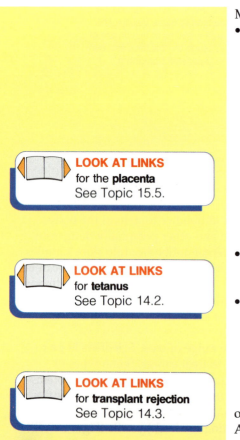

LOOK AT LINKS
for the **placenta**
See Topic 15.5.

LOOK AT LINKS
for **tetanus**
See Topic 14.2.

LOOK AT LINKS
for **transplant rejection**
See Topic 14.3.

Monoclonal antibodies can be put to other uses:
- the urine of pregnant women contains **human chorionic gonadotrophin (hCG).** This hormone, produced by the placenta, can be detected with monoclonal antibodies as early as 12 days after fertilisation. A pregnancy test kit consists of a dipstick impregnated with antibodies. The position of the coloured band after dipping the stick into a urine sample tells the woman whether or not she is pregnant. If hCG is present in the urine, the molecules bind to the antibodies at the end of the dipstick. The hCG/antibody complex diffuses up the dipstick and meets a band of antibodies which also bind hCG. Colour change at this level shows that the woman is pregnant. If there is no hCG in the urine, antibodies alone diffuse further up the dipstick to meet a second band of different antibodies. The antibodies bind together producing a colour change which shows that the woman is not pregnant.
- monoclonal antibodies bind with poisons, inactivating them. Tetanus toxin and overdosage of digoxin (a drug used to treat heart disease) have been inactivated with different types of antibody.
- the success of transplants depends on matching the tissue of donor and patient as nearly as possible. The closer the match, the less chance there is of **tissue rejection**. Monoclonal antibodies, produced against the cell surface proteins of the donor's and patient's tissues, make tissue matching more accurate. *Explain how monoclonal antibodies improve the accuracy of tissue matching.* Reading p. 244 will help you.
Developing different types of monoclonal antibody for medical and other uses is expensive. However, the potential for profit is considerable. As a result, new companies have sprung up in response to the demand.

Antibiotics

Penicillin was the first of a family of antibiotic drugs made by biotechnology. Today hundreds of different antibiotics are used to combat diseases caused by bacteria. Many more substances with antibiotic activity have been discovered. However, most are too expensive for commercial production and/or have harmful side-effects. Nevertheless, some have their uses outside medicine, in research, agriculture and food production.

There are four groups of antibiotic:

- **penicillins** – commercially produced penicillin is a mixture of compounds. The main component is penicillin G which can be converted into other forms, each with a slightly different activity. For example, penicillin G is broken down by stomach acid, and is therefore best given by injection; ampicillin is unaffected by stomach acid and can therefore be given by mouth as tablets or in a syrup – something children may prefer! The range of penicillins allows medical staff to choose which type is the best for treating a particular disease. Choice also helps to combat the development of drug resistance.
- **cephalosporins** – cephalosporins made by the mould *Cephalosporium* were discovered in 1948. All are active against a similar range of bacteria. Newer cephalosporins are effective against bacteria which have developed resistance to penicillin.
- **tetracyclines** – tetracycline is made by the bacterium *Streptomyces aureofaciens*. Its various forms are active against a similar range of bacteria, although the development of widespread resistance has reduced effectiveness. Tetracyclines bind to calcium and are deposited in growing bones and teeth. *Why should tetracyclines not be given to children and pregnant women?*
- **erythromycins** – the activity range of erthromycin is similar to that of penicillin. Erythromycin, therefore, is useful against bacteria resistant to penicillin or where the patient is allergic to penicillin.

Sales of antibiotics approach 16 billion pounds annually world-wide. Some, like penicillin, destroy a limited number of species of bacteria. They are **narrow-spectrum** antibiotics. Others, such as tetracycline, are **broad-spectrum** antibiotics. They act against a wide range of bacteria.

The story of penicillin

In the 1880s, new ways of staining cells and further development of the light microscope boosted research on microorganisms. Bacteria were clearly visible for the first time. The way was open for fresh discoveries, which eventually led to our conquest of many of the diseases caused by bacteria.

At the time, scientists noticed that laboratory cultures of bacteria did not grow on culture plates on which colonies of mould were also growing. Nobody followed up the observations until 1928 when the British bacteriologist Alexander Fleming (Figure 20.5C) noticed that some of his bacteria cultures were contaminated with mould. Figure 20.5D shows what happened. Fleming identified the mould as *Penicillium notatum* (Figure 20.5E) and reasoned that it produced a substance which killed bacteria. He isolated the substance and called it **penicillin**.

Fleming thought that penicillin might help fight disease, but extracting it from the mould was very difficult. There matters rested until 1938 when Howard Florey and Ernst Chain, working at Oxford, made

IT'S A FACT

The word antibiotic means 'against life'. Antibiotic drugs kill bacteria. However, they do not affect viruses.

LOOK AT LINKS
for more diseases caused by bacteria
See Topic 4.3.

LOOK AT LINKS
for the action of antibiotics
See Topic 18.3.

LOOK AT LINKS
for more about **drug resistance**
See Topic 18.3.

IT'S A FACT

Biotechnology improves yields
Penicillium notatum produces very little penicillin. *P. chrysogenum* yields more. In the 1950s, a one-litre batch of mould and nutrient produced 60 mg of penicillin. Nowadays, 20 g of penicillin can be extracted from the same volume of material owing to advances in biotechnology and control of fermentation processes.

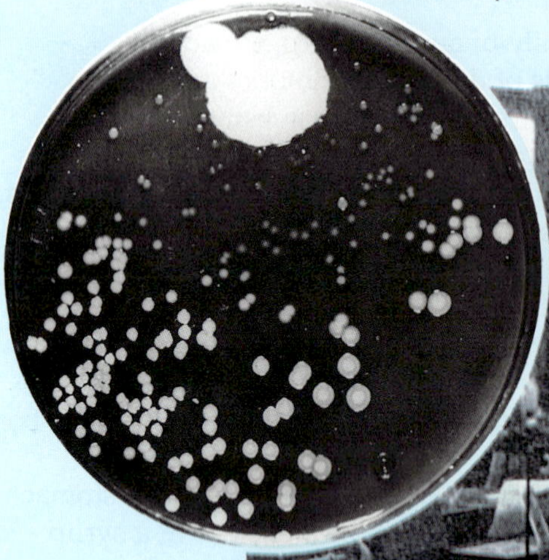

Figure 20.5D ● Colonies of bacteria do not grow near the mould because it secretes penicillin into its surroundings.

Figure 20.5C ● Alexander Fleming (1881–1955) in his laboratory at St Mary's hospital, London. He was awarded the Nobel Prize with Florey and Chain for working on penicillin

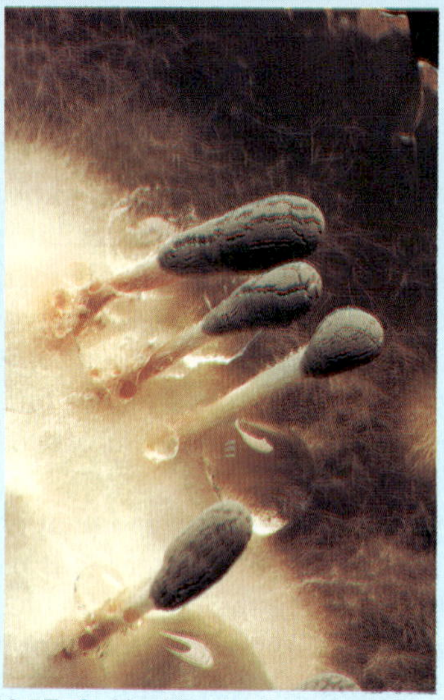

Figure 20.5E ● Hyphae and spore capsules (see p. 17) of *Penicillium*

commercial production of penicillin possible. *Penicillium notatum* was replaced by a better strain, *P. chrysogenum*. The scientists developed methods to produce enough penicillin for clinical trials, using all kinds of vessels (including milk bottles!) to contain the mould–sugar nutrient mixture.

By then (1940) the Second World War (1939–45) had begun and there was an urgent need to treat wounded soldiers. Before penicillin, wounds often became infected with bacteria which caused gangrene, blood poisoning and other fatal diseases. Penicillin prevented infections taking hold and troops could return to duty once they had recovered.

The work moved to the USA where large-scale production techniques were developed, using 'cornsteep liquor' (a waste product of the manufacture of starch from maize) as the nutrient solution for the growth of the penicillin mould. By 1944, enough penicillin was being produced to treat all the British and American casualties during the Normandy landings and invasion of Germany.

Today the massive demand for penicillin world-wide is met by the development of new high-yielding strains of the mould. The yield is improved through genetic manipulation, selection, mutation and protoplast fusion. Huge fermenters holding up to 200 000 litres of mould/nutrient solution produce penicillin by **batch culture**. The mould is then filtered off and the penicillin extracted.

LOOK AT LINKS
for **protoplast fusion**
See Topic 20.3.

Interferons

Have you ever suffered from the symptoms of 'flu? If so, **interferon** was responsible for your shivers, aches, fevers and tiredness. It was produced in response to the virus causing your illness.

Interferons are proteins. Viruses stimulate their production by virus-infected cells. Interferons help healthy cells to break down viruses and prevent them from multiplying. Figure 20.5F shows you the idea.

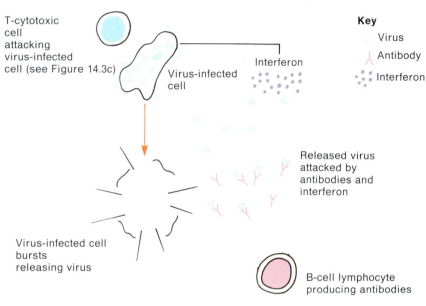

Figure 20.5F ● Interferon does not directly attack viruses, but interacts with host cells, helping them to protect themselves from the virus

Three different types of human interferon have been isolated so far. Other animal interferons have been identified. However, their action is specific. Interferon is only effective in the animal species that produces it.

Originally, interferon was extracted from white blood cells. Thousands of litres of blood were processed but yields of interferon were minute. At a cost of 10 million pounds per gram, research designed to find out how interferon worked was slow and very expensive.

In 1980, the gene for one type of human interferon was isolated and inserted into the bacterium *Escherichia coli*. Prices fell sharply as much more genetically engineered interferon became available at a fraction of the previous production costs.

Hopes have run high that interferon would be the treatment for viral diseases that penicillin has proved to be for diseases caused by bacteria. Hopes remain, but turning interferon into the answer for viral infections is proving very difficult. Although the effect of interferon is as outlined in Figure 20.5F, its actions are very complicated and incompletely known.

Different interferons are available for medical use. Some products seem to be effective against certain types of viral disease. For example, the drug **Alferon**, produced by Interferon Sciences Inc. in the USA, shows promising results against the virus which causes **genital herpes**.

Other products treat different forms of cancer. For example, interferons are used to treat **HIV** infected patients who develop **Kaposi's sarcoma** (Figure 20.5G) *Do viruses cause some types of cancer? How does the effect of interferon on Kaposi's sarcoma help you to answer the question?*

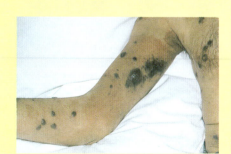

Figure 20.5G ● Kaposi's sarcoma is a type of cancer. It is part of AIDS, as a result of HIV infection

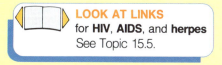

LOOK AT LINKS
for **HIV**, **AIDS**, and **herpes**
See Topic 15.5.

Vaccines

LOOK AT LINKS
for more about **vaccines**
and **immunisation**
See Topic 14.2.

IT'S A FACT

Ten million people world-wide were infected with smallpox in the early 1970s. Since then a mass immunisation programme organised by the World Health Organisation has eliminated the disease.

LOOK AT LINKS
Remind yourself about
antigens and **antibodies**
See Topic 14.2.

LOOK AT LINKS
for **malaria**
See Topic 4.2.

Vaccines made by the methods described on p. 237 protect millions of people world-wide from serious diseases. However, there is a minute risk that:

- microorganisms used to make vaccines may survive the production process
- microorganisms used to make vaccines may regain the ability to cause disease
- patients may have an allergic reaction to the cellular debris that remains from vaccine production despite purification processes
- people working in production may come in contact with the dangerous organisms used to make vaccines despite strict safety precautions.

Genetic engineering is helping to reduce even these minute risks. The principles are:

- isolate genes from the disease organism responsible for producing the antigens that stimulate lymphocytes to produce antibodies
- insert the genes into a less dangerous organism
- culture the engineered organism, which makes antigen in large quantities
- extract the antigen, which is then used to make vaccine.

Difficulties arise in locating the genes which contain the code for the antigens most useful in stimulating lymphocytes to produce antibodies, and in identifying the antigens themselves. However, progress is being made. *Salmonella* bacteria have been engineered to carry the genes responsible for the cell surface proteins of the malaria parasite *Plasmodium*. If the cell surface proteins are effective antigens then a vaccine made from them will be safer than one made from the whole organism.

Hepatitis B virus causes serious liver disease. Medical workers and patients in places where there may be human blood contaminated with the virus and families of hepatitis B carriers are at risk from the disease. They are high-priority groups for hepatitis B vaccine.

The vaccine is made from the blood plasma of people carrying the virus, but genetic engineering is taking over! Genes responsible for the protein coat (antigen) surrounding the virus have been isolated and inserted into yeast cells. Culturing the engineered yeast produces large quantities of virus coat protein which is used to make the vaccine. Figure 20.5H shows the method.

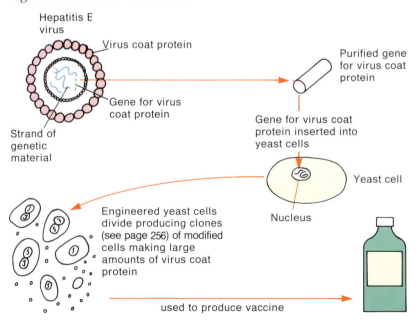

Figure 20.5H ● Genetic engineering makes hepatitis B vaccine – conventional techniques include the genetic material of the virus in vaccine production. The genetic material is responsible for the virus reproducing itself, causing disease. Genetic engineering eliminates the risk

LOOK AT LINKS
for **genetics**
See Topic 19.

LOOK AT LINKS
for **sickle-cell trait** and **malaria**
See Topic 4.2.

LOOK AT LINKS
for **haemoglobin**
See Topic 14.2.

LOOK AT LINKS
for **gene probes**
See Topic 19.3.

IT'S A FACT

People with sickle-cell trait (carriers of the recessive allele for sickle-cell anaemia) have greater resistance to malaria than individuals who are not carriers. This is why the sickle-cell allele persists in human populations living in regions of the world where malaria is endemic.

SUMMARY

Biotechnology is advancing our understanding and treatment of disease. New products and techniques of diagnosis will help to prevent and possibly eliminate many serious diseases.

Gene therapy

First things first! Before continuing you need to know that:
• any particular characteristic (e.g. height of pea plants) is controlled by pairs of genes called **alleles**
• a pair of alleles may be the same (**homozygous**) or different (**heterozygous**)
• a dominant allele masks the effect of its recessive partner.

Check the facts by reading pp. 345–7. They will help you to understand the exciting possibility of treating diseases by gene therapy.

We all carry a few faulty genes. Most of us are unaware of them because their dominant partners mask their harmful effects. Only individuals who are homozygous for faulty alleles show symptoms of disease. For example, **sickle-cell anaemia** is caused by a single mutation on the gene controlling the synthesis of the blood pigment haemoglobin (Figure 20.5I). In different regions of Africa up to 30% of the population may carry the recessive allele for sickle-cell anaemia but its effect is masked by its normal partner. Carriers of the faulty allele are heterozygous and do not develop the disease. However, an individual who inherits the allele for sickle-cell anaemia from each parent is homozygous and soon develops the disease. He or she is unlikely to survive childhood.

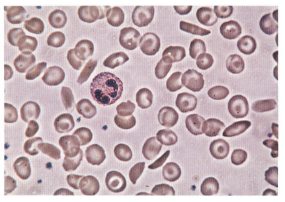

Figure 20.5I ● The disc-shaped red blood cells carry normal haemoglobin. The cells shaped like the blade of a sickle contain abnormal haemoglobin. Sickle cells do not carry as much oxygen as normal cells, so the affected person soon becomes breathless and tires easily

Gene therapy to control sickle-cell anaemia would aim to add normal genes to the patient's genetic make up. Before gene therapy can begin, the faulty gene must be identified. **Gene probes** are powerful tools used in the hunt. They are designed to bind to specific portions of DNA and can distinguish between faulty genes and their normal partners.

CHECKPOINT

❶ What are monoclonal antibodies? Describe how monoclonal antibodies could be used to treat cancer.

❷ Penicillin is an antibiotic drug. Different types of penicillin each have a slightly different activity. Why is it useful for medical staff to be able to choose between different types of penicillin for treating patients?

❸ How does interferon work?

❹ Vaccines protect millions of people world-wide from serious diseases. However, there are minute risks attached to their use. What are the risks? How is genetic engineering helping to reduce the risks?

❺ Briefly discuss the principles behind treating disease by gene therapy.

20.6 Treating sewage

Sewage consists of:
- faeces and urine
- industrial and household wastes
- water, grit and other debris draining from roads and paths

Reading this section will tell you what happens to sewage.

 LOOK AT LINKS

for **cholera** and other diseases
See Topic 4.3;
for **decomposition**
See Topic 2.5;
for the meaning of **aerobic** and **anaerobic**
See Topic 12.1.

Animal dung forms manure, which is a natural fertiliser. Farms which are used solely for rearing animals do not need manure for fertilising crops. Instead the manure is washed into lagoons where it forms a liquid slurry. These lagoons may spring leaks and the slurry may seep into streams and rivers. This causes the water to become richer and richer in nutrients and bacteria to flourish. In doing so, the bacteria use up the oxygen dissolved in the water (biological oxygen demand, BOD). Fish and other wildlife die. Eventually the water becomes foul-smelling and virtually lifeless. The process of enrichment, increasing BOD and elimination of wildlife is called **eutrophication**.

Have you thought where wastes and water go when you flush the toilet, pull the plug after a bath or finish the washing up? Where do you think rain water runs to off the streets? What happens to waste chemicals and other materials from factories? Make a brief summary of your ideas.

Before there were drains and sewage works, sewage went directly into streams or open ditches which emptied into the nearest river. Drinking water supplies were often contaminated with microorganisms causing diseases such as cholera, typhoid and diphtheria. Microorganisms decompose the organic matter in sewage, but in doing so use up the water's oxygen (called the **Biological Oxygen Demand – BOD**) and release poisonous substances like ammonia. Wildlife dies through lack of oxygen and/or poisoning. Anaerobic bacteria, which function without oxygen, continue the decomposition of waste organic matter, blackening the water and producing evil-smelling gases. Figure 20.6A gives you the idea.

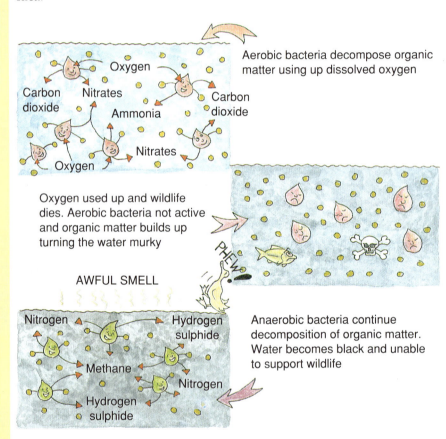

Figure 20.6A ● Aerobic and anaerobic bacteria decomposing waste organic matter

Managing sewage

Raw sewage is a health hazard. Microorganisms help us to manage sewage by breaking it down into harmless and even useful substances. Treatment occurs in sewage works. Biotechnology, engineering and computer control are harnessed to deal with the wastes that pour in from homes, factories and gutters and drains which collect water and rubbish from every street. Figure 20.6B shows what happens.

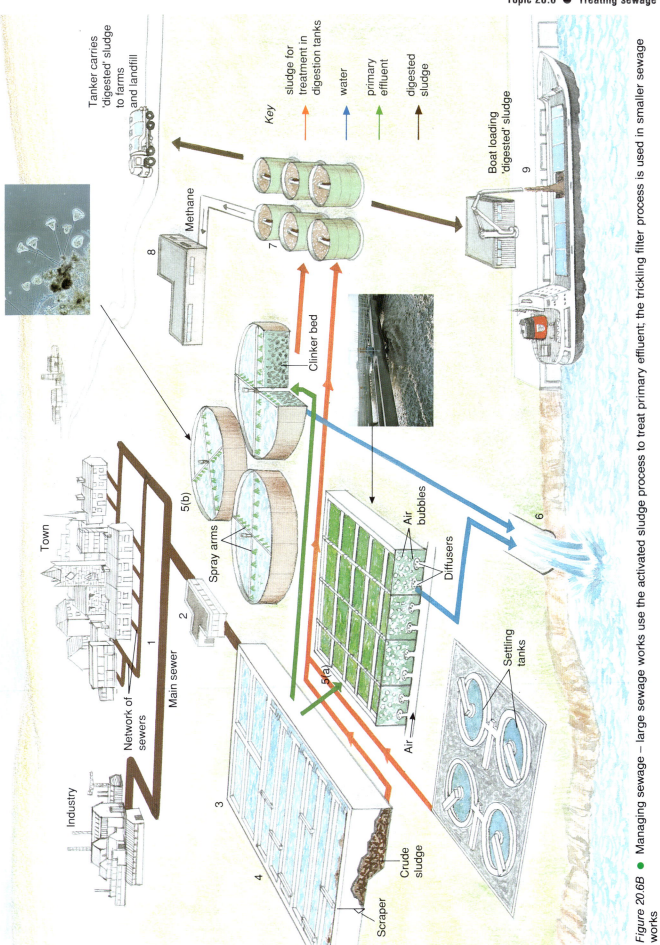

Figure 20.6B ● **Managing sewage – large sewage works use the activated sludge process to treat primary effluent; the trickling filter process is used in smaller sewage works**

Tanker carries 'digested' sludge to farms and landfill

Key

sludge for treatment in digestion tanks

water

primary effluent

digested sludge

Boat loading 'digested' sludge

9

Methane

8

7

Clinker bed

5(b)

Spray arms

Town

1

2

Main sewer

Network of sewers

Industry

Air bubbles

Diffusers

Settling tanks

6

5(a)

Air

3

4

Scraper

Crude sludge

379

In 1858, MPs were driven from the Commons by the awful smell coming from the Thames flowing by the Houses of Parliament. Sheets soaked with disinfectant were hung over the windows to try to clear the air. Parliament authorised Joseph Bazalgette to clean up the Thames. He designed a system of sewers that carried wastes away from central London to huge lagoons further down the river. Untreated sewage was released from the lagoons on the ebb-tide in the hope that it would flow out to sea.
Unfortunately a good deal of the raw sewage returned on the next high tide!

Sewage treatment began in the 1920s but progress was slow. Since 1950, industry, local government and other authorities have co-operated to improve matters. The Thames has lost its smell and wildlife has returned.

Key to Figure 20.6B
1. Sewers – waste from homes, industry and the streets passes into a network of underground sewers which run to the treatment works
2. Pumphouse and screen – sewage passes through closely spaced metal bars which remove rags, paper, wood and large objects that would otherwise damage the machinery
3. Grit removal – sewage trickles through channels. Grit and sand in suspension settle out, are washed, collected and used to fill holes in the ground
4. Primary settling tanks – sewage flows into large tanks where solids settle to the bottom forming **crude sludge**. Electrically powered scapers load crude sludge for transfer to sludge 'digestion' tanks (7)
5. Aerobic processing – liquid (called **primary effluent**) passes from the primary settling tanks to the **secondary treatment plant**. Here a great variety of microorganisms (bacteria, protists, fungi) break down the organic material in the primary effluent into minerals, gases and water. Their activities require a lot of oxygen. The main reactions are:

$$\text{Carbon compounds} \rightarrow \underset{\text{Carbon dioxide}}{CO_2}$$

$$\underset{\text{Hydrogen sulphide}}{H_2S} \rightarrow \underset{\text{Sulphates}}{SO_4^{2-}}$$

$$\underset{\text{Ammonium compounds}}{NH_4^+} \rightarrow \underset{\text{Nitrates}}{NO_3^-}$$

There are two systems:
a) **Activated sludge process** – diffusers bubble air through the primary effluent. It takes about 8 hours for the microorganisms to convert most of the organic material present into simpler substances
b) **Trickling filter process** – primary effluent is sprayed from slowly rotating horizontal arms onto a filter bed of clinker, gravel or moulded plastic contained in a concrete tank. A film of microorganisms covers the filter bed. As effluent slowly trickles over and through the filter bed, the microorganisms break down the organic matter. The reactions are similar to those of the activated sludge process. Protists such as *Vorticella* feed on the bacteria and fungi preventing them from smothering the filter and slowing down the process of breakdown.
Effluent from the activated sludge process or the trickling filter process passes into **settling tanks** where microorganisms and any remaining organic material separate from the water, forming a sludge. Some of the sludge is used for the next batch of aerobic processing, the rest is sent to the sludge 'digestion' tanks (7)
6. Outfall – water from the settling tanks is usually clean enough to go into a river. If the water needs to be especially clean before discharge it is filtered through beds of fine sand and activated charcoal, and treated with chlorine to prevent the growth of any remaining microorganisms
7. Anaerobic processing – sludge from the primary settling tanks and aerobic processing is treated for 2 to 3 weeks in tanks in the absence of oxygen at a temperature between 30° and 40 °C. Anaerobic 'digestion' of the sludge by bacteria such as *Methanobacterium* converts the organic material into methane (CH_4), carbon dioxide (CO_2) and hydrogen (H_2) gases along with water and minerals
8. Powerhouse – methane gas produced during anaerobic processing is burned to heat the 'digestion' tanks and generate electricity, which is used to power machinery and/or supply the National Grid
9. Dumping – 'digested' sludge may be dumped at sea or spread on land as a soil 'conditioner', providing nitrates and phosphates and improving water retention. Sometimes 'digested' sludge is used for landfill. The methane produced is piped to nearby houses as cheap fuel.

Biogas fermenters

Notice in Figure 20.6B that the anaerobic stage in sewage treatment produces the fuel gas **methane**, which is burned and used to generate power to run the sewage works. In China, India and other countries, village communities have used **biogas fermenters** to produce methane for hundreds of years. Human and animal faeces, leafy waste from crops, paper and cardboard are all grist to the bacterial mill inside the fermenter. Figure 20.6C shows one in action.

LOOK AT LINKS
for **the environment** and **conservation**
See Topic 5.

LOOK AT LINKS
for **soil erosion**
See Topic 5.1.

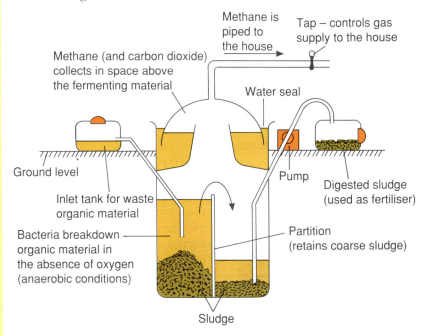

Figure 20.6C ● Biogas fermenter – 'digestion' of organic material by *Methanobacterium* and other anaerobic bacteria produces methane

SUMMARY

Raw sewage is a health hazard. Bacteria break down sewage waste into harmless and even useful substances. Sewage works are designed to deal with large volumes of waste. Biogas fermenters digest small amounts of waste and serve the needs of local communities.

The biogas fermenter is an example of small-scale technology serving local needs. In developing countries where wood is scarce, the methane produced is a valuable fuel for cooking, lighting and powering refrigerators. Left-over sludge is used as a soil conditioner. Biogas fermenters can be expensive. Cheaper designs are being developed so that more people can afford them.

Make the link between these developments and benefits to the environment. For example, how can biogas fermenters help conserve trees and soil? Report your ideas in a way that would help people decide to buy a biogas fermenter for their village.

CHECKPOINT

❶ Discuss how the activities of microorganisms that decompose the organic matter in sewage use up dissolved oxygen.

❷ Summarise the stages in the treatment of sewage in a modern sewage works.

❸ Distinguish between the activated sludge process and the trickling filter process in sewage works.

❹ Describe how a biogas fermenter works.

EVOLUTION

TOPIC 21

21.1 Evolution: the growth of an idea

Species of living organisms change over the course of time. This change is called evolution. How evolution takes place was a puzzle until Charles Darwin discovered the principles of natural selection. In this section you can find out how the idea developed in Darwin's mind.

The variety of life today almost certainly owes its origins to a scenario billions of years ago similar to the one in Figure 21.1B. *How do the events in Earth's history take us from such beginnings to the present day?* Present-day living things are descended from ancestors that have gradually changed through thousands of generations. We call the process **evolution**.

Look at the picture strip of Earth's **geological eras** at the bottom of the page. It is a timetable showing major events during the evolution of life from its beginnings to the appearance of the first human beings approximately 2.5 million years ago.

Charles Darwin

The British naturalist Charles Darwin was the first person to work out a theory explaining **how** species can evolve. In 1831 Darwin was invited to join the company of HMS *Beagle* on a world voyage which included a survey of the South American coast. The journey took five years. During the *Beagle's* time around South America, Darwin took over as the ship's naturalist. He collected fossils and specimens of plants and animals on expeditions inland. He made many observations about the wildlife he encountered.

On one April day in 1832 Darwin collected 68 different species of small beetle from the rain forest around Rio de Janeiro in Brazil. What puzzled him was that there should be so many different species of one kind of insect. *How did such variety come about?*

Figure 21.1A ● Charles Darwin (1809–82)

Geological era	Earth's origins				

Mars

Venus

Mercury

Sun

Seas full of organic molecules

Bacteria-like cells form stony pillars called stromatolites. These form some of Earth's oldest rocks

Earth

Time scale (millions of years ago)	4800	4400	4000	3600	3200	2800

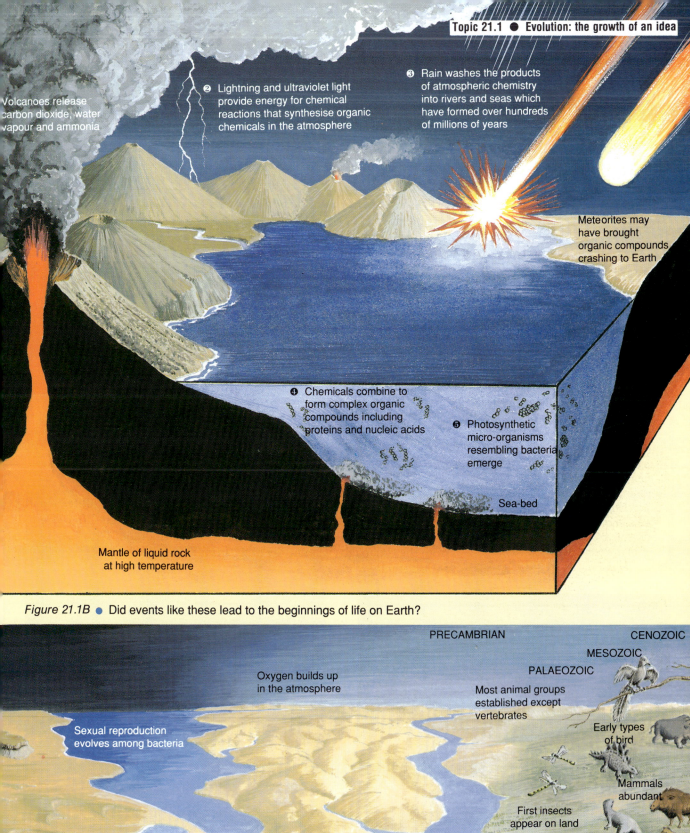

Volcanoes release carbon dioxide, water vapour and ammonia

❷ Lightning and ultraviolet light provide energy for chemical reactions that synthesise organic chemicals in the atmosphere

❸ Rain washes the products of atmospheric chemistry into rivers and seas which have formed over hundreds of millions of years

Meteorites may have brought organic compounds crashing to Earth

❹ Chemicals combine to form complex organic compounds including proteins and nucleic acids

❺ Photosynthetic micro-organisms resembling bacteria emerge

Sea-bed

Mantle of liquid rock at high temperature

Figure 21.1B ● Did events like these lead to the beginnings of life on Earth?

PRECAMBRIAN

CENOZOIC

MESOZOIC

PALAEOZOIC

Oxygen builds up in the atmosphere

Most animal groups established except vertebrates

Early types of bird

Sexual reproduction evolves among bacteria

Mammals abundant

First insects appear on land

Reptiles abundant

Early land plants

First flowering plants

Simple multicellular animals appear such as worms and jellyfish

Humans appear

Fish abundant

383

2400   2000   1600   1200   800   400

IT'S A
FACT

IT'S A
FACT

Space is filled with dust and gases made of carbon, hydrogen and nitrogen; some of the chemicals necessary to build organic compounds. Some scientists believe that dust and gases entering the Earth's atmosphere 4000 million years ago helped to trigger the appearance of life on Earth.

Figure 21.1C ● HMS *Beagle* during her voyage around South America

IT'S A
FACT

Deep-sea vents are places where lava escapes from the Earth's core on to the sea-bed. Today they support thriving communities of animals and bacteria. Some scientists think they may have provided nutrients and suitable environments for the first organisms. *Why are there no plants around deep-sea vents?*

As the *Beagle* sailed up the Pacific coast of South America, Darwin noticed that one type of organism gave way to another. He also noticed that many of the animals were different from those he had seen on the Atlantic coast.

Its South American survey complete, the *Beagle* set sail for the Pacific islands of the Galapagos on 7 September 1835. Once ashore, Darwin observed that the wildlife of the islands was similar to the wildlife he had seen on the South American mainland. However, he noticed differences in detail. The birds shown in Figure 21.1D are cormorants. One comes from the Galapagos islands, the other from Brazil. *What differences can you see between them?*

(a) Galapagos Islands cormorant

(b) Brazilian cormorant

Figure 21.1D ● Flying demands a lot of energy. The absence of predators on the Galapagos Islands means that cormorants do not have to fly away to escape danger. Birds with small wings, therefore, are at an advantage. However, predators threaten Brazilian cormorants. Birds with large wings can fly away and are at an advantage even though they use a lot of energy.

Each of the Galapagos Islands Darwin visited was inhabited by tortoises which were similar to the mainland forms but much larger. He noticed something else: the tortoises on each island were slightly different, so he could tell which island particular tortoises came from (Figure 21.1E). Darwin wondered why the tortoises were different from island to island.

(a) Isla Isabela tortoise

(b) Isla Espanola tortoise

Figure 21.1E ● Tortoises from Isla Isabela (formerly Albemarle) and Isla Espanola (formerly Hood), two of the Galapagos islands. Isla Isabela is well watered and the tortoises wallow in the pools. Isla Espanola is arid and here the tortoises browse on the water-filled stems of cacti and juicy leaves and berries. Notice the shape of the shell of the tortoises from each island. *Can you account for the differences?*

In Darwin's day, most people believed that each species was fixed and unchangeable; but Darwin could not agree. The variety of species discovered on his expeditions in South America and the Galapagos islands convinced him that species are not fixed but change through time; that is, they evolve. *The puzzle was, what mechanism brought about their evolution?*

● *The influence of Lyell and Malthus*

Just before Darwin set sail in the *Beagle*, he was given the first volume of a newly published book *Principles of Geology* by Charles Lyell (Figure 21.1F). The second and third volumes were sent to him during the voyage.

Lyell believed that the Earth's rocks were very old and that natural forces had produced continuous geological change during the course of the Earth's history. Lyell explained that the age of rocks could be estimated from the types of fossil they contained. He also stated that the fossil record was laid down over hundreds of millions of years.

Darwin reasoned from Lyell's work that if rocks and rock formations have changed slowly, over long periods of time, living things might have a similar history.

Another piece of the jigsaw fell into place soon after Darwin's return to England. One day he was reading *An Essay on the Principle of Population* written in 1798 by the Reverend Thomas Malthus (Figure 21.1G). In the essay Malthus stated that a population would increase indefinitely unless it was kept in check by shortages of resources such as food, water and living space.

Darwin understood that living organisms produce more offspring than can normally be expected to survive. A beech tree, for example, produces thousands of seeds each year. Hundreds of seedlings sprout from the seeds but only one or two grow into mature trees (Figure 21.1H). He also understood that the supply of food, amount of space and other resources in the environment are limited. Only those organisms with the structures and the way of life that suited (adapted) them to make best use of these

Figure 21.1F ● Sir Charles Lyell (1797–1875) greatly influenced Darwin's view of Earth and its history. Darwin met Lyell on his return to England and they became great friends.

Figure 21.1G ● Reverend Thomas Malthus

385

LOOK AT LINKS
Characteristics of the sexually reproduced members of a generation vary because each member inherits different combinations of alleles from its parents.
See Topic 19.2.

limited resources survived long enough to reproduce (see Figures 21.1D and 21.1E). Less well-suited organisms left fewer offspring or did not survive to reproduce at all. Therefore, variations which were favourable for survival accumulated from one generation to the next, while less favourable variations died out (became extinct). Darwin called this process natural selection. He realised that this is the mechanism of evolution. At last he understood how species change through time.

Figure 21.1H ● Hundreds of seedlings are growing in an area about 10 m². Very few will grow to maturity. Most perish, crowded out by the two or three that grow the fastest.

21.2 Artificial selection

FIRST THOUGHTS

By choosing particular characteristics, breeders have developed new varieties of plants and animals. The examples described in this section illustrate the process of breeding and its importance in support of Darwin's idea.

Darwin was a cautious man. He knew that other people who had proposed theories of evolution before him had been ignored and disliked. After the *Beagle's* voyage, he spent the next twenty years gathering more evidence to support his ideas. His search for evidence led him to investigate the work of breeders of animals and plants.

Figure 21.1A shows two very different breeds of dog. Centuries of selecting dogs with particular characteristics such as size, colour, length of coat and shape of ear have resulted in a wide variety of breeds. The corgi, spaniel, labrador and bulldog are just a few examples. *How many other breeds can you think of?*

Dog breeders have taken advantage of the large amount of variation in the characteristics of the members of generations of dogs and chosen the characteristics which they want to be passed on to the next generation. The dogs with these characteristics have been allowed to reproduce and so the genes controlling the selected characteristics have been passed to their offspring. This process of choosing which characteristics should pass to the next generation and which should not is called **artificial selection**. (See Figure 21.2B.)

Figure 21.2A ● The same species?

Italian spaniel

Individuals with desirable characteristics are selected for breeding. The genes for the characteristics are passed on through a number of generations. Eventually a new breed is established.

Norfolk spaniel

Individuals with desirable characteristics are selected for breeding. More new breeds are established

Cocker spaniel

Springer spaniel

Figure 21.2B ● The artificial selection of desired characteristics produces breeds of dog that look very different from their ancestors

Generations of artificial selection have resulted in the great variety of dog breeds we know today. All dogs belong to the same species, *Canis familiaris*, and are descended from the same ancestral species even though they look very different from one another.

Figure 21.2C shows another example of artificial selection. Varieties of vegetable have been produced by choosing different characteristics which make good eating and breeding for each one. All the plants belong to the same species, *Brassica oleracea* (a relative of the mustard plant), and are descended from the same ancestral species.

Darwin added the experience of breeding pigeons to his observations of other people's work. He saw that by mating pigeons with different characteristics, new varieties of pigeons could be produced. All the pigeons in Figure 21.2D belong to the species *Columba livia*.

Darwin reasoned that if artificial selection produced change in domestic animals and plants, natural selection should have the same effect on wildlife. The only difference was that artificial selection by human beings produced new varieties in a relatively short time, selection by nature took much longer.

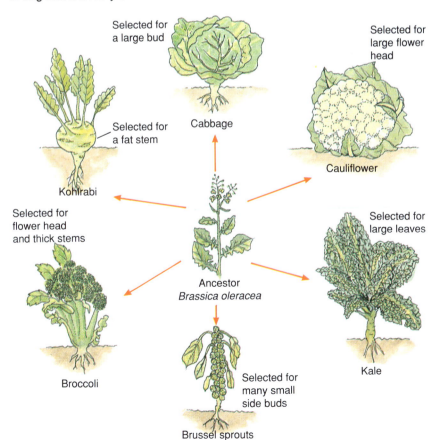

Selected for a large bud

Selected for large flower head

Selected for a fat stem

Cabbage

Cauliflower

Kohlrabi

Selected for flower head and thick stems

Selected for large leaves

Ancestor *Brassica oleracea*

Broccoli

Selected for many small side buds

Kale

Brussel sprouts

Figure 21.2C ● A variety of vegetables bred from a single ancestral species

Figure 21.2D ● The many varieties of domestic pigeon

21.3 The Origin of Species

Darwin hesitated from publishing his ideas on natural selection, but in 1858 he had an unpleasant surprise. He received a letter and an essay from Alfred Russel Wallace who was a naturalist living and working in Malaya (Figure 21.3A). In the letter, Wallace asked for Darwin's opinion as a respected naturalist on the essay which set out the principles of natural selection as the mechanism for evolutionary change. This was a terrible blow to Darwin who had worked patiently on the same ideas for 20 years.

Like Darwin, Wallace was led to the idea of natural selection through reading Malthus' essay on population. Over the years, Darwin had discussed natural selection with only a few close friends and fellow scientists. They had urged him to publish his ideas. When Wallace's letter arrived, Darwin was stung into action.

Both scientists recognised the other's contribution and acted most generously over the matter. They agreed to the joint publication of their ideas which appeared in the *Journal of the Linnean Society* for 1858.

Offspring reproduced asexually are genetically identical with each other and with their parent. Mutation (see p. 148) is the only source of genetic variation. Sexually reproduced offspring inherit genetic material from both parents. They are genetically different from each other (except in the case of identical twins) and from their parents.

Natural selection works on genetic variation selecting the variations which enable organisms to adapt to a changing environment. In stable environments the selection pressure for organisms to adapt is low. Asexual reproduction, therefore, has advantages. One parent reproducing identical copies of itself can more easily populate and exploit the environment than two parents reproducing sexually.

In rapidly changing environments, however, the balance tips in favour of sexual reproduction. Selection pressure is high and favours organisms which can change in response to changing circumstances. Only sexual reproduction is able to produce enormous amounts of genetic variation. Natural selection can then select the change which enables organisms to adapt to the new environment.

Figure 21.3A ● Alfred Russel Wallace

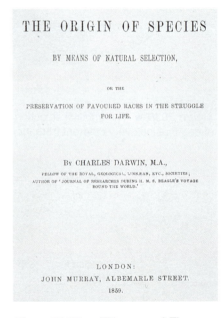

THE ORIGIN OF SPECIES

BY MEANS OF NATURAL SELECTION,

OR THE

PRESERVATION OF FAVOURED RACES IN THE STRUGGLE FOR LIFE.

By CHARLES DARWIN, M.A.,

FELLOW OF THE ROYAL, GEOLOGICAL, LINNÆAN, ETC., SOCIETIES; AUTHOR OF 'JOURNAL OF RESEARCHES DURING H. M. S. BEAGLE'S VOYAGE ROUND THE WORLD.'

LONDON:
JOHN MURRAY, ALBEMARLE STREET.
1859.

Figure 21.3B ● Title page of *The Origin of Species*

Although Wallace was a distinguished naturalist in his own right, he realised that Darwin's ideas were more thoroughly worked out than his own. For the rest of his life Wallace insisted that Darwin take most of the credit. After the publication of the joint paper, Darwin hastily prepared a fuller account of his work. *The Origin of Species* was published in 1859 (Figure 21.3B).

In the book Darwin brought together so much evidence that evolution was established as a viable theory. However, the position of natural selection was less secure. Many scientists thought it likely that natural selection was the mechanism of evolution but wanted more evidence. Not until the development of modern genetics, beginning with the discovery of Mendel's work, was natural selection generally accepted as the mechanism for evolutionary change. Understanding genetics allows us to understand why organisms vary. Natural selection works on these variations to produce evolutionary change.

IT'S A FACT

Although Mendel published his theories of heredity in 1866, Darwin never knew about them. It was only after their deaths that Darwin's work on natural selection and Mendel's work on genetics were brought together to give an explanation of evolutionary change.

❶ Briefly explain your understanding of evolution.

❷ Why do you think studying life in the past helps us to understand biology today?

❸ What is an ancestor?

❹ What is a descendant?

❺ What is meant by the term 'natural selection'?

FIRST THOUGHTS

21.4 Evolution in action

In this section you can read about examples of evolutionary change.

LOOK AT LINKS
for **populations**
See Topic 2.

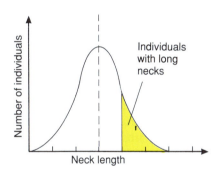

Individuals with long necks

Neck length

Number of individuals

A long neck enables giraffes to browse on the leaves of trees. This is an advantage during the dry season when grass and other food plants are in short supply. This passes more genes for long necks on to the next generation

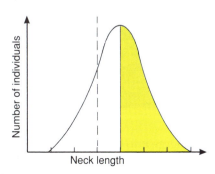

Neck length

Number of individuals

Figure 21.4A ● Passing on favourable characteristics

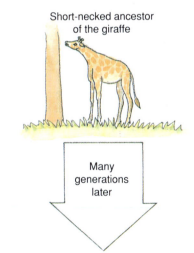

Short-necked ancestor of the giraffe

Many generations later

Long-necked descendants

Members of a species do not live in isolation. They are part of a **population**. No two individuals of a population are the same genetically (except identical twins). Some individuals have genes controlling characteristics which are better suited for survival than the genes of other members of the population. The individuals with the favourable characteristics are more likely to reproduce successfully and have offspring more likely to survive than the individuals with less favourable characteristics. The result is evolution (see Figure 21.4A).

Evolution is still happening

When a new *insecticide* is used against an insect pest, a small amount is enough to kill most of the population. However, a few individuals are not affected because they have genes which code for enzymes that help to break down the insecticide into harmless substances. These individuals are resistant to the insecticide. These are the insects that survive and reproduce. The majority of their offspring inherit the genes for the enzymes that break down insecticide. In this way the number of resistant insects in the population increases.

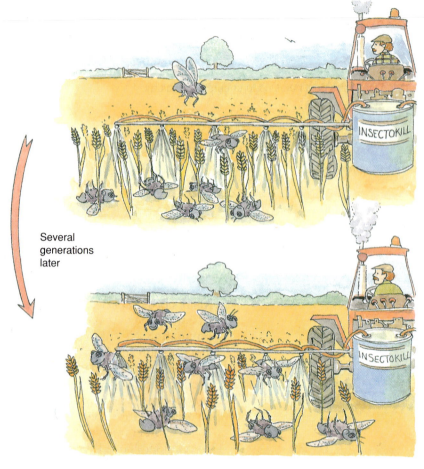

Several generations later

Figure 21.4B ● Resistance of insect pests to insecticide

Insects reproduce quickly so resistance spreads quickly through the population. More and more insecticide has to be used until it becomes so inefficient and so expensive that an alternative insecticide has to be found. Even then, the effect is only temporary because the insect pest soon develops resistance to the new insecticide (Figure 21.4B). The insecticide acts as the agent of selection.

Resistance to antibiotics has evolved in bacteria in much the same way as resistance to insecticides has evolved in insects. Populations of bacteria always contain a few individuals with genes that make them resistant to an antibiotic. These individuals survive and reproduce. The new generation inherits the genes for resistance. Resistance develops rapidly in bacteria because they reproduce rapidly (in some species a new generation is produced every 20 minutes). The antibiotic acts as the agent of selection.

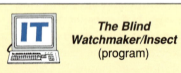

The Blind Watchmaker/Insect (program)

Using *The Blind Watchmaker* or *Insect* you can select which offspring survive and produce the next generation. Mutation occurs, allowing you to design your own evolving creature.

IT'S A FACT

The insecticide DDT was used to control insect pests during the Second World War. In 1947 the first cases of resistance to the new compound were reported in houseflies. Since then resistance in houseflies has developed so that some varieties are resistant to one hundred times the dose of DDT needed to kill unresistant flies. Today, hundreds of insect species are resistant to a range of insecticides.

Figure 21.4C shows two forms of the peppered moth *Biston betularia*. The speckled, pale moth (called the peppered variety) is found most often in the countryside where it merges into the light-coloured background of lichen covered trees and rocks. In towns and cities air pollution kills the lichens leaving trees and other surfaces bare and sooty. The black moth (called the melanic variety) is the most common form in this environment where it merges into the dark background.

Figure 21.4C ● Can you see the moths?

Birds eat moths whose colour does not merge into the background because they are more easily seen. Natural selection (in the form of moth-eating birds) favours the melanic variety in urban areas and the peppered variety in the countryside (Figure 21.4D).

Resistance in bacteria means that antibiotics become less effective for the treatment of disease. Doctors try to get round the problem by using different types of antibiotic and by making sure that patients are given an antibiotic only if it is absolutely necessary. In hospitals, the pressure of selection by antibiotics is so intense that strains of bacteria resistant to a range of antibiotics rapidly evolve.

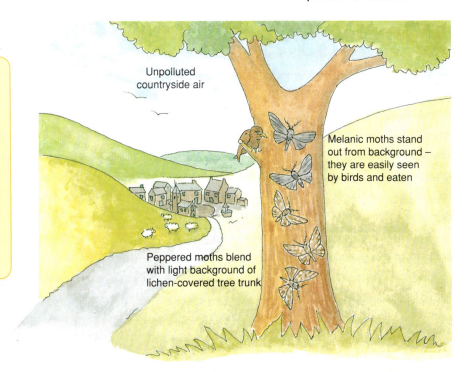

Unpolluted countryside air

Melanic moths stand out from background – they are easily seen by birds and eaten

Peppered moths blend with light background of lichen-covered tree trunk

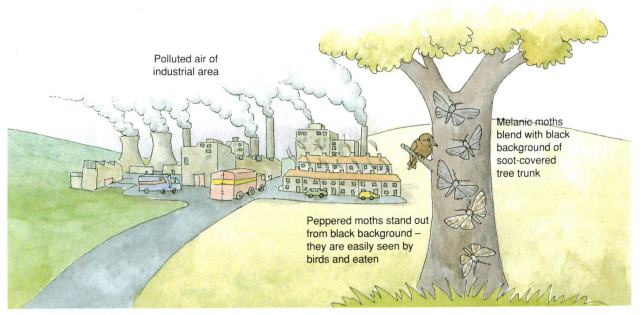

Polluted air of industrial area

Melanic moths blend with black background of soot-covered tree trunk

Peppered moths stand out from black background – they are easily seen by birds and eaten

Figure 21.4D ● Maintaining the balance between the peppered and melanic varieties of the moth *Biston betularia*

CHECKPOINT

❶ Look at Figures 21.1D and 21.1E on pages 384 and 385. Explain in your own words the evolution of flightless cormorants and different varieties of tortoise on the Galapagos Islands. In your explanation mention variation, favourable characteristics, ancestors and descendants.

❷ Look at Figure 21.4A. The graphs show the variation in length of neck in giraffes and their ancestors. What do the graphs tell you about the process of evolutionary change? Why do long necks in giraffes favour survival?

❸ Briefly summarise the reasons for the differences in abundance of peppered moths and melanic moths in the countryside and industrial areas.

❹ Why is the bird in Figure 21.4D an agent of natural selection?

21.5 The vertebrate story

FIRST
THOUGHTS

Fish, amphibians, reptiles, birds and mammals are vertebrates. Their ancestors lived in the sea about 500 million years ago. This section traces the evolution of the vertebrates.

LOOK AT LINKS
for **vertebrates**
See Topic 1.5.

The Earth's surface consists of layers of rock. The more recently formed a layer is, the nearer it is to the Earth's surface. Undisturbed rock layers follow on like the chapters of a book. The fossils in each rock layer are a record of life on Earth at the time when the layer was formed. The sequence of fossils traces the history of life on Earth.

The rock layers of the Grand Canyon have lain undisturbed for hundreds of millions of years (Figure 21.5A). They have been exposed by the Colorado River eroding the soft rock. Today the river flows through a gorge 1.6 kilometres deep. The layers of rock at the bottom of the canyon are 2000 million years old. To travel down one of the trails from the rim to the bottom is to travel back in time. The fossils in each layer represent a particular stage in the evolution of life (Figure 21.5B).

Rocks are 200 million years old

Rocks are 400 million years old

Rocks are 500 million years old

Figure 21.5A ● The Grand Canyon. The rocks are mostly sandstones and limestones deposited in shallow seas that once covered the western USA

Figure 21.5B ● Animals representative of those living at the time when the rock layers were deposited, and which have been exposed in the strata by the Colorado River

LOOK AT LINKS
Rock layers can be deformed. Sometimes they can be vertical or even upside down.
See *KS: Chemistry*, Topic 2.2.

IT'S A FACT

Everyone believed that the coelocanth (a type of fish) had been extinct for 200 million years. Then in 1938 a fisherman caught one in the Indian Ocean. Since then a number of specimens have been caught. The fin bones and muscles of the coelocanth differ from those of other fish living today. Coelocanths resemble the fossil remains of *Eusthenopteron*.

The vertebrate story begins in the rock layers dated at about 400 million years old. The fossil bones of ancient types of fish are found here. In the younger rocks near the surface, fossils of reptiles appear for the first time. In the deeper, older layers there are no fossil bones at all although there are shells and impressions of invertebrate animals that lived in the sea 500 million years ago or more.

What do the rocks of the Grand Canyon tell us about the evolution of life? Careful study of the fossils in successive layers of rock reveals that:

- Representatives of the major invertebrate groups were in existence by 500 million years ago.
- Fish were the first vertebrates to appear about 400 million years ago.
- Reptiles appeared about 200 million years after the fish.
- Birds and mammals appeared more recently, too late to be represented as fossils in the rocks of the Grand Canyon.

The sequence of fossils also tells us that the older the fossils, the more they differ from present-day living things. From the fossil record we can conclude that organisms change through time. In other words, present-day living things are a product of evolution.

The invasion of land

Eusthenopteron was a type of fish that lived about 375 million years ago. These fish had lungs and were able to gulp air. Breathing air enabled them to survive periods of drought when the lakes and marshes in which they lived dried out. They also had strong fins which allowed them to move on land. There was abundant invertebrate food on the banks of the pools in which they lived; insects and plants had moved onto land about 25 million years before the appearance of *Eusthenopteron*.

As *Eusthenopteron*, and other fish like it, could live for part of the time out of water they were more likely to survive and leave behind new generations of air-breathing descendants. The vertebrate invasion of land had begun.

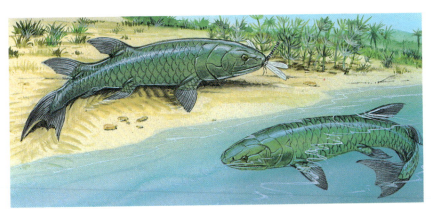

Figure 21.5C ● Eusthenopteron

Amphibians

Frogs, newts and salamanders and all other amphibians are descendants of the earliest land dwellers (Figure 21.5D). The adults usually live on land, but need water for breeding. The female lays her eggs in water. They are fertilised in water and hatch into tadpoles which swim and have gills. The tadpole grows: legs develop, it loses its tail and its gills are replaced by lungs. The change from tadpole to adult is called a metamorphosis.

Figure 21.5D ● Ichthyostega – an early type of amphibian that lived about 350 million years ago

The amphibian body is covered with soft skin through which the body quickly loses water in a dry atmosphere. This is why you find frogs and toads on land only where it is damp.

Reptiles

Reptiles, descended from an ancient type of amphibian, completed the move from water to land around 200 million years ago. Over the next 130 million years an enormous variety of reptiles appeared: land giants, agile runners, fliers, swimmers. They dominated the environment (Figure 21.5E).

Reptiles are true land animals. The most important features which allow them to be so are:

- Horny scales which waterproof the skin.
- Well developed legs.
- An egg which is surrounded by a hard impermeable shell. The liquid inside the egg provides the developing embryo with its own 'private pool'. This allows reptiles to breed away from water.

About 70 million years ago many species of reptile became extinct. Today tortoises, turtles, snakes, lizards and crocodiles are virtually all that remain. Why so many species of reptile became extinct is not clear. However, before they became extinct two quite different groups of reptile had given rise to the ancestors of birds and mammals.

IT'S A FACT

The extinction of most reptiles occurred in a relatively short period of time. The extinction of dinosaurs in North America, for example, happened over 12 million years. This seems like a very long time to us but it is an instant compared with the 4000 million years of life on Earth. Think of it like this: if the 4000 million years of life on Earth occupied one day, the reptiles would have died out in the space of five minutes.

Dimorphodon – although the dinosaur probably lacked the muscles for flapping flight, it had a wingspan of two metres and probably glided

Megalosaurus – a large carnivorous dinosaur

Scelidosaurus – herbivorous dinosaur

Compsognathus – chicken-sized dinosaur with powerful legs for running

Plesiosaurus – long necked reptile well adapted for swimming in the sea

Ichthyosaurus – reptile adapted for swimming. Specimens have been found with the skeletons of embryos inside them. It seems possible that the female Icthyosaurus gave birth to live young

Figure 21.5E ● 155 million years ago reptiles dominated the land

Warm-bloodedness and cold-bloodedness

The **metabolism** of cells releases heat which warms the animal body. Most enzymes which control metabolism work best at a temperature of about 37 °C. If an animal can maintain its body temperature at around 37 °C, even when the temperature outside changes, then metabolism will be at its most efficient. Birds and mammals have evolved ways of keeping the temperature at the centre of the body (the core temperature) at around 37 °C. This is why they are described as **warm-blooded** even though the temperature at the body's surface may fluctuate a little with changes in the temperature outside.

Being warm-blooded means that birds and mammals can live in places where other animals would soon perish because they cannot maintain a steady core temperature. Animals that cannot maintain a steady core temperature are called **cold-blooded**, not because they are 'cold' but because their body temperature fluctuates with the temperature of the environment. Their body temperature drops when the temperature of the environment drops and rises when the temperature of the environment rises (Figure 21.5F). Fish, amphibians and all invertebrates are cold-blooded. Many reptiles are cold-blooded but some species are able to achieve limited control of body temperature (Figure 21.5G).

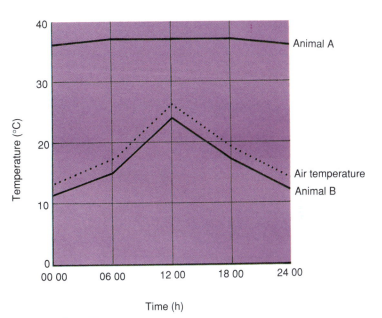

Figure 21.5F ● The body temperatures of two animals and the air temperature recorded over 24 hours. *Which animal is warm-blooded and which cold-blooded?*

LOOK AT LINKS
for **metabolism**
See Topic 10.2;
for **homoeostasis**
See Topic 13.1 and
Topic 16.5

Camels are adapted to survive in hot deserts:
- **fat** insulates surfaces exposed to the sun, restricting heat flow into the body. The underparts have much less insulation so heat can radiate out to the ground
- **flattened nostrils** reduce water loss
- **core body temperature** can fluctuate by 6 °C with body temperature rising in the daytime and dropping during the night. This saves the camel around 5 litres of water a day.

Polar bears are adapted to survive Arctic cold:
- their **fur** does not mat when wet, is easily shaken dry and stays erect trapping a layer of air
- **fat** forms a thick layer of blubber, which insulates the body, under the skin.

IT'S A FACT

The Earth's climate cooled around 70 million years ago. Perhaps this is one reason why many species of reptile became extinct. However, like lizards today, it seems likely that some ancient species of reptile could regulate body temperature and so offset the effects of cold weather.

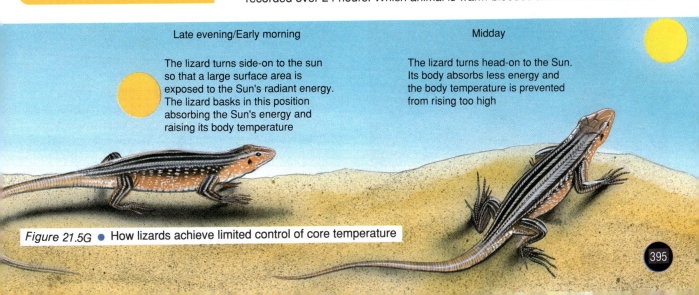

Late evening/Early morning

The lizard turns side-on to the sun so that a large surface area is exposed to the Sun's radiant energy. The lizard basks in this position absorbing the Sun's energy and raising its body temperature

Midday

The lizard turns head-on to the Sun. Its body absorbs less energy and the body temperature is prevented from rising too high

Figure 21.5G ● How lizards achieve limited control of core temperature

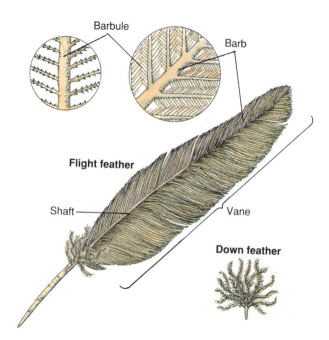

Figure 21.5H ● Feathers

Figure 21.5I ● *Archaeopteryx* was about the size of a crow. Notice the impressions of feathers (characteristic of birds). *Archaeopteryx* also had teeth in its beak (characteristic of reptiles)

Figure 21.5J ● Hoatzin chick

Birds

A bird's body is covered with feathers. Its forelimbs are modified for flying. Feathers make flight possible and help to retain body heat (Figure 21.5H).

Flight feathers give the body shape and lift. Each one has a central shaft. Either side of the shaft are hundreds of parallel filaments called barbs which together make the vane. Each barb has several hundred tiny barbules along it, each carrying very small hooks which fasten on to the hooks of the barbules above. This arrangement gives flight feathers their shape. Down feathers are fluffy because they do not have hooks on their barbules. They provide insulation.

The colour of a bird's feathers also helps to attract mates and provide camouflage.

Birds spend a great deal of time preening themselves to keep their feathers in good condition. Feathers are fluffed out and the bird nibbles them, drawing each one through its beak to remove fleas, lice and mites. The preen gland (called the parson's nose) lies just above the tail. It produces oil which the bird works into the feathers to waterproof them. Preening also repairs feathers and puts them back into place. When the fossil remains of an early type of bird called *Archaeopteryx* were found in 1861, it might have been mistaken for a reptile if it had not been for the beautifully preserved impressions of feathers which surrounded the specimen (Figure 21.5I).

● Flying

How did flight evolve? One idea is that the reptile ancestors of birds lived in trees and that the evolution of feathers enabled them to glide from branch to branch. In Figure 21.5I you can see a claw on the leading edge of the wing. Scientists think that this helped *Archaeopteryx* to clamber along the branches. The chicks of the hoatzin have claws like this and lead a similar sort of life in the tangled branches of the South American rain forest (Figure 21.5J).

Another idea suggests that the evolution of flight began with a feathered carnivorous reptile running after its prey in hops and bounds, helped by its feathered arms. According to this theory, flight developed from long leaps after food.

No one knows the origins of flight. All we do know is that some long extinct group of reptiles gave rise to the feathered warm-blooded creatures we call birds, and that *Archaeopteryx* is an early stage in their evolution.

LOOK AT LINKS
for **flight**
See Topic 17.2.

(a) Duck-billed platypus

(b) Spiny anteater

Figure 21.5K ●

Mammals

One of the early groups of reptile which lived 200 million years ago was the ancestor of mammals. The fossil evidence shows that at first the mammal line flourished but later it nearly died out in the 'age of reptiles' that followed. However, mammals became abundant after most of the reptile species had become extinct 70 million years ago.

Of present-day mammals, the duck-billed platypus and spiny anteater most resemble the animals that were the ancestors of the mammal line (Figure 21.5K). Both animals have a mixture of reptile and mammal features. For example, they lay eggs as reptiles do. However, they are covered with hair and suckle their young with milk as mammals do.

● *Skin, hair and control of body temperature*

Hair is made of the protein keratin. It grows from pockets of cells in the skin called **hair follicles**. Figure 21.5L shows hair follicles and other structures in a section of human skin. Although parts of the human body appear hairless, we can see where the hair follicles are when we are cold because the follicle muscles contract, covering the skin with small bumps which we call 'goose pimples'.

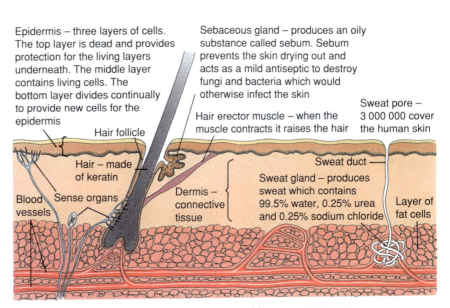

Epidermis – three layers of cells. The top layer is dead and provides protection for the living layers underneath. The middle layer contains living cells. The bottom layer divides continually to provide new cells for the epidermis

Sebaceous gland – produces an oily substance called sebum. Sebum prevents the skin drying out and acts as a mild antiseptic to destroy fungi and bacteria which would otherwise infect the skin

Hair follicle

Hair – made of keratin

Hair erector muscle – when the muscle contracts it raises the hair

Sweat pore – 3 000 000 cover the human skin

Blood vessels

Sense organs

Dermis – connective tissue

Sweat duct

Sweat gland – produces sweat which contains 99.5% water, 0.25% urea and 0.25% sodium chloride

Layer of fat cells

Figure 21.5L ● Section through the human skin

Hair and other mechanisms help to control the body's temperature:

- Hairs raised by the erector muscles trap a layer of air which insulates the body in cold weather (air is a poor conductor of heat). In warm weather the hair is lowered and no air is trapped.
- Fat insulates the body and reduces heat loss.
- Sweat cools the body because it carries heat energy away from the body as it evaporates.
- Millions of temperature-sensitive sense receptors cover the skin. Nerves connect them to the brain which controls the body's response to changes in temperature in the environment.
- When it is warm, blood vessels in the skin dilate. More blood flows through the vessels in the skin and loses heat to the environment. In cold weather, the blood vessels in the skin constrict and less heat is lost to the environment.

Chemical reactions (metabolism – see page 156) in the liver release a lot of heat energy, which is distributed all over the body by blood. Birds and mammals have a high metabolic rate. The large amount of heat released from the chemical reactions explains why they are warm blooded.

Figure 21.5M ● A grey kangaroo carrying her offspring in her marsupium

● Present-day mammals

Apart from the duck-billed platypus and the spiny anteater, the class mammals includes two groups: the marsupials and the placentals. Marsupial mammals include the kangaroo, wallaby and koala bear. They give birth to offspring which are at an early stage of development. The offspring crawl into their mother's pouch (called the marsupium) and attach themselves to the nipples of her milk glands. Inside the pouch they grow and develop until they can look after themselves (Figure 21.5M).

Placental mammals include moles, bats, rodents, whales, monkeys and humans (Figure 21.5N). They give birth to offspring which are at a later stage of development than marsupials. The embryo is attached to the wall of the uterus by the placenta through which it obtains nourishment.

Mammals are successful animals. They colonise every environment: from polar ice caps to tropical forests to hot deserts. They fill the air, land and sea.

Humans are relative newcomers to the scene. Our ancestors first appeared approximately 2.5 million years ago. They were as we are – curious. We use words to communicate our curiosity, construct codes of moral behaviour and ask questions about our relationship with the world of nature. We have even invented a way of investigating such questions. It is called Science!

Figure 21.5N ● Some of the many placental mammals

SUMMARY

It is possible that there is life somewhere else in the Universe, but we do not know of it yet. For the time being, therefore, Earth is unique. It is filled with life – a product of 4000 million years of evolution.

● *Topic 19*

1 The diagram shows some of the features of the inheritance of sex.

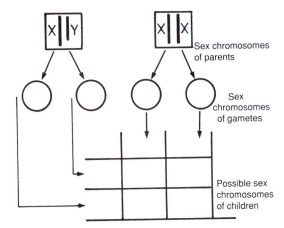

(a) Copy the diagram and indicate the sex of each parent.
(b) Complete your copy of the diagram by drawing in the spaces provided, the chromosomes present in the gametes and in the children.
(c) The diagram shows the chromosomes taken from a cell of an adult human.

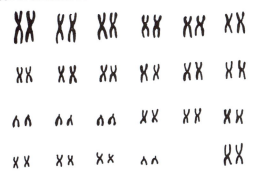

  (i) State whether the chromosomes are from a male or female.
  (ii) Give one reason for your answer.   (WJEC)

2 Diagram A shows the chromosomes of a body cell from a female insect. Diagram B shows the chromosomes of a body cell from a male insect of the same species.

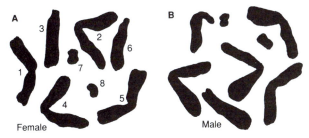

(a) Using the numbers, arrange the chromosomes of the female into groups of matching pairs. The first matching pair is shown.

    4 matches with 2

(b) Carefully draw the pair of chromosomes in the male which do not match.
(c) Name the structures concerned with heredity which are found arranged along chromosomes.
(d) How many chromosomes would be found in a sperm produced by this insect?

  (WJEC)

3 The diagram below shows the offspring of crosses between pure bred Aberdeen Angus bulls, which are black, and pure Redpoll cows, which are red. The ratio of the colours of the offspring of the first generation is also shown. Coat colour is controlled by a single gene which has two forms (alleles): one for black and one for red coat colour.

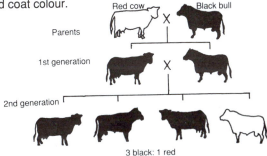

3 black: 1 red

(a) What letters are suitable to represent the two forms (alleles) of the gene?
(b) Make a rough copy of the diagram.
  (i) Draw a circle around each animal in the diagram which is definitely homozygous for the gene for coat colour.
  (ii) Draw a square around each animal in the diagram which is definitely heterozygous for the gene for coat colour.
(c) Explain why some of the animals in the diagram could be either homozygous or heterozygous for the gene for coat colour.

  (LEAG)

● *Topic 20*

4 The table shows the results obtained when **three** different types of yeast were each allowed to completely ferment a standard sugar solution at 16 °C. The results show six separate occasions when this was done for each of the three types of yeast. In each case the time taken, to the nearest hour, from the beginning to the end of fermentation was recorded.

Type of yeast	Time taken for complete fermentation of a standard sugar solution at 16 °C/h					
Type A (S. carlsbergensis)	48	47	41	38	36	42
Type B (S. cerevisiae)	15	30	24	12	16	15
Type C (S. ellipsoideus)	24	29	30	32	36	33

(a) Using the data in the table calculate the mean time for Type A yeast to complete fermentation. Show your working.

(b) Suggest which **one** of the yeasts, Type A, Type B or Type C, would be most suitable for use in bread making. Explain your answer.

(c) Explain the role of yeast in bread making.

5 The diagram shows stages in a biotechnological process in which a human gene that codes for making the hormone insulin is inserted into a piece of circular DNA obtained from a bacterial cell. The process results in large amounts of human insulin being made in a short period of time.

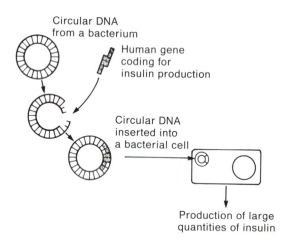

Circular DNA from a bacterium

Human gene coding for insulin production

Circular DNA inserted into a bacterial cell

Production of large quantities of insulin

(a) Of what type of biotechnological process is the diagram an example?

(b) Using the information given above and in the diagram and what you know about bacterial cells describe the stages in the process shown by the arrows.

(c) Explain why bacteria are ideal organisms for making large amounts of human insulin in a relatively short time.

(d) People suffering from diabetes may be given human insulin in order to stay alive. Why is the bodily function of a diabetic not normal and how does the supplying of insulin restore the body function to normal?

(e) Until recently diabetics were supplied with insulin obtained from sheep or pigs. Discuss the advantages and possible disadvantages of treating diabetics with insulin produced by the process shown in the diagram rather than using insulin obtained from sheep or pigs.

6 The diagram shows the main stages of a bio-technological process which uses a fermenter. The process was designed to produce single-cell protein

(SCP) from bacteria which grow on methanol. The bacteria and raw materials are put into the fermenter which is kept at a constant temperature of 40 °C.

(a) Using **only** the information provided in the diagram, answer the following questions:

(i) What other raw materials, in addition to methanol, are used in the process?

(ii) What process ensures that the raw materials entering the fermenter are free of disease organisms?

(iii) Why is the process in (ii) necessary?

(iv) What substance is recycled into the fermenter after being separated from the bacteria?

(v) State **two** forms in which SCP is sold.

(b) Explain the importance of keeping the fermenter at a constant temperature of 40 °C.

(c) State **three** advantages of using bacteria to produce protein food.

● **Topic 21**

7 Snails are eaten by thrushes. Some snails have shells that are very striped, others unstriped. Every September for several years a scientist counted all the snails he could find in an area of grassland. Here are the results:

Year	% covered by grass	Number of snails with ... very striped shells	unstriped shells
1971	98	58	13
1972	25	24	22
1973	5	2	33
1974	97	34	10
1975	96		
1976		9	43
1977	98	68	13

(a) State:

(i) a probable number of snails you would have expected to find in 1975,

(ii) a probable percentage cover of grass in 1976.

(b) (i) All the snails were of the same type (species). During the seven years of study a single specimen was found with a completely black shell. What word could be used to describe this unusual form of the species?

(ii) Choose the best answer. The term which best **explains** the results in the table is

1 heredity   3 conservation

2 natural selection   4 artificial selection.

(WJEC)

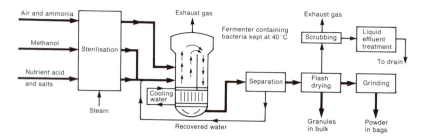

Air and ammonia

Exhaust gas

Exhaust gas

Fermenter containing bacteria kept at 40 °C

Scrubbing

Liquid effluent treatment

Methanol

Sterilisation

To drain

Nutrient acid and salts

Cooling water

Separation

Flash drying

Grinding

Steam

Recovered water

Granules in bulk

Powder in bags